Differential Geometry
Proceedings of the VIII International Colloquium

Differential Geometry

Proceedings of the VIII International Colloquium

Santiago de Compostela, Spain   7 – 11 July 2008

editors

Jesús A Álvarez López
Eduardo García-Río
University of Santiago de Compostela, Spain

World Scientific

NEW JERSEY · LONDON · SINGAPORE · BEIJING · SHANGHAI · HONG KONG · TAIPEI · CHENNAI

Published by

World Scientific Publishing Co. Pte. Ltd.
5 Toh Tuck Link, Singapore 596224
USA office: 27 Warren Street, Suite 401-402, Hackensack, NJ 07601
UK office: 57 Shelton Street, Covent Garden, London WC2H 9HE

British Library Cataloguing-in-Publication Data
A catalogue record for this book is available from the British Library.

Cover image: Self-portrait of Enrique Vidal Abascal

DIFFERENTIAL GEOMETRY
Proceedings of the VIII International Colloquium

Copyright © 2009 by World Scientific Publishing Co. Pte. Ltd.

All rights reserved. This book, or parts thereof, may not be reproduced in any form or by any means, electronic or mechanical, including photocopying, recording or any information storage and retrieval system now known or to be invented, without written permission from the Publisher.

For photocopying of material in this volume, please pay a copying fee through the Copyright Clearance Center, Inc., 222 Rosewood Drive, Danvers, MA 01923, USA. In this case permission to photocopy is not required from the publisher.

ISBN-13 978-981-4261-16-6
ISBN-10 981-4261-16-5

Printed in Singapore.

PREFACE

This volume contains the papers contributed by the speakers of the VIII International Colloquium on Differential Geometry, which was held in Santiago de Compostela (Spain) during July 7–11, 2008.

The Colloquium was organized by all members of the Department of Geometry and Topology of the University of Santiago de Compostela, and M. Ángeles de Prada of the University of País Vasco (Spain). The chairman was Luis Á. Cordero Rego. There were about 135 participants from many countries.

We are most thankful to the following institutions that have cosponsored the Colloquium: the Department of Geometry and Topology, the Institute of Mathematics and the Faculty of Mathematics of the University of Santiago de Compostela, the "Ministerio de Educación y Ciencia" (projects MTM2006-01432 and MTM2007-30162-E), the "Ministerio de Ciencia y Tecnología" (project MTM2008-02640), the "Consellería de Educación e Ordenación Universitaria da Xunta de Galicia", the "Consellería de Innovación e Industria da Xunta de Galicia" (project PGIDIT06PXIB207054PR), and the "Consejo Superior de Investigaciones Científicas" of Spain. Their support was essential to carry out our program.

The papers of this volume were mostly from the lectures presented at the Colloquium, which was designed with two main parts: Foliation Theory and Riemannian Geometry. All contributed papers were refereed, and we believe that they are of high quality, and make a significant progress on these subjects.

Finally, we express our deep gratitude to all participants, and particularly to the contributors of the papers of this volume.

Editors
J. A. Álvarez López
Eduardo García-Río

ORGANIZING COMMITTEES

VIII International Colloquium on Differential Geometry

ORGANIZING COMMITTEE

L. A. Cordero (Chairman)	– University of Santiago de Compostela, Spain
M. A. de Prada	– University of the Bask Country, Spain
Members of the Department of Geometry and Topology	– University of Santiago de Compostela, Spain

HONORARY COMMITTEE

Emilio Pérez Touriño	President of the Galician Government
Laura E. Sánchez Piñón	Galician Minister for Education.
Fernando X. Blanco Álvarez	Galician Minister for Innovation and Industry
Senén Barro Ameneiro	Rector of the University of Santiago de Compostela
Xosé A. Sánchez Bugallo	Mayor of Santiago de Compostela
Ernesto Viéitez Cortizo	President of the Royal Academy of Sciences of Galicia
Juan M. Viaño Rey	Dean of the USC Faculty of Mathematics
Juan J. Nieto Roig	Director of the USC Institute of Mathematics

INTERNATIONAL ADVISORY COMMITTEE

Jesús A. Álvarez López	– University of Santiago de Compostela, Spain
Alberto Candel	– SISSA, California State University, USA
Eduardo García-Río	– University of Santiago de Compostela, Spain
Peter B. Gilkey	– University of Oregon, USA
Gilbert Hector	– University Claude Bernard, Lyon 1, France
Stefan Ivanov	– University of Sofia, Bulgaria
Manuel de León	– ICMAT-CSIC, Spain
Shigenori Matsumoto	– Nihon University, Japan
Antonio M. Naveira	– University of Valencia, Spain
Paul A. Schweitzer	– Catholic University of Rio de Janeiro, Brazil
Pawel G. Walczak	– Lódz University, Poland

CONTENTS

Preface	v
Organizing Committees	vii
A brief portrait of the life and work of Professor Enrique Vidal Abascal *L. A. Cordero*	1

Part A   Foliation theory — 9

Characteristic classes for Riemannian foliations *S. Hurder*	11
Non unique-ergodicity of harmonic measures: Smoothing Samuel Petite's examples *B. Deroin*	36
On the uniform simplicity of diffeomorphism groups *T. Tsuboi*	43
On Bennequin's isotopy lemma and Thurston's inequality *Y. Mitsumatsu*	56
On the Julia sets of complex codimension-one transversally holomorphic foliations *T. Asuke*	65
Singular Riemannian foliations on spaces without conjugate points *A. Lytchak*	75

Variational formulae for the total mean curvatures of a
codimension-one distribution 83
V. Rovenski and P. Walczak

On a Weitzenböck-like formula for Riemannian foliations 94
V. Slesar

Duality and minimality for Riemannian foliations on open
manifolds 102
X. M. Masa

Open problems on foliations 104

Part B Riemannian geometry 109

Graphs with prescribed mean curvature 111
M. Dajczer

Genuine isometric and conformal deformations of submanifolds 120
R. Tojeiro

Totally geodesic submanifolds in Riemannian symmetric spaces 136
S. Klein

The orbits of cohomogeneity one actions on complex
hyperbolic spaces 146
J. C. Díaz-Ramos

Rigidity results for geodesic spheres in space forms 156
J. Roth

Mean curvature flow and Bernstein-Calabi results for
spacelike graphs 164
G. Li and I. M. C. Salavessa

Riemannian geometric realizations for Ricci tensors of
generalized algebraic curvature operators 175
P. Gilkey, S. Nikčević and D. Westerman

Conformally Osserman multiply warped product structures
in the Riemannian setting   185
   M. Brozos-Vázquez, M. E. Vázquez-Abal and
   R. Vázquez-Lorenzo

Riemannian Γ-symmetric spaces   195
   M. Goze and E. Remm

Methods for solving the Jacobi equation. Constant osculating
rank vs. constant Jacobi osculating rank   207
   T. Arias-Marco

On the reparametrization of affine homogeneous geodesics   217
   Z. Dušek

Conjugate connections and differential equations on infinite
dimensional manifolds   227
   M. Aghasi, C. T. J. Dodson, G. N. Galanis and A. Suri

Totally biharmonic submanifolds   237
   D. Impera and S. Montaldo

The biharmonicity of unit vector fields on the Poincaré half-
space H^n 247
   M. K. Markellos

Perspectives on biharmonic maps and submanifolds   257
   A. Balmuş

Contact pair structures and associated metrics   266
   G. Bande and A. Hadjar

Paraquaternionic manifolds and mixed 3-structures   276
   S. Ianuş and G. E. Vîlcu

On topological obstruction of compact positively Ricci
curved manifolds   286
   W.-H. Chen

Gray curvature conditions and the Tanaka-Webster connection   291
R. Mocanu

Riemannian structures on higher order frame bundles from
classical linear connections   296
J. Kurek and W. M. Mikulski

Distributions on the cotangent bundle from torsion-free
connections   301
J. Kurek and W. M. Mikulski

On the geodesics of the rotational surfaces in the Bianchi-
Cartan-Vranceanu spaces   306
P. Piu and M. M. Profir

Cotangent bundles with general natural Kähler structures of
quasi-constant holomorphic sectional curvatures   311
S. L. Druţă

Polynomial translation Weingarten surfaces in 3-dimensional
Euclidean space   316
M. I. Munteanu and A. I. Nistor

G-structures defined on pseudo-Riemannian manifolds   321
I. Sánchez-Rodríguez

List of Participants   327

A BRIEF PORTRAIT OF THE LIFE AND WORK OF PROFESSOR ENRIQUE VIDAL ABASCAL

Luis A. Cordero

Department of Geometry and Topology, University of Santiago de Compostela
15782 Santiago de Compostela, Spain
E-mail: luisangel.cordero@usc.es

The "VIIIth International Colloquium on Differential Geometry (E. Vidal Abascal Centennial Congress)" held in Santiago de Compostela (Spain), 7-11 July 2008, has been a part of the celebration of the one hundreth anniversary of the birth of Prof. Enrique Vidal Abascal. In what follows you will find a brief summary of his life and work; if you are interested in knowing more about this Spanish mathematician, you can find more details about him at the webpage

ENRIQUE VIDAL ABASCAL

http://xtsunxet.usc.es/icdg2008/evidala.htm.

Some Biographical Data

Enrique Vidal Abascal was born in Oviedo (Spain) in October 12, 1908, and died in Santiago de Compostela (Spain) in October 31, 1994.

Graduated in Mathematics at the Complutense University of Madrid in 1931, Vidal became immediately involved in the teaching of mathematics at the college level from 1932 until 1955, although his work at the university started in 1941 at the University of Santiago, sharing his position at a college in Santiago with a nomination as Assistant Professor at the university. In 1955 he was nominated full professor to occupy a chair on Differential

Geometry at the University of Santiago, staying in this position until his academic retirement in 1978.

In 1944 he obtained the Title of Doctor in Exact Sciences by the Complutense University of Madrid; his doctoral thesis was written under the guidance of Ramón María Aller Ulloa, being his advisor Esteban Terradas (Complutense University), and it was devoted to the geometric study of the calculation of the orbits of double stars.

During his academic life at the university he has been engaged at the following positions: Vice Dean of the Faculty of Sciences of the University of Santiago during the periods 1960-61, 1966-69 and 1976-78; Dean of the Faculty of Mathematics of the same University from January to October of 1978, date of his retirement; Vice President of the Royal Spanish Mathematical Society in the periods 1963-66 and 1973-78; Head of the Mathematical Section of the Astronomical Observatory of the University of Santiago, associated to the "Consejo Superior de Investigaciones Científicas" (Spain), from 1942; and Director of the Seminary of Mathematics of the University of Santiago, center co-ordinated with the Institute "Jorge Juan" of the "Consejo Superior de Investigaciones Científicas", between 1967 and 1978.

Vidal was also Member of the Royal Academy of Galicia from 1971, and the creation of the Royal Academy of Sciences of Galicia, in 1978, is owed to his personal initiative and management. He was the first President of this Academy between 1978 and 1982, when he resigned and then being nominated as Honorary President of the Academy. Member of the "American Mathematical Society", the "Círcolo Matemático di Palermo" and the "Royal Spanish Mathematical Society", he was also reviewer of "Mathematical Reviews" and of "Zentralblatt für Mathematik".

In 1952 he was granted for one stay at the "Bureau International d'Education" in Geneva (Switzerland), for visiting training centers in Switzerland in order to know the problematic of the teaching of Mathematics at the secondary level in that country; in 1953 he traveled again to Switzerland, staying for three months in Lausanne working with Georges de Rham on the foundations of Integral Geometry, and being invited to give a conference in the Polytechnical School of this city.

In 1963, 1966, 1973 and 1977 he was invited to pronounce conferences and to lecture courses on the subjects of his research at the University of Paris VI and at the "College de France". Also, he was invited to participate in numerous seminaries and scientific meetings at the Universities of Paris and Strasbourg, and in the research centers of Oberwolfach and Brussels. He also participated in the International Congresses of Mathematicians celebrated in Edinburg (1958), Moscow (1966) and Nize (1970), in numerous "Annual Meetings of Spanish Mathematicians", and he was one of the invited lecturers in the "Reunión de Matemáticos de Expresión Latina" held in Palma de Mallorca (Spain) in 1977.

The Scientific Work of E. Vidal Abascal

The scientific interests of the Prof. Vidal were centered in three great areas: Astronomy, in particular in the calculation of orbits of double stars; Classical Differential Geometry (curves and surfaces) and Integral Geometry; and Differential Geometry of Manifolds, and in this particular context in Foliation Theory, Almost-Product Structures and Hermitian Geometries.

An objective proof of the interest and importance of his research and results is the fact that many of their articles were published in important world-wide distributed journals of recognized prestige; among them: *Astronomical Journal* (Yale, USA), *Journal of Differential Geometry* (Leigh Univ., USA), *Proceedings of the American Mathematical Society* (USA), *Bulletin of the American Mathematical Society* (USA), *Annals de l'Institut Fourier* (Grenoble, France), *Comptes Rendu de l'Académie des Sciences* (Paris, France), *Rendiconti dei Círcolo Matemático di Palermo* (Palermo, Italy), or *Tensor N.S.* (Japan)

He was a very prolific author, and in the list of his publications, probably incomplete, there are a total of 112 publications that can be grouped in the following form: 3 monographs and 13 articles on Astronomy; 1 book, 5 monographs and 43 articles on Differential and Integral Geometry, and 47 publications more: books on general Mathematics, discourses, books of essay, and articles of divulgation in prestigious magazines, like "Revista de Occidente" (Spain) for example.

Vidal was awarded with several prizes for his scientific work: by the "Consejo Superior de Investigaciones Científicas", Madrid (Spain) in 1949; by the "Royal Academy of Exact, Physical and Natural Sciences of Spain" in 1953 and 1959; and by the Galician Government in 1989, by the whole of his scientific work. He also received the following medals: "Officier dans l'Ordre des Palmes Académiques", granted by the French Government, in 1974, and the "Medalla Castelao", granted by the Galician Government in 1986, as a recognition of a whole life dedicated to Galicia.

The interest of Vidal in his studies of Astronomy was mainly centered in the calculation of orbits of double stars, the subject of his Ph.D. thesis. When commenting his works on this subject, Prof. Baize, of the Astronomical Observatory of Paris (France), one of the maximum authorities of the world on this topic at that time, wrote in 1980 the following:

"Prof. Vidal is first of all a mathematician, his works on Differential Geometry universally well-known and appreciated are an evidence of that. He has not observed personally double stars, but he has been interested in the problems that arise in the calculation of its orbits, calculation for which there exist numerous methods, of an unequal practical value, and that can be classified in two groups; graphical methods and analytical methods. The method imagined by Vidal belongs to the second group, but it solves the problem by completely new routes, giving an elegant proof and applying it to the star calculations ... The whole of his research, as much on the non-elliptic orbits as on the elliptic ones, has been condensed by Vidal in a very important work, published in 1953, entitled "Calculation of Apparent Orbits of Double Stars", book that constitutes without a doubt, as I already wrote in the moment of its appearance, the most remarkable contribution in our time to the study of the orbits of double stars. On the other hand, Vidal did not limit himself only to the theory, he was also interested in the practical application of his methods, inventing and making construct by the prestigious Swiss company Coradi his ingenious "Orbígrafo", a device that allows to directly draw up on the paper the curve that represents the angles of position as a function of the distances, respecting rigorously the law of areas. This instrument is still used by numerous researchers, mainly in the Astronomical Observatory of the University of Santiago."

The first works of Vidal on Differential Geometry go back to 1943-47. His studies on parallel curves on surfaces of constant curvature are specially outstanding, because they lead to a generalization of the classical formulas of Steiner for parallel connected curves in the plane; the methods used by Vidal in this study were used later by the C.B. Allendoerfer on the spheres. His works on these subjects, that belong to what at that time was known as Differential Geometry "in the large", led him in a natural way to consider the study of some problems of Integral Geometry on surfaces. His numerous contributions in this area have not been out of phase with the passage of time, and it is not difficult to find references to them in recent articles dedicated, for example, to the study of the volume of geodesic tubes in Riemannian manifolds of arbitrary curvature.

The study of the integral invariants of geodesics led Vidal to consider its generalization and, in the last term, to the study of the measures in foliated manifolds, via by which Vidal introduces himself in a new subject of research, very novel at that time, subject of which he has been the pioneer in Spain. Once again, the results obtained by Vidal between 1964 and 1967 continue being mentioned in recent publications on the topic.

From 1966, the year in which the first doctoral thesis directed by him appears, Vidal stops being an isolated researcher and begins to form a compact team of researchers constituted by young people graduated in the Section of Mathematics of the Faculty of Sciences of the University of Santiago. The numerous subjects of thesis that Vidal proposes to his students extend his personal interests in research to other subjects, such as almost-product or almost-hermitian structures. Between 1966 and 1978 Vidal directed a total of fifteen doctoral theses and, to an age in which a certain diminution in his research activity would be logical, he published five articles on Differential Geometry, gave four communications in Congresses and wrote three monographs. In his articles of these last years there are again very remarkable contributions, such as the notion of almost foliated metric for almost-product structures, or the definition and characterization of two new families of almost-hermitian structures, whose scientific name universally accepted and adopted is the one of "geometries G_1 and G_2", being the "G" by the adjective "Gallegas" as it was indicated specifically in the note published in the "Comptes Rendus" of the Academy of Sciences of Paris in which they appeared in 1976.

The Creative Restlessness of E. Vidal Abascal

When Vidal arrives at the university, first as student and years later as professor, mathematics in Spanish was totally out of phase and almost isolated with respect to the currents of study followed in the most important research centers of abroad. The first trips that he made abroad (to Switzerland in 1952 and 1953, and to the ICM in Edinburg in 1958) allowed him to state this reality, and he became aware of the urgent and inexcusable necessity of putting a remedy to this situation. Vidal did not content himself with speaking or writing about what was precise to do, like many of his contemporaries in Spain did, instead, he acted, and as a result of his actions Vidal became a pioneer in opening the mathematical research in Galician universities, or even more, it should be said, in the Spanish universities, to the international mathematical community, in particular to Europe.

Once his academic situation at the university became steady, and in spite of the difficulties and obstacles that existed in Spain at that time, Vidal manages the first visits of prestigious foreign mathematicians to Santiago de Compostela. Between 1960 and 1978 more than forty foreign professors visited Vidal's department, coming from Brazil, Belgium, England, France, Germany, Israel, Portugal, Romania, Switzerland and the USA; here, they lectured graduate courses, gave conferences, or participated in the International Colloquia organized by Vidal. Among those mathematicians one can find the names of some of the most prestigious geometers of those decades. A fact to be specially remarked is that Prof. René Deheuvels, from the University of Paris VII (France), was nominated as Visiting Professor of the University of Santiago to lecture, during two consecutive academic years, courses of doctorate with full academic validity, a singular case in the Spanish mathematics of that time.

Moreover, Vidal organized three International Colloquia on Differential Geometry, held at the University of Santiago in 1963, 1967 and 1972, the first of them being, in fact, the first international congress of mathematics celebrated in Spain. In 1978 a fourth International Colloquium specially dedicated to Vidal was held as a tribute in the occasion of his academic retirement. Later four new Colloquia, celebrated in 1984, 1988, 1994 and 2008, gave a continuity to the series initiated under the direction of Vidal.

With these two actions, the visits of foreign mathematicians, and the organization of international meetings, Vidal reached the first of his objectives: for his students to know some of the most actual lines of research on differential geometry of that time.

But Vidal had a second objective in mind: to get his students traveling abroad to improve his training as researchers. And he also accomplished this objective. Some of his students went, for long stays, to France (in Paris and Strasbourg), to England (in Durhan) and to the USA (in Harvard and Maryland).

The profits of these two actions for the Spanish mathematics were clear: between 1965 and 1978 Vidal directed 15 doctoral thesis, and one can find, even today, the traces of his work through the work of his students, and not only in the University of Santiago but also in the Universities of Sevilla, Granada, Valencia, Bilbao, or La Laguna (Canary Islands), for example.

Prof. Luis A. Santaló (University of Buenos Aires, Argentine), when commenting the set of the work of Vidal Abascal, wrote in 1980:

"It is fundamental to remark that the importance of his works is even, if not surpassed, by the fact of having created in Santiago de Compostela a School of Geometry from where some brilliant students have already went to another Spanish universities and there they are having an outstanding behavior. He knew the way to form a School. He knew how to create and to direct, for long years, a center with its own publications, a place for important national and international congresses and meetings, obliged visit for the most outstanding figures of the time, who came there to lecture courses and conferences with the certainty that their teachings will felt in a land that has been diligently prepared by Vidal Abascal, who knew how to made of his Institute a warm and welcoming complement for the own beauties of Santiago. His work has been persistent and always directed with tenacity, intelligence and love. Students and colleagues know well of his extraordinary charming, the smoothness in his behavior, the nobility in his relationship, and his cleverness in the direction."

These efforts and initiatives of Vidal Abascal were not recognized at Spanish level, as it had been of justice. Nevertheless, the French Government compensated his efforts granting to him the medal of "Officier dans l'Ordre des Palmes Académiques" in 1974. This medal was, at that time, rarely granted outside France, and with it the exceptional personality of Prof. Vidal Abascal was recognized. Prof. Deheuvels in the act of imposition of this medal said:

"... He knew how to stimulate in fifteen years the mathematical activity in the Spanish Universities... He got, in few years, to make Santiago well known in the scientific world, not only by the outstanding Colloquia that he organized but also by his own scientific works or by the works of his students... the concession by the French Government of this medal shows that his reputation has exceeded widely the frontiers."

Vidal Abascal essayist, painter, etc.

As much in his speech of entrance as Member of the Royal Academy of Galicia ("The Crisis of the European University", A Coruña, 1971), like in his essay "Science and the Socialized University" (Ed. Dossat, Madrid 1972), or in his numerous journalistic collaborations, Vidal showed clearly his restlessness before the crisis by which the Spanish university was passing through, and in particular the crisis of the Galician university, that at that time was reduced to the University of Santiago. Through his writings his will of a liberal and progressive man is clear, deeply worried about a university unable to give a suitable answer to the social demands of that time.

Also in Vidal's writings the poverty of the resources destined to promote research, basic or applied, were an object of reflection as well as the necessity of making a suitable and long-term planning that allowed to obtain a greater profitability of the limited existing resources, or the deficiencies in the system of access to the university, or in the systems of training and promotion of its teaching staffs.

Vidal was also a painter. His taste for painting came from his youth, although he confessed that painting always had been for him only a "hobby". He made a first showing of his paintings in A Coruña, 1947; this was followed by others in Santiago de Compostela (1948, 1975 and 1978), Vigo (1950 and

1976), Pontevedra (1974, 1975 and 1980), Ourense (1977), Madrid (1979) and Barcelona (1983). Vidal also published a few articles about painting and was a critic of art; in 1972 he pronounced a conference in the Galician Center of Buenos Aires (Argentine) about "The Galician Painting School", and in 1979 he published a book entitled "On the University and the Galician Painting" (Univ. of Santiago, 1979). Many of his pictorial works appear at the present in museums and particular collections.

PART A
Foliation theory

CHARACTERISTIC CLASSES FOR RIEMANNIAN FOLIATIONS

Steven Hurder

Department of Mathematics (m/c 249), University of Illinois at Chicago
Chicago, IL 60607-7045, USA
E-mail: hurder@uic.edu

The purpose of this paper is to both survey and offer some new results on the non-triviality of the characteristic classes of Riemannian foliations. We give examples where the primary Pontrjagin classes are all linearly independent. The independence of the secondary classes is also discussed, along with their total variation. Finally, we give a negative solution of a conjecture that the map of classifying spaces $FR\Gamma_q \to F\Gamma_q$ is trivial for codimension $q > 1$.

Keywords: Riemannian foliation, characteristic classes, secondary classes, Chern-Simons classes

1. Introduction

The Chern-Simons class[9] of a closed 3-manifold M, considered as foliated by its points, is the most well-known of the secondary classes for Riemannian foliations. Foliations with leaves of positive dimension offer a much richer class to study, and the values of their secondary classes reflect both geometric (metric) and dynamical properties of the foliations. It is known that all of these classes can be realized independently for explicit examples (Theorem 4.3), but there remain a number of open problems to study. In this note, we survey the known results, highlight some of the open problems, and provide a negative answer to an outstanding conjecture.

Let M be a smooth manifold of dimension n, and let $\mathcal{F}$ be a smooth foliation of codimension q. We say that $\mathcal{F}$ is a *Riemannian foliation* if there is a smooth Riemannian metric g on TM which is *projectable* with respect to $\mathcal{F}$. Identify the normal bundle Q with the orthogonal space $T\mathcal{F}^\perp$, and let Q have the restricted Riemannian metric $g_Q = g|Q$. For a vector $X \in T_xM$ let $X^\perp \in Q_x$ denote its orthogonal projection. Given a leafwise path γ between points x, y on a leaf L, the transverse holonomy h_γ along

γ induces a linear transformation $dh_x[\gamma] : Q_x \to Q_y$. The fact that the Riemannian metric g on TM is projectable is equivalent to the fact that the transverse linear holonomy transformation $dh_x[\gamma]$ is an isometry for all such paths.[16,17,39–41,49]

There are a large variety of examples of Riemannian foliations which arise naturally in geometry. Given a smooth fibration $\pi \colon M \to B$, the connected components of the fibers of π define the leaves of a foliation $\mathcal{F}$ of M. A Riemannian metric g on TM is projectable if there is a Riemannian metric g_B on TB such that the restriction of g to the normal bundle $Q \equiv T\mathcal{F}^\perp$ is the lift of the metric g_B. The pair $(\pi \colon M \to B, g)$ is said to be a *Riemannian submersion*. Such foliations provide the most basic examples of Riemannian foliations.

Suspensions of isometric actions of finitely generated groups provide another canonical class of examples of Riemannian foliations. The celebrated Molino Structure Theory for Riemannian foliations of compact manifolds reduces, in a broad sense, the study of the geometry of Riemannian foliations to a mélange of these two types of examples – a combination of fibrations and group actions; see Theorems 6.2 and 6.3 below. When the dimension of M is at most 4, the Molino approach yields a "classification" of all Riemannian foliations. However, in general the structure theory is too rich and subtle to effect a classification for codimension $q \geq 3$ and leaf dimensions $p \geq 2$. The survey by Ghys, Appendix E of[41] gives an overview of the classification problem circa 1988.

The secondary characteristic classes of Riemannian foliations give another approach to a broad classification scheme. Their study focuses attention on various classes of Riemannian foliations, which are investigated in terms of known examples and their Molino Structure Theory, and the values of their characteristic classes, often leading to new insights.

The characteristic classes of a Riemannian foliation are divided into three types: the primary classes, given by the ring generated by the Euler and Pontrjagin classes; the secondary classes; and the blend of these two as defined by the Cheeger-Simons differential characters. Each of these types of invariants have been more or less extensively studied, as discussed below. The paper also includes various new results and unpublished observations, some of which were presented in the author's talk.[22]

The main new result of this paper uses characteristic classes to give a negative answer to Conjecture 3 of the Ghys survey [*op. cit.*]. The proof of the following is given in §3.

Theorem 1.1. *For $q \geq 2$, the map $H_{4k-1}(FR\Gamma_q; \mathbb{Z}) \to H_{4k-1}(F\Gamma_q; \mathbb{Z})$ has* *infinite-dimensional image for all degrees* $4k - 1 \geq 2q$.

This paper is an expanded version of a talk given at the joint AMS-RSME Meeting in Seville, Spain in June 2003. The talk was dedicated to the memory of Connor Lazarov, who passed away on February 27, 2003. We dedicate this work to his memory, and especially his fun-loving approach to all things, including his mathematics, which contributed so much to the field of Riemannian foliations.

This work was supported in part by NSF grant DMS-0406254.

2. Classifying spaces

The universal Riemannian groupoid $R\Gamma_q$ is generated by the collection of all local isometries $\gamma\colon (U_\gamma, g'_\gamma) \to (V_\gamma, g''_\gamma)$ where g'_γ and g''_γ are complete Riemannian metrics on $\mathbb{R}^q$, and $U_\gamma, V_\gamma \subset \mathbb{R}^q$ are open subsets. Let $BR\Gamma_q$ denote the classifying space of the groupoid $R\Gamma_q$. The Hausdorff topological space $BR\Gamma_q$ is well-defined up to weak-homotopy equivalence.[14,15] If we restrict to orientation-preserving maps of $\mathbb{R}^q$, then we obtain the groupoid denoted by $R\Gamma_q^+$ with classifying space $BR\Gamma_q^+$.

The universal groupoid Γ_q of $\mathbb{R}^q$ is that generated by the collection of all local diffeomorphisms $\gamma\colon U_\gamma \to V_\gamma$ where $U_\gamma, V_\gamma \subset \mathbb{R}^q$ are open subsets. The realization of the groupoid Γ_q is a non-Hausdorff topological space $B\Gamma_q$, which is well-defined up to weak-homotopy equivalence.

An $R\Gamma_q$-structure on M is an open covering $\mathcal{U} = \{U_\alpha \mid \alpha \in \mathcal{A}\}$ of M and for each $\alpha \in \mathcal{A}$, there is given

- a smooth map $f_\alpha\colon U_\alpha \to V_\alpha \subset \mathbb{R}^q$
- a Riemannian metric g'_α on $\mathbb{R}^q$

such that the pull-backs $f_\alpha^{-1}(T\mathbb{R}^q) \to U_\alpha$ define a smooth vector bundle $Q \to M$ with Riemannian metric $g|Q = g_\alpha = f_\alpha^* g'_\alpha$. An $R\Gamma_q$-structure on M determines a continuous map $M \to |\mathcal{U}| \to BR\Gamma_q$.

Foliations $\mathcal{F}_0$ and $\mathcal{F}_1$ of codimension q of M are *integrably homotopic* if there is a foliation $\mathcal{F}$ of $M \times \mathbb{R}$ of codimension q such that $\mathcal{F}$ is everywhere transverse to the slices $M \times \{t\}$, so defines a foliation $\mathcal{F}_t$ of codimension-q of $M \times \{t\}$, and $\mathcal{F}_t$ of M_t agrees with $\mathcal{F}_t$ of M for $t = 0, 1$. This notion extends to Riemannian foliations, where we require that $\mathcal{F}$ defines a Riemannian foliation of codimension-q of $M \times \mathbb{R}$.

Theorem 2.1 (Haefliger[14,15]). *A Riemannian foliation* $(\mathcal{F}, g)$ *of* M *with oriented normal bundle defines an* $R\Gamma_q^+$*-structure on* M. *The* homotopy class of the composition $h_{\mathcal{F},g}\colon M \to B R \Gamma_q^+$ depends only on the integrable homotopy class of $(\mathcal{F}, g)$.

The derivative of a local isometry $\gamma\colon (U_\gamma, g'_\gamma) \to (V_\gamma, g''_\gamma)$ takes values in $\mathbf{SO}(q)$, and is functorial, so induces a classifying map $\nu\colon B R \Gamma_q^+ \to B\mathbf{SO}(q)$. The homotopy fiber of ν is denoted by $F R \Gamma_q$. The space $F R \Gamma_q$ classifies $R \Gamma_q^+$-structures with a (homotopy class of) framing for Q. Let $P \to M$ be the bundle of oriented orthonormal frames of $Q \to M$, and $s\colon M \to P$ a choice of framing of Q. Then we have the commutative diagram:

$$
\begin{array}{ccccc}
\mathbf{SO}(q) & = & \mathbf{SO}(q) & = & \mathbf{SO}(q) \\
\downarrow & & \downarrow & & \downarrow \\
P & \xrightarrow{h^s_{\mathcal{F},g}} & F R \Gamma_q & \xrightarrow{f} & F \Gamma_q \\
s \uparrow \downarrow & & \downarrow & & \downarrow \\
M & \xrightarrow{h_{\mathcal{F},g}} & B R \Gamma_q^+ & \xrightarrow{f} & B \Gamma_q^+ \\
& & \downarrow \nu & & \downarrow \nu \\
& & B\mathbf{SO}(q) & = & B\mathbf{SO}(q)
\end{array}
$$

where the right-hand column is the sequence of classifying spaces for the groupoid defined by the germs of local diffeomorphisms of $\mathbb{R}^q$. The natural maps $f\colon F R \Gamma_q \to F \Gamma_q$ and $f\colon B R \Gamma_q^+ \to B \Gamma_q^+$ are induced by the natural transformation which "forgets" the normal Riemannian metric data.

The approach to classifying foliations initiated by Haefliger in [14,15] is based on the study of the homotopy classes of maps $[M, B R \Gamma_q^+]$ from a manifold M without boundary to $B R \Gamma_q^+$. Given a homotopy class of an embedding of an oriented subbundle $Q \subset TM$ of dimension q, one studies the homotopy classes of maps $h_{\mathcal{F},g}\colon M \to B R \Gamma_q^+$ such that the composition $\nu \circ h_{\mathcal{F},g}\colon M \to B\mathbf{SO}(q)$ classifies the homotopy type of the subbundle Q. The "Haefliger classification" of Riemannian foliations is thus based on the study of the homotopy types of the spaces $B R \Gamma_q^+$ and $F R \Gamma_q$.

In the case of codimension-one, a Riemannian foliation with oriented normal bundle of M is equivalent to specifying a closed, non-vanishing 1-form ω on M. As $\mathbf{SO}(1)$ is the trivial group, $F R \Gamma_1^+ = B R \Gamma_1^+$, and the classifying map $M \to B R \Gamma_1^+$ is determined by the real cohomology class of ω, which follows from the following result of Joel Pasternack. Let $\mathbb{R}_\delta$ denote the real line, considered as a *discrete* group, and $B \mathbb{R}_\delta$ its classifying space.

Theorem 2.2 (Pasternack[45]). *There is a natural homotopy equivalence* $BR\Gamma_1^+ \simeq FR\Gamma_1 \simeq B\mathbb{R}_\delta$.

For codimension $q \geq 2$, $\mathbf{SO}(q)$ is not contractible, and the homotopy types of $BR\Gamma_q^+$ and $FR\Gamma_q$ are related by the above fibration sequence. For the space $FR\Gamma_q$ there is a partial generalization of Pasternack's Theorem.

Theorem 2.3 (Hurder[18,19]). *The space $FR\Gamma_q$ is $(q-1)$-connected. That is, $\pi_\ell(FR\Gamma_q) = \{0\}$ for $0 \leq \ell < q$. Moreover, the volume form associated to the transverse metric defines a surjection* vol: $\pi_q(FR\Gamma_q) \to \mathbb{R}$.

Proof. We just give a sketch; see[19] for details. Following a remark by Milnor, one observes that by the Phillips Immersion Theorem,[46-48] an $FR\Gamma_q$-structure on $\mathbb{S}^\ell$ for $0 < \ell < q$ corresponds to a Riemannian metric defined on an open neighborhood retract of the ℓ-sphere, $\mathbb{S}^\ell \subset U \subset \mathbb{R}^q$.

Given an $FR\Gamma_q$-structure on the open set $U \subset \mathbb{R}^q$ – which is equivalent to specifying a Riemannian metric on TU – one then constructs an explicit integrable homotopy through framed $R\Gamma_q$-structures on a smaller open neighborhood $\mathbb{S}^\ell \subset V \subset U$. The integrable homotopy starts with the given Riemannian metric on TV, and ends with the standard Euclidean metric on TV, which represents the "trivial" $FR\Gamma_q$-structure on $\mathbb{S}^\ell$. Thus, every $FR\Gamma_q$-structure on $\mathbb{S}^\ell$ is homotopic to the trivial structure.

The surjection vol: $\pi_q(FR\Gamma_q) \to \mathbb{R}$ is well-known, and is realized by varying the total volume of a Riemannian metric on $\mathbb{S}^q$, considered as foliated by points. □

Associated to the classifying map $\nu\colon BR\Gamma_q^+ \to B\mathbf{SO}(q)$ is the Puppe sequence
$$\cdots \longrightarrow \Omega FR\Gamma_q \xrightarrow{\Omega \nu} \Omega BR\Gamma_q^+ \longrightarrow \mathbf{SO}(q) \xrightarrow{\delta} FR\Gamma_q \longrightarrow BR\Gamma_q^+ \xrightarrow{\nu} B\mathbf{SO}(q) \quad (1)$$
In the case of codimension $q = 2$, $\mathbf{SO}(2) = \mathbb{S}^1$ and $FR\Gamma_2$ is 1-connected, so the map $\delta\colon \mathbf{SO}(2) \to FR\Gamma_2$ is contractible. This yields as an immediate consequence:

Theorem 2.4 (Hurder[21]). $\Omega BR\Gamma_2^+ \cong \mathbf{SO}(2) \times \Omega FR\Gamma_2$.

It is noted in[21] that the homotopy equivalence in Theorem 2.4 is not an H-space equivalence, as this would imply that map $\nu^*\colon H^*(B\mathbf{SO}(2); \mathbb{R}) \to H^*(BR\Gamma_2; \mathbb{R})$ is an injection, which is false. In contrast, we have the following result:

Theorem 2.5. *The connecting map* $\delta\colon \mathbf{SO}(q) \to F R\Gamma_q$ *in (1) is not homotopic to a constant for $q \geq 3$.*

Note that the map $\delta\colon \mathbf{SO}(q) \to F R\Gamma_q$ classifies the Riemannian foliation with standard framed normal bundle on $\mathbf{SO}(q) \times \mathbb{R}^q$, obtained via the pullback of the standard product foliation of $\mathbf{SO}(q) \times \mathbb{R}^q$ via the action of $\mathbf{SO}(q)$ on $\mathbb{R}^q$. Theorem 2.5 asserts that the canonical twisted foliation of $\mathbf{SO}(q) \times \mathbb{R}^q$ is not integrably homotopic through framed Riemannian foliations to the standard product foliation. This will be proven in section 4, using basic properties of the secondary classes for Riemannian foliations. For the non-Riemannian case, it is conjectured that the connecting map $\delta\colon \mathbf{SO}(q) \to F\Gamma_q$ is homotopic to the constant map.[23]

To close this discussion of general properties of the classifying spaces of Riemannian foliations, we pose a problem particular to codimension two:

Problem 2.1. *Prove that the map induced by the volume form* $\mathrm{vol}\colon \pi_2(FR\Gamma_2) \to \mathbb{R}$ *is an isomorphism. That is, given two $R\Gamma_2$-structures $\mathcal{F}_0$ and $\mathcal{F}_1$ on $M = \mathbb{R}^3 - \{0\}$, with homotopic normal bundles, prove that $\mathcal{F}_0$ and $\mathcal{F}_1$ are homotopic as $R\Gamma_2$-structures if and only if they have cohomologous transverse volume forms.*

One can view this as asking for a "transverse uniformization theorem" for Riemannian foliations of codimension two. Note that Example 5.2 below shows the conclusion of Problem 2.1 is false for $q = 3$.

3. Primary classes

The primary classes of a Riemannian foliation are those obtained from the cohomology of the classifying space of the normal bundle $Q \to M$, pulled-back via the classifying map $\nu\colon M \to B\mathbf{SO}(q)$. Recall[38] that the cohomology groups of $\mathbf{SO}(q)$ are isomorphic to free polynomial ring:

$$H^*(B\mathbf{SO}(2); \mathbb{Z}) \cong \mathbb{Z}[E_1]$$
$$H^*(B\mathbf{SO}(q); \mathbb{Z}) \cong \mathbb{Z}[E_m, P_1, \ldots, P_{m-1}]\ ,\ q = 2m \geq 4$$
$$H^*(B\mathbf{SO}(q); \mathbb{Z}) \cong \mathbb{Z}[P_1, \ldots, P_m]\ ,\ q = 2m+1 \geq 3$$

As usual, P_j denotes the Pontrjagin cohomology class of degree $4j$, E_m denotes the Euler class of degree $2m$, and the square $E_m^2 = P_m$ is the top degree generator of the Pontrjagin ring.

There are three main results concerning the universal map $\nu^*\colon H^\ell(B\mathbf{SO}(q); \mathcal{R}) \to H^\ell(BR\Gamma_q^+; \mathcal{R})$, where $\mathcal{R}$ is a coefficient ring, which we discuss in detail below.

Theorem 3.1 (Pasternack[45]). $\nu^* \colon H^\ell(B\mathbf{SO}(q); \mathbb{R}) \to H^\ell(B R\Gamma_q^+; \mathbb{R})$ is trivial for $\ell > q$.

Theorem 3.2 (Bott, Heitsch[5]). $\nu^* \colon H^*(B\mathbf{SO}(q); \mathbb{Z}) \to H^*(B R\Gamma_q^+; \mathbb{Z})$ is injective.

Theorem 3.3 (Hurder[19,22]). $\nu^* \colon H^\ell(B\mathbf{SO}(q); \mathbb{R}) \to H^\ell(B R\Gamma_q^+; \mathbb{R})$ is injective for $\ell \leq q$.

The contrast between Theorems 3.1 and 3.2 is one of the themes of this section, while the proof of Theorem 3.3 is based on an observation.

Let ∇_g denote the Levi-Civita connection on $Q \to M$ associated to the projectable metric g for $\mathcal{F}$. The Chern-Weil construction associates to each universal class P_j the closed Pontrjagin form $p_j(\nabla_g) \in \Omega^{4j}(M; \mathbb{R})$. For $q = 2m$, as Q is assumed to be oriented, there is also the Euler form $e_m(\nabla_g) \in \Omega^{2m}(M; \mathbb{R})$ whose square $e_m(\nabla_g)^2 = p_m(\nabla_g)$. The universal map $\nu^* \colon H^*(B\mathbf{SO}(q); \mathbb{R}) \to H^*(B R\Gamma_q^+; \mathbb{R})$ is defined by its values on foliated manifolds, where $\nu^*(P_j) = [p_j(\nabla_g)] \in H^{4j}(M; \mathbb{R})$, and $[\beta]$ represents the de Rham cohomology class of a closed form β.

Let m be the least integer such that $q \leq 2m + 2$. Given $J = (j_1, j_2, \ldots, j_m)$ with each $j_\ell \geq 0$, set $p_J = p_1^{j_1} \cdot p_2^{j_2} \cdots p_m^{j_m}$, which has degree $4|J| = 4(j_1 + \cdots + j_m)$. Let $\mathcal{P}$ denote a basis monomial: for $q = 2m+1$, it has the form $\mathcal{P} = p_J$ with $\deg(\mathcal{P}) = 4|J|$. For $q = 2m$, either $\mathcal{P} = p_J$ with $\deg(\mathcal{P}) = 4|J|$, or $\mathcal{P} = e_m \cdot p_J$ with $\deg(\mathcal{P}) = 4|J| + 2m$.

Pasternack[44] first observed in his thesis that the proof of the Bott Vanishing Theorem[4] can be strengthened in the case of Riemannian foliations, as the adapted metric ∇_g is projectable. He showed that on the level of differential forms, an analogue of the Bott Vanishing Theorem holds.

Theorem 3.4 (Pasternack[44,45]). If $\deg(\mathcal{P}) > q$ then $\mathcal{P}(\nabla_g) = 0$.

Theorem 3.1 follows immediately. Today, this result is considered "obvious", but that is due to the later extensive development of this field in the 1970's.

Next consider the injectivity of $\nu^* \colon H^k(B\mathbf{SO}(q); \mathbb{R}) \to H^k(B R\Gamma_q^+; \mathbb{R})$. We recall a basic observation of Thom.[38]

Theorem 3.5. *There is a compact, orientable Riemannian manifold B of dimension q such that all of the Pontrjagin and Euler classes up to degree q are independent in $H^*(B; \mathbb{R})$. If q is odd, then B can be chosen to be a connected manifold.*

Proof. For q even, let B equal the disjoint union of all products of the form $\mathbf{CP}^{i_1} \times \cdots \times \mathbf{CP}^{i_k} \times S^1 \times \cdots \times S^1$ with dimension q. For q odd, B is the connected sum of all products of the form $\mathbf{CP}^{i_1} \times \cdots \times \mathbf{CP}^{i_k} \times S^1 \times \cdots \times S^1$ with dimension q. The claim then follows by the Splitting Principle[38] for the Pontrjagin classes. □

Proof of Theorem 3.3. This now follows from the universal properties of $B R\Gamma_q^+$, as we endow the manifold B with the foliation $\mathcal{F}$ by points, with the standard Riemannian metric on B. □

The proof of Theorem 3.3 in[19] used the fact that $\nu\colon B R\Gamma_q^+ \to BSO(q)$ is q-connected.

Next, we discuss the results of Bott and Heitsch.[5] Let $K \subset \mathbf{SO}(q)$ be a closed Lie subgroup, and let $\Gamma \subset K$ be a finitely-generated subgroup. Suppose that B is a closed connected manifold, with basepoint $b_0 \in B$. Assume there is a surjection $\rho\colon \Lambda = \pi_1(B, b_0) \to \Gamma \subset K \subset \mathbf{SO}(q)$. Then via the natural action of $\mathbf{SO}(q)$ on $\mathbb{R}^q$ we obtain an action of Λ on $\mathbb{R}^q$. Let $\widetilde{B} \to B$ denote the universal covering of B, equipped with the right action of Λ by deck transformations. Then form the flat bundle

$$\mathbb{E}_\rho = \widetilde{B} \times \mathbb{R}^q / (b \cdot \gamma, \vec{v}) \sim (b, \rho(\gamma) \cdot \vec{v}) \xrightarrow{\pi} \widetilde{B}/\Lambda = B \qquad (2)$$

As the action of Λ on $\mathbb{R}^q$ preserves the standard Riemannian metric, we obtain a Riemannian foliation $\mathcal{F}_\rho$ on $\mathbb{E}_\rho$ whose leaves are the integral manifolds of the flat structure. The classifying map of the foliation $\mathcal{F}_\rho$ is given by the composition of maps

$$\mathbb{E}_\rho \to B\Lambda \to B(K_\delta) \to B(\mathbf{SO}(q)_\delta) \to B R\Gamma_q^+ \qquad (3)$$

where K_δ and $\mathbf{SO}(q)_\delta$ denotes the corresponding Lie groups considered with the discrete topology, and $B(K_\delta)$ and $B(\mathbf{SO}(q)_\delta)$ are the corresponding classifying spaces.

The Bott-Heitsch examples take K to be a maximal torus, so that for $q = 2m$ or $q = 2m+1$, we have $K = \mathbb{T}^m = \mathbf{SO}(2) \times \cdots \times \mathbf{SO}(2)$ with m factors. Consider first the case $q = 2$. For an odd prime p, let $\Gamma = \mathbb{Z}/p\mathbb{Z}$, embedded as the p-th roots of unity in $K = \mathbf{SO}(2)$. Let $B = \mathbb{S}^{2\ell+1}/\Gamma$ be the quotient of the standard odd-dimensional sphere, and consider the composition

$$\nu \circ \rho\colon B \to \mathbb{E}_\rho \to B(\mathbb{Z}/p\mathbb{Z}) \to B(\mathbf{SO}(2)_\delta) \to B R\Gamma_2^+ \xrightarrow{\nu} BSO(2) \qquad (4)$$

The composition $\nu \circ \rho$ classifies the Euler class of the flat bundle $\mathbb{E}_\rho \to B$, which is torsion. The map in cohomology with $\mathbb{Z}/p\mathbb{Z}$-coefficients,

$$(\nu \circ \rho)^* \colon H^*(B\mathbf{SO}(2); \mathbb{Z}/p\mathbb{Z}) \to H^*(B; \mathbb{Z}/p\mathbb{Z}) \tag{5}$$

is injective for $* \leq 2\ell$. It follows that the map

$$\nu^* \colon H^*(B\mathbf{SO}(2); \mathbb{Z}/p\mathbb{Z}) \to H^*(B R\Gamma_2^+; \mathbb{Z}/p\mathbb{Z}) \tag{6}$$

is injective in all degrees. As this holds true for all odd primes, it is also injective for integral cohomology. □

Theorem 3.2 is a striking result, as Theorem 3.1 implies that $\nu^* \colon H^*(B\mathbf{SO}(2); \mathbb{Q}) \to H^*(B R\Gamma_2^+; \mathbb{Q})$ is the trivial map for $* > 2$. One thus concludes from the Universal Coefficient Theorem for cohomology[5] that the homology groups $H_*(B R\Gamma_2^+; \mathbb{Z})$ cannot be finitely generated in all odd degrees $* \geq 3$.

The treatment of the cases where $q = 2m > 2$ and $q = 2m+1 > 2$ follows similarly, where one takes $\Gamma = (\mathbb{Z}/p\mathbb{Z})^m \subset \mathbb{T}^m \subset \mathbf{SO}(q)$, and let $p \to \infty$. An application of the splitting theorem for the Pontrjagin classes of vector bundles then yields Theorem 3.2.

The fibration sequence $FR\Gamma_q \to BR\Gamma_q^+ \to B\mathbf{SO}(q)$ yields a spectral sequence converging to the homology groups $H^*(BR\Gamma_q^+; \mathbb{Z})$ with E^2-term

$$E^2_{r,s} \cong H_r(B\mathbf{SO}(q); H_s(FR\Gamma_q; \mathbb{Z})) \tag{7}$$

It follows that the groups $H_s(FR\Gamma_q; \mathbb{Z})$ cannot all be finitely generated for odd degrees $* \geq q$. In fact, we will see that this follows from the results of Pasternack and Lazarov discussed in the next section on secondary classes, but the homology classes being detected via the torsion classes above seem to be of a different "sort" than those detected via the secondary classes.

Recall that the universal classifying map $f \colon FR\Gamma_q \to F\Gamma_q$ "forgets" the added structure of a holonomy-invariant transverse Riemannian metric for the foliation. It has been conjectured (see page 308[41]) that this map induces the trivial map in homotopy.

Conjecture 3.1. $f_\# \colon \pi_k(FR\Gamma_q) \to \pi_k(F\Gamma_q)$ *is trivial for all* $k > 0$.

The ideas of the proof of Theorem 3.2 imply that Conjecture 3.1 is false.

Theorem 3.6. *The image of* $f_* \colon H_{4k-1}(BR\Gamma_q; \mathbb{Z}) \to H_{4k-1}(B\Gamma_q; \mathbb{Z})$ *is infinite-dimensional for* $4k > 2q \geq 4$.

Proof. Our approach uses the homological methods of the proof of Theorem 3.2 in[5] and especially the commutative diagram from page 144.

Let $\mathcal{P} \in H^{4k}(BSO(q); \mathbb{Z})$ for $4k > q$ be a generating monomial. The Bott-Heitsch Theorem 3.2 implies that the image $f^* \circ \nu^*(\mathcal{P}) \in H^{4k}(BR\Gamma_q^+; \mathbb{Z})$ is not a torsion class under the composition

$$H^{4k}(BSO(q); \mathbb{Z}) \xrightarrow{\nu^*} H^{4k}(B\Gamma_q^+; \mathbb{Z}) \xrightarrow{f^*} H^{4k}(BR\Gamma_q^+; \mathbb{Z}) \quad (8)$$

Let $\mathcal{A}_{4k-1} = \text{image}\{H_{4k-1}(BR\Gamma_q^+; \mathbb{Z}) \to H_{4k-1}(B\Gamma_q^+; \mathbb{Z})\}$. Suppose that $\mathcal{A}_{4k-1}$ is finite-dimensional, then $\text{Ext}(\mathcal{A}_{4k-1}, \mathbb{Z})$ is a torsion group. Consider the commutative diagram:

$$
\begin{array}{ccccc}
& & H^{4k}(BSO(q); \mathbb{Z}) & & \\
& & \downarrow \nu^* & & \\
\text{Ext}(H_{4k-1}(B\Gamma_q^+; \mathbb{Z}), \mathbb{Z}) & \xrightarrow{\tau} & H^{4k}(B\Gamma_q^+; \mathbb{Z}) & \xrightarrow{e} & \text{Hom}(H_{4k}(B\Gamma_q^+; \mathbb{Z}), \mathbb{Z}) \\
\downarrow \iota^* & & & & \\
\text{Ext}(\mathcal{A}_{4k-1}, \mathbb{Z}) & & \downarrow f^* & & \downarrow f^* \\
\downarrow \sigma^* & & & & \\
\text{Ext}(H_{4k-1}(BR\Gamma_q^+; \mathbb{Z}), \mathbb{Z}) & \xrightarrow{\tau} & H^{4k}(BR\Gamma_q^+; \mathbb{Z}) & \xrightarrow{e} & \text{Hom}(H_{4k}(BR\Gamma_q^+; \mathbb{Z}), \mathbb{Z})
\end{array}
$$

In the diagram, e is the evaluation map of cohomology on homology, and τ maps onto its kernel. The inclusion $\iota \colon \mathcal{A}_{4k-1} \subset H_{4k-1}(B\Gamma_q^+; \mathbb{Z})$ induces the map ι^*, and the surjection $\sigma \colon H_{4k-1}(BR\Gamma_q^+; \mathbb{Z}) \to \mathcal{A}_{4k-1}$ induces σ^*.

The Bott Vanishing Theorem implies that the class

$$e \circ \nu^*(\mathcal{P}) \in \text{Hom}(H_{4k}(B\Gamma_q^+; \mathbb{Z}), \mathbb{Z}) \subset \text{Hom}(H_{4k}(B\Gamma_q^+; \mathbb{Z}), \mathbb{Q})$$

is trivial for $\deg(\mathcal{P}) > 2q$. Thus, there exists $\mathcal{P}_\tau \in \text{Ext}(H_{4k-1}(B\Gamma_q^+; \mathbb{Z}), \mathbb{Z})$ such that $\tau(\mathcal{P}_\tau) = \nu^*(\mathcal{P})$. The class $\iota^*(\mathcal{P}_\tau) \in \text{Ext}(\mathcal{A}_{4k-1}, \mathbb{Z})$ is torsion, by the assumption on $\mathcal{A}_{4k-1}$. Thus, $f^* \circ \nu^*(\mathcal{P}) = \tau \circ \sigma^* \circ \iota^*(\mathcal{P}_\tau)$ is a torsion class, which contradicts the Bott-Heitsch results. Thus, $\mathcal{A}_{4k-1}$ cannot be finite-dimensional for $4k > 2q$. $\square$

Corollary 3.1. *The image of $f_* \colon H_{4k-1}(FR\Gamma_q; \mathbb{Z}) \to H_{4k-1}(F\Gamma_q; \mathbb{Z})$ is infinite-dimensional for $4k > 2q \geq 4$.*

Proof. This follows from the commutative diagram following Theorem 2.1, the functorial properties of the spectral sequence (7), the fact that $H_*(BSO(q); \mathbb{Z})$ is finitely generated in all degrees, and Theorem 3.6. $\square$

Problem 3.1. *Find geometric interpretations of the cycles in the image of the map* $f_*\colon H_{4k-1}(F R\Gamma_q;\mathbb{Z}) \to H_{4k-1}(F\Gamma_q;\mathbb{Z})$.

The construction of foliations with solenoidal minimal sets in[10,26] give one realization of some of the classes in the image of this map, as discussed in the talk by the author[25] at the conference of these Proceedings. Neither these examples,[26] nor the situation overall, is understood in sufficient depth.

4. Secondary classes

Assume that $(\mathcal{F}, g)$ is a Riemannian foliation of codimension q. We also assume that there exists a framing $s\colon M \to P$ of the normal bundle. Then the data $(\mathcal{F}, g, s)$ yields a classifying map $h^s_{\mathcal{F},g}\colon M \to FR\Gamma_q$. In this section, we discuss the construction of the secondary characteristic classes of such foliations, constructed using the Chern-Weil method,[8] and some of the results about these classes.

Recall that ∇_g denotes the Levi-Civita connection of the projectable metric g on Q.

Let $\mathcal{I}(\mathbf{SO}(q))$ denote the ring of Ad-invariant polynomials on the Lie algebra $\mathfrak{so}(q)$ of $\mathbf{SO}(q)$. Then we have

$$\mathcal{I}(\mathbf{SO}(2)) \cong \mathbb{R}[e_m]$$
$$\mathcal{I}(\mathbf{SO}(2m)) \cong \mathbb{R}[e_m, p_1, \ldots, p_{m-1}] \,, \ q = 2m \geq 4$$
$$\mathcal{I}(\mathbf{SO}(2m+1)) \cong \mathbb{R}[p_1, \ldots, p_m] \,, \ q = 2m+1 \geq 3$$

where the p_j are the Pontrjagin polynomials, and e_m is the Euler polynomial defined for q even.

The symmetric polynomials p_j evaluated on the curvature matrix of 2-forms associated to the connection ∇_g yields closed forms $\Delta_{\mathcal{F},g}(p_j) = p_\ell(\nabla_g) \in \Omega^{4j}(M)$. Then $\Delta_{\mathcal{F}}[p_j] = [p_j(\nabla_g)] \in H^{4j}(M;\mathbb{R})$ represents the Pontrjagin class $P_j(Q)$. The Euler form $\Delta_{\mathcal{F},g}(e_m) = e_m(\nabla_g) \in \Omega^{2m}(M)$ and the Euler class $\Delta_{\mathcal{F},g}[e_m] \in H^{2m}(M,\mathbb{R})$ are similarly defined when $q = 2m$. We thus obtain a multiplicative homomorphism

$$\Delta_{\mathcal{F},g}\colon \mathcal{I}(\mathbf{SO}(q)) \to H^*(M;\mathbb{R})$$

As noted in Theorem 3.4, Pasternack first observed that for ∇_g the adapted connection to a Riemannian foliation, the map $\Delta_{\mathcal{F},g}$ vanishes identically in degrees greater than q.

Definition 4.1. For $q = 2m$, set

$$\mathcal{I}(\mathbf{SO}(q))_{2m} \equiv \mathbb{R}[e_m, p_1, p_2, \ldots, p_{m-1}]/(\mathcal{P} \mid \deg(\mathcal{P}) > q)$$

For $q = 2m + 1$, set

$$\mathcal{I}(\mathbf{SO}(q))_{2m+1} \equiv \mathbb{R}[p_1, p_2, \ldots, p_m]/(\mathcal{P} \mid \deg(\mathcal{P}) > q)$$

Corollary 4.1 (Pasternack). *Let $(\mathcal{F}, g)$ be a Riemannian foliation of M with codimension q. Then there is a characteristic homomorphism $\Delta_{\mathcal{F},g}: \mathcal{I}(\mathbf{SO}(q))_q \to H^*(M; \mathbb{R})$, which is functorial for transversal maps between foliated manifolds.*

Of course, if we assume that the normal bundle Q is trivial, then this map is zero in cohomology. The point of the construction of secondary classes is to obtain geometric information from the forms $p_j(\nabla_g) \in \Omega^{4j}(M)$, even if they are exact. If we do not assume that Q is trivial, then one still knows that the cohomology classes $[p_j(\nabla_g)] \in H^{4j}(M; \mathbb{R})$ lie in the image of the integral cohomology, $H^*(M; \mathbb{Z}) \to H^*(M; \mathbb{R})$ so that one can use the construction of Cheeger-Simons differential characters as in[7,9,31,52] to define secondary invariants in the groups $H^{4j-1}(M; \mathbb{R}/\mathbb{Z})$. These classes are closely related to the Bott-Heitsch examples above, and to the secondary classes constructed below.

Given a trivialization $s: M \to P$, let ∇_s be the flat connection on Q for which s is parallel. Set $\nabla_t = t\nabla_g + (1-t)\nabla_s$, which we consider as a connection on the bundle Q extended as product over $M \times \mathbb{R}$. Then the Pontrjagin forms for ∇_t yield closed forms $p_j(\nabla_t) \in \Omega^{4j}(M \times \mathbb{R})$. Define the $4j - 1$ degree transgression form

$$h_j = h_j(\nabla_g, s) = \int_0^1 \{\iota(\partial/\partial t) p_j(\nabla_t)\} \wedge dt \ \in \Omega^{4j-1}(M) \qquad (9)$$

which satisfies the coboundary relation on forms:

$$dh_j(\nabla_g, s) = p_j(\nabla_g) - p_j(\nabla_s) = p_j(\nabla_g)$$

For $q = 2m$ we also introduce the transgression of the Euler form,

$$\chi_m = \chi_m(\nabla_g, s) = \int_0^1 \{\iota(\partial/\partial t) e_m(\nabla_t)\} \wedge dt \ \in \Omega^{q-1}(M) \qquad (10)$$

which satisfies the coboundary equation $d\chi_m = e_m(\nabla_g)$. Note that if $4j > q$, then the form $p_j(\nabla_g) = 0$, so the transgression form h_j is closed. The cohomology class $[h_j] = \Delta^s_{\mathcal{F},g}(h_j) \in H^{4j-1}(M; \mathbb{R})$ is said to be a *secondary cohomology class*. In general, introduce the graded differential complexes:

$$\begin{aligned} RW_{2m} &= \Lambda(h_1, \ldots, h_{m-1}, \chi_m) \otimes \mathcal{I}(\mathbf{SO}(q))_{2m} \\ RW_{2m+1} &= \Lambda(h_1, \ldots, h_m) \otimes \mathcal{I}(\mathbf{SO}(q))_{2m+1} \end{aligned}$$

where $d_W(h_j \otimes 1) = 1 \otimes p_j$ and $d_W(\chi_m \otimes 1) = e_m \otimes 1$. For $I = (i_1 < \cdots < i_\ell)$ and $J = (j_1 \leq \cdots \leq j_k)$ set

$$h_I \otimes p_J = h_{i_1} \wedge \cdots \wedge h_{i_\ell} \otimes p_{j_1} \wedge \cdots \wedge p_{j_k} \qquad (11)$$

Note that $\deg(h_I \otimes p_J) = 4(|I|+|J|)-\ell$, and that $d_W(h_I \otimes p_J) = 0$ exactly when $4i_1 + 4|J| > q$. In the following, the expression $h_I \otimes p_J$ will always assume that the indexing sets I and J are ordered as above.

Theorem 4.1 (Lazarov - Pasternack[34]). *Let $(\mathcal{F},g)$ be a Riemannian foliation of codimension $q \geq 2$ of a manifold M without boundary, and assume that there is given a framing of the normal bundle, $s\colon M \to P$. Then the above constructions yield a map of differential graded algebras*

$$\Delta^s_{\mathcal{F},g}\colon RW_q \to \Omega^*(M) \qquad (12)$$

such that the induced map on cohomology, $\Delta^s_{\mathcal{F},g}\colon H^*(RW_q) \to H^*(M;\mathbb{R})$, is independent of the choice of basic connection ∇_g, and depends only on the integrable homotopy class of $\mathcal{F}$ as a Riemannian foliation and the homotopy class of the framing s.

This construction can also be recovered from the method of truncated Weil algebras applied to the Lie algebra $\mathfrak{so}(q)$ (see Kamber and Tondeur[29,30]). The functoriality of the construction of $\Delta^s_{\mathcal{F},g}$ implies, in the usual way:[32,34]

Corollary 4.2. *There exists a universal characteristic homomorphism*

$$\Delta\colon H^*(RW_q, d_W) \to H^*(F R \Gamma_q; \mathbb{R}) \qquad (13)$$

There are many natural questions about how the values of these secondary classes are related to the geometry and dynamical properties of the foliation $(\mathcal{F}, g, s)$. We discuss some known results in the following.

First, consider the role of the section $s\colon M \to P$. Given any smooth map $\varphi\colon M \to \mathbf{SO}(q)$, we obtain a new framing $s' = s \cdot \varphi \colon M \to P$ by setting $s'(x) = s(x) \cdot \varphi(x)$. Thus, φ can be thought of as a gauge transformation of the normal bundle $Q \to M$.

The cohomology of the Lie algebra $\mathfrak{so}(q)$ is isomorphic to an exterior algebra, generated by the cohomology classes of left-invariant closed forms $\tau_j \in \Lambda^{4j-1}(\mathfrak{so}(q))$ for $j < q/2$, and the Euler form $\chi_m \in \Lambda^{2m-1}(\mathfrak{so}(q))$ when $q = 2m$. The map φ pulls these back to closed forms $\varphi^*(\tau_j) \in \Omega^{4j-1}(M)$.

Theorem 4.2 (Lazarov[33,34]). *Suppose that two framings s, s' of Q are related by a gauge transformation $\varphi\colon M \to \mathbf{SO}(q)$, $s' = s \cdot \varphi$. Then on the level of forms,*

$$\Delta^{s'}_{\mathcal{F},g}(h_j) = \Delta^s_{\mathcal{F},g}(h_j) + \varphi^*(\tau_j) \qquad (14)$$

In particular, for $j > q/4$, we have the relation in cohomology

$$\Delta_{\mathcal{F},g}^{s'}[h_j] = \Delta_{\mathcal{F},g}^{s}[h_j] + \varphi^*[\tau_j] \in H^{4j-1}(M;\mathbb{R}) \qquad (15)$$

The relation (14) can be used to easily calculate exactly how the cohomology classes $\Delta_{\mathcal{F},g}^{s}[h_I \otimes p_J]$ and $\Delta_{\mathcal{F},g}^{s'}[h_I \otimes p_J]$ associated to framings s, s' are related. (See §4,[34] and[33] for details.) Here is one simple application of Theorem 4.2:

Proof of Theorem 2.5. For the product foliation of $\mathbf{SO}(q) \times \mathbb{R}^q$ we have a natural identification of the transverse orthogonal frame bundle $P = \mathbf{SO}(q) \times \mathbf{SO}(q)$. Let $s: \mathbf{SO}(q) \to P$ be the map $s(x) = x \times \{Id\}$, called the product framing. Then the map $\Delta_{\mathcal{F},g}^{s}: RW_q \to \Omega^*(M)$ is identically zero.

On the other hand, the connecting map $\delta: \mathbf{SO}(q) \to F R \Gamma_q$ in (1) classifies the Riemannian foliation $\mathcal{F}_\delta$ of $\mathbf{SO}(q) \times \mathbb{R}^q$, obtained via the pull-back of the standard product foliation of $\mathbf{SO}(q) \times \mathbb{R}^q$ via the action of $\mathbf{SO}(q)$ on $\mathbb{R}^q$. However, the normal framing of $\mathcal{F}_\delta$ is the product framing on $\mathbf{SO}(q) \times \mathbb{R}^q$. Let $\varphi: \mathbf{SO}(q) \to \mathbf{SO}(q)$ be defined by $\varphi(x) = x^{-1}$ for $x \in \mathbf{SO}(q)$. Then $\mathcal{F}_\delta$ is diffeomorphic to the product foliation of $\mathbf{SO}(q) \times \mathbb{R}^q$ with the framing defined by the gauge action of φ.

It follows from Theorem 4.2 that for $j > q/4$, $\Delta_{\mathcal{F},g}^{s'}[h_j] = \varphi^*[\tau_j] = \pm \tau_j \in H^{4j-1}(\mathbf{SO}(q);\mathbb{R})$ is a generator. Hence, the connecting map $\delta: \mathbf{SO}(q) \to F R \Gamma_q$ cannot be homotopic to the identity if there exists $j > q/4$ such that $\tau_j \in H^{4j-1}(\mathbf{SO}(q);\mathbb{R})$ is non-zero. This is the case for all $q > 2$. □

The original Chern-Simons invariants of 3-manifolds[9] can be considered as examples of the above constructions. Let M be a closed oriented, connected 3-manifold with Riemannian metric g. Consider M as foliated by points, then we obtain a Riemannian foliation of codimension 3. Choose an oriented framing $s: M \times \mathbb{R}^3 \to TM$, then the transgression form $\Delta_{\mathcal{F},g}^{s}(h_1) \in H^3(M;\mathbb{R}) \cong \mathbb{R}$ is well-defined. Note that by formula (15), the mod $\mathbb{Z}$-reduction $\overline{\Delta_{\mathcal{F},g}^{s}(h_1)} \in H^3(M;\mathbb{R}/\mathbb{Z}) \cong \mathbb{R}/\mathbb{Z}$ is then independent of the choice of framing. This invariant of the metric is just the Chern-Simons invariant.[9] for (M,g). On the other hand, Atiyah showed[1] that for a 3-manifold, there is a "canonical" choice of framing s_0 for TM, so that there is a canonical $\mathbb{R}$-valued Chern-Simons invariant, $\Delta_{\mathcal{F},g}^{s_0}(h_1) \in \mathbb{R}$.

Chern and Simons[9] also show that the values of $\overline{\Delta_{\mathcal{F},g}^{s}(h_1)} \in \mathbb{R}/\mathbb{Z}$ can vary non-trivially with the choice of Riemannian metric.

One of the standard problems in foliation theory, is to determine whether the universal characteristic map is injective. For the classifying space $B\Gamma_q$ of smooth foliations, this remains one of the outstanding open problems.[24] In contrast, for Riemannian foliations, the universal map (13) is injective. We present here a new proof of this, based on Theorem 3.5.

Theorem 4.3 (Hurder[19]). *There exists a compact manifold M and a Riemannian foliation $\mathcal{F}$ of M with trivial normal bundle, such that $\mathcal{F}$ is defined by a fibration over a compact manifold of dimension q, and the characteristic map $\Delta^s_{\mathcal{F},g}\colon H^*(RW_q) \to H^*(M)$ is injective. Moreover, if q is odd, then M can be chosen to be connected.*

Proof. Let B be the compact, oriented Riemannian manifold defined in the proof Theorem 3.5. Let M be the bundle of oriented orthonormal frames for TB. The basepoint map $\pi\colon M \to B$ defines a fibration $\mathbf{SO}(q) \to M \to B$, whose fiber $L_x = \pi^{-1}(x)$ over $x \in B$ is the group $\mathbf{SO}(q)$ of oriented orthonormal frames in T_xB. Let $\mathcal{F}$ be the foliation defined by the fibration. The Riemannian metric on B lifts to the transverse metric on the normal bundle $Q = \pi^*TB$. The bundle Q has a canonical framing s, where for $b \in B$ and $A \in \mathbf{SO}(q)$ the framing of $Q_{x,A}$ is that defined by the matrix A.

The normal bundle restricted to L_x is trivial, as it is just the constant lift of T_xB. That is, $Q|L_x \cong \pi^*(T_xB) \cong L_x \times \mathbb{R}^q$. The basic connection ∇_g restricted to $Q|L_x$ is the connection associated to the product bundle $L_x \times \mathbb{R}^q$. However, the canonical framing of $Q \to M$ restricted to $Q|L_x$ is twisted by $\mathbf{SO}(q)$. Thus, the connection ∇_s on Q for which the canonical framing is parallel, restricts to the Maurer-Cartan form on $\mathbf{SO}(q) \times \mathbb{R}^q$ along each fiber L_x.

By Chern-Weil theory, the forms $\Delta^s_{\mathcal{F},g}(h_j) = h_j(\nabla_g, s)$ restricted to $L_x = \mathbf{SO}(q)$ are closed, and their classes in cohomology define the free exterior generators for the cohomology $H^*(\mathbf{SO}(q);\mathbb{R})$. (In the even case $q = 2m$, one must include the Euler class χ_m as well.)

Give the algebra RW_q the basic filtration by the degree in $\mathcal{I}(\mathbf{SO}(q))_q$, and the forms in $\Omega^*(M)$ the basic filtration by their degree in $\pi^*\Omega^*(B)$. (See[30] for example.) The characteristic map $\Delta^s_{\mathcal{F},g}$ preserves the filtrations, hence induces a map of the associated Leray-Hirsch spectral sequences,

$$\Delta^{*,*}_r\colon E^{*,*}_r(RW_q, d_W) \to E^{*,*}_r(M, d_r)$$

For $r = 2$, we then have

$$\Delta^{*,*}_2\colon E^{*,*}_2(RW_q) \cong (RW_q, d_W) \to E^{*,*}_r(M, d_2) \cong H^*(\mathbf{SO}(q);\mathbb{R}) \otimes H^*(B;\mathbb{R})$$

which is injective by the remark above. Pass to the E_∞-limit to obtain that $\Delta^s_{\mathcal{F},g}\colon H^*(RW_q) \to H^*(M)$ induces an injective map of associated graded algebras, hence is injective. □

It seems to be an artifact of the proof that for $q \geq 4$ even, the manifold M we obtain is not connected.

Problem 4.1. *For $q \geq 4$ even, does there exists a closed, connected manifold M and a Riemannian foliation $\mathcal{F}$ of M of codimension-q and trivial normal bundle, such that the secondary characteristic map $\Delta^s_{\mathcal{F},g} \colon H^*(RW_q) \to H^*(M)$ injects? Is there a cohomological obstruction to the existence of such an example?*

Note that in the examples constructed in the proof of Theorem 4.3, the image of the monomials $h_I \otimes p_J$ for $4i_1 + 4|J| > q$ (and hence $h_I \otimes p_J$ is d_W-closed) are integral:

$$\Delta^s_{\mathcal{F},g}[h_I \otimes p_J] \in \text{Image} \{H^*(M, \mathbb{Z}) \to H^*(M, \mathbb{R})\}$$

This follows since the restriction of the forms $\Delta^s_{\mathcal{F},g}(h_I)$ to the leaves of $\mathcal{F}$ are integral cohomology classes. In general, one cannot expect a similar integrality result to hold for examples with all leaves compact, as is shown by the Chern-Simons example previously mentioned. However, a more restricted statement holds.

Definition 4.2. A foliation $\mathcal{F}$ of a manifold M is *compact Hausdorff* if every leaf of $\mathcal{F}$ is a compact manifold, and the leaf space $M/\mathcal{F}$ is a Hausdorff space.

Theorem 4.4 (Epstein,[12] Millett[37]). *A compact Hausdorff foliation $\mathcal{F}$ admits a holonomy-invariant Riemannian metric on its normal bundle Q.*

In the next section, we discuss the division of the secondary classes into "rigid" and "variable" classes. One can show the following:

Theorem 4.5. *Let $\mathcal{F}$ be a compact Hausdorff foliation of codimension q of M with trivial normal bundle. If $h_I \otimes p_J$ is a rigid class, then*

$$\Delta^s_{\mathcal{F},g}[h_I \otimes p_J] \in \text{Image} \{H^*(M, \mathbb{Q}) \to H^*(M, \mathbb{R})\}$$

This follows for the case when the leaf space $M/\mathcal{F}$ is a smooth manifold from[18] whose methods extend to this more general situation.

It is an interesting problem to find geometric conditions on a Riemannian foliation which imply the rationality of the secondary classes.[11] Rationality should be associated to rigidity properties for the global holonomy of the leaf closures, one of the fundamental geometric concepts in the Molino Structure theory discussed in §6. One expects rationality results for the secondary classes analogous to the celebrated results of Reznikov,[50,51] possibly with some additional assumptions on the geometry of the leaves.

5. Variation of secondary classes

The secondary classes of a foliation are divided into two types, the "rigid" and the "variable" classes. Examples show that the variable classes are sensitive to both the geometry and dynamical properties of the foliation, while the rigid classes seem to be topological in nature.

A monomial $h_I \otimes p_J \in RW_q$ is said to be *rigid* if $\deg(p_{i_1} \wedge p_J) > q + 2$. Note that if $4i_1 + 4|J| > q$, then this condition is automatically satisfied when $q = 4k$ or $q = 4k + 1$. Here is the key property of the rigid classes:

Theorem 5.1 (Lazarov and Pasternack, Theorem 5.5[34]). *Let $(\mathcal{F}_t, g_t, s_t)$ be a smooth 1-parameter family of framed Riemannian foliations. Let $h_I \otimes p_J \in RW_q$ be a rigid class. Then*

$$\Delta^{s_0}_{\mathcal{F}_0, g_0}[h_I \otimes p_J] = \Delta^{s_1}_{\mathcal{F}_1, g_1}[h_I \otimes p_J] \in H^*(M; \mathbb{R})$$

Note that the family $\{(\mathcal{F}_t, g_t) \mid 0 \le t \le 1\}$ need not be a Riemannian foliation of codimension-q of $M \times [0, 1]$.

For the special case where $q = 4k - 2 \ge 6$, a stronger form of the above result is true:

Theorem 5.2 (Lazarov and Pasternack, Theorem 5.6[34]).
Let $(\mathcal{F}, g_t, s_t)$ be a smooth 1-parameter family, where $\mathcal{F}$ is a fixed foliation of codimension q, each g_t is a holonomy invariant Riemannian metric on Q, and s_t is a smooth family of framings on Q. Let $h_I \otimes p_J \in RW_q$ satisfy $\deg(p_{i_1} \wedge p_J) > q + 1$. Then

$$\Delta^{s_0}_{\mathcal{F}, g_0}[h_I \otimes p_J] = \Delta^{s_1}_{\mathcal{F}, g_1}[h_I \otimes p_J] \in H^*(M; \mathbb{R})$$

We say that these classes are metric rigid. *Thus, the classes $[h_I \otimes p_J] \in H^*(RW_q)$ are metric rigid when $\deg(h_{i_1} \otimes p_J) > q$, and rigid under all deformations when $\deg(h_{i_1} \otimes p_J) > q + 1$.*

A closed monomial $h_I \otimes p_J$ which is not rigid, is said to be *variable*. In the special case $q = 2$, the class $[\chi_1 \otimes e_1] \in H^3(RW_2)$ is variable. For $q > 2$, neither the Euler class e_m or its transgression χ_m can occur in a variable class, so for $q = 4k - 2$ or $q = 4k - 1$, the variable classes are spanned by the closed monomials

$$\mathcal{V}_q = \{h_I \otimes p_J \mid 4i_1 + 4|J| = 4k\} \tag{16}$$

Let v_q^k denote the dimension of the subspace of $H^k(RW_q)$ spanned by the variable monomials.

Theorem 5.2 implies that for codimension $q = 4k - 2 \geq 6$, in order to continuously vary the value of a variable class $h_I \otimes p_J$ it is necessary to deform the underlying foliation. For $q = 4k - 1$, the value of variable class may (possibly) be continuously varied by simply changing the transverse metric for the foliation. We illustrate this with two examples.

Example 5.1 (Chern-Simons, Example 2 in §6[9]). *Consider* $\mathbb{S}^3$ *as the Lie group* **SU**(2) *with Lie algebra spanned by*

$$X = \begin{bmatrix} i & 0 \\ 0 & -i \end{bmatrix}, Y = \begin{bmatrix} 0 & i \\ i & 0 \end{bmatrix}, Z = \begin{bmatrix} 0 & -1 \\ 1 & 0 \end{bmatrix}$$

which gives a framing s of $T\mathbb{S}^3$. Let g_u be the Riemannian metric on $\mathbb{S}^3$ for which the parallel Lie vector fields $\{u \cdot X, Y, Z\}$ are an orthonormal basis. Let $\mathcal{F}$ denote the point-foliation of $\mathbb{S}^3$. Then $[h_1] \in H^3(RW_3)$ and for each $u > 0$, we have $\Delta^s_{\mathcal{F},g_u}[h_1] \in H^3(\mathbb{S}^3;\mathbb{R}) \cong \mathbb{R}$.

Theorem 5.3 (Theorem 6.9[9]). $\frac{d}{du}|_{u=1}\left(\Delta^s_{\mathcal{F},g_u}[h_1]\right) \neq 0.$

One expects similar results also hold for other compact Lie groups of dimension $4k - 1 \geq 7$, although the author does not know of a published calculation of this.

Chern and Simons also prove a fundamental fact about the conformal rigidity of the transgression classes, and as their calculations are all local, the result carries over to Riemannian foliations:

Theorem 5.4 (Theorem 4.5[9]). *The rigid secondary classes in codimension $q = 4k-1$ are conformal invariants. That is, let $(\mathcal{F}, g)$ be a Riemannian foliation of codimension $q = 4k - 1$ of the closed manifold M. Let s be a framing of the normal bundle Q. Let $\mu \colon M \to \mathbb{R}$ be a smooth function, which is constant along the leaves of $\mathcal{F}$. Define a conformal deformation of g by setting $g_t = \exp(\mu(t)) \cdot g$. Then for all $[h_I \otimes p_J] \in H^*(RW_q, d_W)$ with $4i_1 + 4|J| = q + 1$,*

$$\Delta^s_{\mathcal{F},g_t}[h_I \otimes p_J] = \Delta^s_{\mathcal{F},g}[h_I \otimes p_J] \in H^*(M;\mathbb{R})$$

Combining Theorems 5.1, 5.2 and 5.4 we obtain:

Corollary 5.1. *The secondary classes of Riemannian foliations are conformal invariants.*

A modification of the original examples of Bott[3] and Baum-Cheeger[2] show that all of the variable secondary classes vary independently, by a suitable variation of foliations.

Example 5.2 (Lazarov-Pasternack[35]). *Let $\alpha = (\alpha_1, \ldots, \alpha_{2k}) \in \mathbb{R}^{2k}$. Let $(x_1, y_1, x_2, y_2, \ldots, x_{2k}, y_{2k})$ denote coordinates on $\mathbb{R}^{4k}$, and define a Killing vector field X_α on $\mathbb{R}^{4k}$ by*

$$X_\alpha = \sum_{i=1}^{k} \alpha_i \{ x_i \partial/\partial y_i - y_i \partial/\partial x_i \}$$

Let $\phi_t^\alpha \colon \mathbb{R}^{4k} \to \mathbb{R}^{4k}$ be the isometric flow of X_α, which restricts to an isometric flow on the unit sphere $\mathbb{S}^{4k-1}$, so defines a Riemannian foliation $\mathcal{F}_\alpha$ of codimension $q = 4k - 2$ of $\mathbb{S}^{4k-1}$.

Let $h_i \otimes p_J$ satisfy $4i + 4|J| = 4k$. Associated to $p_i \wedge p_J$ is an Ad-invariant polynomial $\varphi_{i,J}$ on $\mathfrak{so}(4k)$ of degree $2k$. Let $M \to \mathbb{S}^{4k-1}$ denote the bundle of orthonormal frames for the normal bundle to $\mathcal{F}_\alpha$, for α near $0 \in \mathbb{R}^{2k}$. The spectral sequence for $\mathbf{SO}(4k-2) \to M \to \mathbb{S}^{4k-1}$ collapses at the $E_2^{r,s}$-term, hence $H^*(M; \mathbb{R}) \cong H^*(\mathbb{S}^{4k-1}, \mathbb{R}) \otimes H^*(\mathbf{SO}(4k-2); \mathbb{R})$. Let $[C] \in H^{4k-1}(M, \mathbb{R})$ correspond to the fundamental class of the base.

Theorem 5.5 (§§2 & 3[35]). *There exists $\lambda \neq 0$ independent of the choice of $p_i \wedge p_J$ such that*

$$\langle \Delta^s_{\mathcal{F}_\alpha, g}[h_i \otimes p_J], [C] \rangle = \lambda \cdot \frac{\varphi_{i,J}(\alpha_1, \ldots, \alpha_{2k})}{\alpha_1 \cdots \alpha_{2k}} \qquad (17)$$

These examples are for $q = 4k - 2$. Multiplying by a factor of $\mathbb{S}^1$ in the transverse direction yields examples with codimension $4k - 1$, and the same secondary invariants. Hence, we have the following corollary, due to Lazarov and Pasternack:

Corollary 5.2 (Theorem 3.6[35]). *Let $q = 4k - 2$ or $4k - 1$. Evaluation on a basis of $H^{4k-1}(RW_q; d_W)$ defines a surjective map*

$$\pi_{4k-1}(B R \Gamma_q^+) \to \mathbf{R}^{v_q^{4k-1}} \qquad (18)$$

In particular, all of the variable secondary classes in degree $4k - 1$ vary independently.

Although not stated by Lazarov and Pasternack,[35] these examples also imply that all of the variable secondary classes for Riemannian foliations vary independently, as stated in Theorem 4[20].

The papers[19,30,34–36,53,54] contain a more extensive collection of examples of the calculation of the secondary classes for Riemannian foliations.

We mention also the very interesting work of Morita[43] which shows there is an extended set of secondary invariants, beyond those described above. This paper uses the Chern-Weil approach of Kamber and Tondeur to extend the construction of secondary invariants for Riemannian foliations to include an affine factor in its transverse holonomy group. Moreover, Morita gives examples to show these additional classes are non-zero for a natural sets of examples, and hence for $q > 2$, give further non-triviality results for the homotopy type of $F R \Gamma_q$.

6. Molino Structure Theory

The values of certain of the secondary classes for Riemannian foliations can vary under an appropriate deformation of the underlying Riemannian foliation. This raises the question, exactly what aspects of the dynamics of $\mathcal{F}$ contributes to this variation? Molino's Structure Theory for Riemannian foliations provides a framework for studying this problem, as highlighted in Molino's survey.[42] We recall below some of the main results of this theory, in order to formulate some of the open questions. The reader can consult Molino,[40,41] Haefliger,[16,17] or Moerdijk and Mrčun[39] for further details.

Recall that we assume M is a closed, connected smooth manifold, $(\mathcal{F}, g)$ is a smooth Riemannian foliation of codimension q with tangential distribution $F = T\mathcal{F}$, and that the normal bundle $Q \to M$ to $\mathcal{F}$ is oriented.

Let $\pi : \widehat{M} \to M$ be the bundle of oriented orthonormal frames for Q. For $x \in M$, the fiber $\pi^{-1}(x) = \mathbf{Fr}^+(Q_x)$ is the space of orthogonal frames of Q_x with positive orientation. The manifold $\widehat{M}$ is a principal right $\mathbf{SO}(q)$-bundle. Set $\widehat{x} = (x, e) \in \widehat{M}$ for $e \in \mathbf{Fr}^+(Q_x)$.

The manifold $\widehat{M}$ has a Riemannian foliation $\widehat{\mathcal{F}}$, whose leaves are the holonomy coverings of the leaves of $\mathcal{F}$. The definition of $\widehat{\mathcal{F}}$ can be found in the sources cited above, but there is an easy intuitive definition. Let X denote a vector field on M which is everywhere tangent to the leaves of $\mathcal{F}$, so that its flow $\varphi_t : M \to M$ defines $\mathcal{F}$-preserving diffeomorphisms. For each $x \in M$, $t \mapsto \varphi_t(x)$ defines a path in the leaf L_x^h through x. The differential of these maps induce transverse isometries $D_x\varphi_t : Q_x \to Q_{\varphi_t(x)}$ which act on the oriented frames of Q, hence define paths in $\widehat{M}$. Given $\widehat{x} = (x, e) \in \widehat{M}$, the leaf $\widehat{L}_{\widehat{x}}^h$ is defined by declaring that the path $t \mapsto D_x\varphi_t(e)$ is tangent to $\widehat{L}_{\widehat{x}}^h$. It follows from the construction that the restriction $\pi : \widehat{L}_{\widehat{x}}^h \to L_x^h$ of the projection π to each leaf of $\widehat{\mathcal{F}}$ is a covering map.

There is an $\mathbf{SO}(q)$-invariant Riemannian metric $\widehat{g}$ on $T\widehat{M}$ such that $\widehat{\mathcal{F}}$ is Riemannian. The metric $\widehat{g}$ satisfies $d\pi : T\widehat{\mathcal{F}} \to T\mathcal{F}$ is an isometry, and the restriction of $\widehat{g}$ to the tangent space $T\pi$ of the fibers of π is induced from the natural bi-invariant metric on $\mathbf{SO}(q)$. Then $d\pi$ restricted to the orthogonal complement $(T\widehat{\mathcal{F}} \oplus T\pi)^\perp$ is a Riemannian submersion to Q.

A fundamental observation is that $\widehat{\mathcal{F}}$ is *Transversally Parallelizable* (TP). Let $\mathbf{Diff}(\widehat{M}, \widehat{\mathcal{F}})$ denote the subgroup of diffeomorphisms of $\widehat{M}$ which map leaves to leaves for $\widehat{\mathcal{F}}$, not necessarily taking a leaf to itself. The TP condition is that $\mathbf{Diff}(\widehat{M}, \widehat{\mathcal{F}})$ acts transitively on $\widehat{M}$.

Given $\widehat{x} = (x, e) \in \widehat{M}$, let $\overline{L_x^h}$ denote the closure of the leaf L_x in M, and let $\overline{L_{\widehat{x}}^h}$ denote the closure of the leaf $\widehat{L}_{\widehat{x}}^h$ in $\widehat{M}$. For notational convenience, we set $N_x = \overline{L_x^h}$ and $N_{\widehat{x}} = \overline{L_{\widehat{x}}^h}$. Note that the distinction between $N_x \subset M$ and $N_{\widehat{x}} \subset \widehat{M}$ is indicated by the basepoint.

Theorem 6.1. *Given any pair of points $\widehat{x}, \widehat{y} \in \widehat{M}$, there is a diffeomorphism $\Phi \in \mathbf{Diff}(\widehat{M}, \widehat{\mathcal{F}})$ which restricts to a foliated diffeomorphism, $\Phi \colon N_{\widehat{x}} \to N_{\widehat{y}}$. Hence, given any pair of points $x, y \in M$, the universal coverings of the leaves L_x^h and L_y^h of $\mathcal{F}$ are diffeomorphic and quasi-isometric.*

This is a key property of Riemannian foliations, and is used to establish the general Molino Structure Theory, which gives a description of the closures of the leaves of $\mathcal{F}$ and $\widehat{\mathcal{F}}$.

Theorem 6.2 (Molino[40,41]). *Let M be a closed, connected smooth manifold, and $(\mathcal{F}, g)$ a smooth Riemannian foliation of codimension q of M. Let $W = M/\overline{\mathcal{F}}$ be the quotient of M by the closures of the leaves of $\mathcal{F}$, and $\Upsilon \colon M \to W$ the quotient map.*

(1) For each $\widehat{x} \in \widehat{M}$, the closure $N_{\widehat{x}}$ of $\widehat{L}_{\widehat{x}}^h$ is a submanifold of $\widehat{M}$.
(2) The set of all leaf closures $N_{\widehat{x}}$ defines a foliation $\widehat{\mathcal{E}}$ of $\widehat{M}$ with all leaves compact without holonomy.
(3) The quotient leaf space $\widehat{W}$ is a closed manifold with an induced right $\mathbf{SO}(q)$-action, and the induced fibration $\widehat{\Upsilon} \colon \widehat{M} \to \widehat{W}$ is $\mathbf{SO}(q)$-equivariant.
(4) W is a Hausdorff space, and there is an $\mathbf{SO}(q)$-equivariant commutative diagram:

$$\begin{array}{ccc} \mathbf{SO}(q) & = & \mathbf{SO}(q) \\ \downarrow & & \downarrow \\ \widehat{M} & \xrightarrow{\widehat{\Upsilon}} & \widehat{W} \\ \pi \downarrow & & \downarrow \widehat{\pi} \\ M & \xrightarrow{\Upsilon} & W \end{array}$$

The second main result of the structure theory provides a description of the closures of the leaves of $\mathcal{F}$ and $\widehat{\mathcal{F}}$, and the structure of $\widehat{\mathcal{F}}|N_{\widehat{x}}$.

Theorem 6.3 (Molino[40,41]). *Let M be a closed, connected smooth manifold, and $(\mathcal{F}, g)$ a smooth Riemannian foliation of codimension q of M.*

(1) There exists a simply connected Lie group G, whose Lie algebra $\mathfrak{g}$ is spanned by the holonomy-invariant vector fields on $N_{\widehat{x}}$ transverse to $\widehat{\mathcal{F}}$, such that the restricted foliation $\widehat{\mathcal{F}}$ of $N_{\widehat{x}}$ is a Lie G-foliation with all leaves dense, defined by a Maurer-Cartan connection 1-form $\omega_{\mathfrak{g}}^{\widehat{x}} : TN_{\widehat{x}} \longrightarrow \mathfrak{g}$.

(2) Let $\rho_{\widehat{x}} : \pi_1(N_{\widehat{x}}, \widehat{x}) \to G$ be the global holonomy map of the flat connection $\omega_{\mathfrak{g}}^{\widehat{x}}$. Then the image $\widehat{\mathcal{N}}_{\widehat{x}} \subset G$ of $\rho_{\widehat{x}}$ is dense in G.

7. Some open problems

Theorems 6.2 and 6.3 suggest a number of questions about the secondary classes of Riemannian foliations. It is worth recalling that for the example constructed in the proof of Theorem 4.3 of a Riemannian foliation for which the characteristic map is injective, all of its leaves are compact, and so the structural Lie group G of Theorem 6.3 reduces to the trivial group. For this example, all of the secondary classes are integral.

The first two problems invoke the structure of the quotient manifold $\widehat{W} = \widehat{M}/\mathcal{E}$ and space $W = M/\overline{\mathcal{F}}$.

Problem 7.1. *Suppose that foliation $\overline{\mathcal{F}}$ of M by the leaf closures of $\mathcal{F}$ is a non-singular foliation. Show that all secondary classes of $\mathcal{F}$ are rational. In the case where every leaf of $\mathcal{F}$ is dense in M, so W reduces to a point, are the secondary classes necessarily integral?*

In all examples where there exists a family of foliations for which the secondary classes vary non-trivially, the quotient space W is singular, hence the action of $\mathbf{SO}(q)$ on $\widehat{W}$ has singular orbits. The action of $\mathbf{SO}(q)$ thus defines a stratification of $\widehat{W}$. (See[28] for a discussion of the stratifications associated to a Riemannian foliation, and some of their properties.)

Problem 7.2. *How do the values of the secondary classes for a Riemannian foliation depend upon the $\mathbf{SO}(q)$-stratification of $\widehat{W}$? Are there conditions on the structure of the stratification which are sufficient to imply that the secondary classes are rational?*

The next problems concern the role of the structural Lie group G of a Riemannian foliation $\mathcal{F}$.

Problem 7.3. *Suppose the structural Lie group G is nilpotent. For example, if all leaves of $\mathcal{F}$ have polynomial growth, the G must be nilpotent.*[6,56] *Show that all rigid secondary classes of $\mathcal{F}$ are rational.*

All of the known examples of families of Riemannian foliations for which the secondary classes vary non-trivially are obtained by the action of an abelian group $\mathbb{R}^p$, and so the structural Lie group G is necessarily abelian. In contrast, one can ask whether there is a generalization to the secondary classes of Riemannian foliations of the results of Reznikov that the rigid secondary classes of flat bundles must be rational.[50,51]

Problem 7.4. *Suppose the structural Lie group G is semi-simple with real rank at least 2, without any factors of $\mathbb{R}$. Must the values of the secondary classes be rigid under deformation? Are all of the characteristic classes of $\mathcal{F}$ are rational?*

Problem 7.5. *Assume the leaves of $\mathcal{F}$ admit a Riemannian metric for which they are Riemannian locally symmetric spaces of higher rank.*[55,57] *Must all of the characteristic classes of $\mathcal{F}$ be rational?*

The final question is more global in nature, as it asks how the topology of the ambient manifold M influences the values of the secondary classes for a Riemannian foliation $\mathcal{F}$ of M. Of course, one influence might be that the cohomology group $H^\ell(M;\mathbb{R}) = \{0\}$ where $\ell = \deg(h_I \otimes p_J)$, and then $\Delta_{\mathcal{F}}(h_I \otimes p_J) = 0$ is rather immediate. Are there more subtle influences, such as whether particular restrictions on the fundamental group $\pi_1(M)$ restrict the values of the secondary classes for Riemannian foliations of M?

Problem 7.6. *How does the topology of a compact manifold M influence the secondary classes for a Riemannian foliation $(\mathcal{F}, g)$ with normal framing s of M?*

There are various partial results for Problem 7.6 in the literature,[34,36,54] but no systematic treatment. It seems likely that an analysis such as in Ghys[13] for Riemannian foliations of simply connected manifolds would yield new results in the direction of this question.

References

1. M. Atiyah, **Topology**, 29:1–7, 1990.
2. P. Baum and J. Cheeger, **Topology**, 8:173–193, 1969.
3. R. Bott, **Michigan Math. J.**, 14:231–244, 1967.

4. R. Bott, **Global Analysis (Proc. Sympos. Pure Math., Vol. XVI, Berkeley, Calif., 1968)**, Amer. Math. Soc., Providence, R.I., 1970:127–131.
5. R. Bott and J. Heitsch, **Topology**, 11:141-146, 1972.
6. Y. Carrière, **Comment. Math. Helv.** 63:1–20, 1988.
7. J. Cheeger and J. Simons, **Geometry and Topology (College Park, Md., 1983/84)**, Lecture Notes in Math. Vol., 1167, Springer, Berlin, 1985: 50–80.
8. S.S. Chern, **Complex manifolds without potential theory**, Second Edition. With an appendix on the geometry of characteristic classes, Springer-Verlag, New York, 1979.
9. S.S. Chern and J. Simons, **Ann. of Math.** (2), 99:48–69, 1974.
10. A. Clark and S. Hurder, *Solenoidal minimal sets for foliations*, submitted, December 2008.
11. J.L. Dupont and F.W. Kamber, **Math. Ann.**, 295:449–468, 1993.
12. D.B.A. Epstein, **Ann. Inst. Fourier**, 26:265–282, 1976.
13. É. Ghys, **Ann. Inst. Fourier**, 34:203–223, 1984.
14. A. Haefliger, **Topology**, 9:183–194, 1970.
15. A. Haefliger, **Manifolds–Amsterdam 1970 (Proc. Nuffic Summer School)**, Lect. Notes in Math. Vol. 197, Springer–Verlag, Berlin, 1971:133–163.
16. A. Haefliger, **Differential geometry (Santiago de Compostela, 1984)**, Res. Notes in Math., Vol. 131:174–197, Pitman, Boston, MS, 1985.
17. A. Haefliger, **Séminaire Bourbaki, Vol. 1988/89**, Asterisque, 177-178, Société Math. de France, 1989, 183–197.
18. S. Hurder, **Dual homotopy invariants of G-foliations**, Thesis, University of Illinois Urbana-Champaign, 1980.
19. S. Hurder, **Proc. Amer. Math. Soc.**, 81:485–489, 1981.
20. S. Hurder, **Comment. Math. Helv.**, 56:307–326, 1981.
21. S. Hurder, **Topology App.**, 50:81-86, 1993.
22. S. Hurder, *Characteristic classes of Riemannian foliations*, talk, Joint International Meeting AMS–RSME, Sevilla, 2003: http://www.math.uic.edu/~ hurder/talks/Seville2003.pdf
23. S. Hurder, *Foliation Problem Set*, preprint, September 2003: http://www.math.uic.edu/~hurder/papers/58manuscript.pdf
24. S. Hurder, *Classifying foliations*, in press, **Foliations, Topology and Geometry**, Contemp. Math., American Math. Soc., 2009.
25. S. Hurder, *Dynamics and Cohomology of Foliations*, talk, VIII International Colloquium on Differential Geometry, Santiago de Compostela, July 2008: http://www.math.uic.edu/~hurder/talks/Santiago2008np.pdf
26. S. Hurder, *Essential solenoids, ghost cycles, and $F\Gamma_q^r$*, **preprint**, 2008.
27. S. Hurder and D. Töben, **Trans. Amer. Math. Soc.**, to appear.
28. S. Hurder and D. Töben, *Equivariant basic cohomology and residues of Riemannian foliations*, **in preparation**, 2008.
29. F.W. Kamber and Ph. Tondeur, **Ann. Sci. École Norm. Sup.**, 8:433–486, 1975.
30. F.W. Kamber and Ph. Tondeur, **Foliated bundles and characteristic**

classes, Lect. Notes in Math. Vol. 493, Springer-Verlag, Berlin, 1975.
31. J.-L. Koszul, *Travaux de S. S. Chern et J. Simons sur les classes caractéristiques*, Séminaire Bourbaki, Vol. 1973/1974, Exp. No. 440, Lect. Notes in Math., Vol. 431, Springer-Verlag, Berlin, 1975, pages 69–88.
32. H.B. Lawson, Jr., **The Quantitative Theory of Foliations**, NSF Regional Conf. Board Math. Sci., Vol. 27, 1975.
33. C. Lazarov, **J. Differential Geometry**, 14:475–486 (1980), 1979.
34. C. Lazarov and J. Pasternack, **J. Differential Geometry**, 11:365–385, 1976.
35. C. Lazarov and J. Pasternack, **J. Differential Geometry**, 11:599–612, 1976.
36. X.-M. Mei, **Proc. Amer. Math. Soc.** 89:359–366, 1983.
37. K. Millett, **Foliations: Dijon 1974**, Lect. Notes in Math. Vol. 484, Springer–Verlag, New York and Berlin, 277–287,1975.
38. J. Milnor and J. Stasheff, **Characteristic classes**, Annals of Mathematics Studies, No. 76., Princeton University Press, Princeton, N. J. 1974.
39. I. Moerdijk and J. Mrčun, **Introduction to foliations and Lie groupoids**, Cambridge Studies in Advanced Mathematics, Vol. 91, 2003.
40. P. Molino, **Nederl. Akad. Wetensch. Indag. Math.** 44:45–76, 1982.
41. P. Molino, **Riemannian foliations**, Translated from the French by Grant Cairns, with appendices by Cairns, Y. Carrière, É. Ghys, E. Salem and V. Sergiescu, Birkhäuser Boston Inc., Boston, MA, 1988.
42. P. Molino, **Geometric Study of Foliations, Tokyo 1993** (eds. Mizutani et al), World Scientific Publishing Co. Inc., River Edge, N.J., 1994, 97–119.
43. S. Morita, **Osaka J. Math.** 16:161–172, 1979.
44. J. Pasternack, **Topological obstructions to integrability and Riemannian geometry of foliations**, Thesis, Princeton University, 1970.
45. J. Pasternack, **Differential geometry (Proc. Sympos. Pure Math., Vol. XXVII, Part 1, Stanford Univ., Stanford, Calif., 1973)**, Amer. Math. Soc., Providence, R.I., 1975:303–310.
46. A. Phillips, **Topology** 6:171–206, 1967.
47. A. Phillips, **Comment. Math. Helv.** 43:204–211, 1968.
48. A. Phillips, **Comment. Math. Helv.** 44:367–370, 1969.
49. B.L. Reinhart, **Differential geometry of foliations**, Ergebnisse der Mathematik und ihrer Grenzgebiete vol. 99, Springer-Verlag, Berlin, 1983.
50. A. Reznikov, **Ann. of Math.** (2), 141:373–386, 1995.
51. A. Reznikov, **J. Differential Geom.**, 43:674–692, 1996.
52. J. Simons, *Characteristic forms and transgression II: Characters associated to a connection*, **preprint, SUNY Stony Brook**, 1972.
53. K. Yamato, **C. R. Acad. Sci. Paris Sér. A-B**, 289:A537–A540, 1979.
54. K. Yamato, **Japan. J. Math. (N.S.)**, 7:227–256, 1981.
55. R.J. Zimmer, **Inst. Hautes Études Sci. Publ. Math.**, 55:37–62, 1982.
56. R.J. Zimmer, **J. Funct. Anal.**, 72:58–64, 1987.
57. R.J. Zimmer, **J. Amer. Math. Soc.**, 1:35–58, 1988.

NON UNIQUE-ERGODICITY OF HARMONIC MEASURES: SMOOTHING SAMUEL PETITE'S EXAMPLES

Bertrand Deroin

CNRS UMR 8628
Département de mathématiques d'Orsay
Université Paris 11 Bât. 425 Orsay Cedex, France
E-mail: Bertrand.Deroin@math.u-psud.fr

We give an example of a smooth foliation of a compact manifold, with all leaves dense, without any transverse invariant measure, but with uncountably many ergodic harmonic measures in the sense of Garnett.[9] We exhibit holomorphic foliations of compact complex manifolds with a similar behaviour.

Keywords: Ergodic theory, foliations

Mathematics Subject Classification 2000: Primary 37A25. 37C85

1. Introduction

A smooth foliation is a partition of a smooth manifold M as the union of immersed p-dimensional submanifolds called *leaves*; it is defined by a cover of M by open balls U_i and diffeomorphisms $\varphi_i : U_i \to \mathbf{B}^p \times \mathbf{B}^q$ such that every leaf intersects U_i in a countable union of sets of the form $\varphi_i^{-1}(\mathbf{B}^p \times \{\cdot\})$ called *plaques*. We denote by

$$\varphi_j \circ \varphi_i^{-1}(x_i, t_i) = (x_j = x_{ji}(x_i, t_i), t_j = t_{ji}(t_i))$$

the change of coordinates.

Garnett initiated the theory of *harmonic measures*[9] (see also[2,3]) on a foliation equipped with a smooth Riemannian metric ds^2 on its tangent bundle. Denote by $\Delta_{\mathcal{F}}$ the Laplacian associated to ds^2 along the leaves of $\mathcal{F}$: a harmonic measure is a probability measure m on M verifying $\Delta_{\mathcal{F}} m = 0$ in the weak sense, i.e. $\int \Delta_{\mathcal{F}} f \, dm = 0$ for any smooth function f on M. Garnett proved that such measures always exist, and moreover, she developped an ergodic theory for such measures, in connexion with the foliated Brownian motion.

In this work[a], a foliation will be called *uniquely ergodic* if on any minimal closed $\mathcal{F}$-saturated subset is supported a unique harmonic measure. In the sixties, Furstenberg[7] gave an example of a diffeomorphism of the 2-torus, with dense orbits, and with uncountably many invariant measures. By suspending such a diffeomorphism, one obtains an example of a foliation with dense leaves and uncountably many transverse invariant measures.

A *transverse invariant measure* for $\mathcal{F}$ is a family of Radon measures ν_i on the ball B^q, such that $(t_{ji})_* \nu_i = \nu_j$ where t_{ji} is defined. This notion has been introduced[12] by Plante. Such a family of measures gives rise to a harmonic measure by forming their product with the leafwise volume. The harmonic measures obtained by this contruction are called *totally invariant*. For instance, by suspending Furtenberg's example of a minimal non-uniquely ergodic diffeomorphism, all the harmonic measures are totally invariant.

When a foliation does not carry any transverse invariant measure, the transverse dynamics is richer, and non unique-ergodicity is much less likely to happen. For instance, unique ergodicity is known in the following cases:

- transversally conformal foliations without any invariant measure, see.[4]
- homogeneous foliations whose leaves are the orbit of a locally free action of the affine group (i.e. $M = G/\Gamma$ where Γ is a lattice of a Lie group, and the foliation is given by the orbit of the left action of a copy of the affine group in G). This fact is a consequence of a combination of the work of Ratner[13] and of Petite.[11]
- the suspension of a representation of the fundamental group of a compact manifold in $SL(d+1, \mathbf{R})$ acting by projective transformations are uniquely ergodic,[10] as soon as the representation is contracting and irreducible.

Petite,[11] however, has constructed minimal laminations with several non totally invariant harmonic measures using tilings of the hyperbolic plane. Our goal is to provide smooth examples of this kind: we construct a smooth foliation by Riemannian surfaces of a 5-dimensional manifold, with dense leaves, and with uncountably many non totally invariant ergodic harmonic measures.

We also provide examples in the holomorphic world: we construct a non uniquely-ergodic (but not minimal) holomorphic foliation by curves of

---

[a]Usually, unique ergodicity means that there is a unique harmonic measure. However we prefer to allow the existence of several minimal sets.

a compact complex manifold, without any transverse invariant measure. These examples contrast with the unique ergodicity phenomenom occuring for (complex) codimension 1 foliations (see[4,6]).

2. The construction

Let $M \in SL(2, \mathbf{Z})$ be a hyperbolic matrix, i.e. it has two eigenvalues $0 < \alpha < 1 < \beta$. Let (a_1, a_2) and (b_1, b_2) be eigenvectors corresponding to α and β, i.e. we have the relations

$$aa_1 + ba_2 = \alpha a_1, \quad ca_1 + da_2 = \alpha a_2 ,$$

$$ab_1 + bb_2 = \beta b_1, \quad cb_1 + db_2 = \beta b_2 ,$$

where a, b, c, d are the coefficient of M. Let $\mathbf{H}$ be the upper half-plane, whose elements are the complex numbers w with positive imaginary part. Consider the group G generated by the following diffeomorphisms of the product $\mathbf{H} \times \mathbf{R}$:

$$g_0(w, z) = (\alpha w, \beta z) ,$$

$$g_i(w, z) = (w + a_i, z + b_i), \quad i = 1, 2 .$$

The subgroup G' of G generated by g_1 and g_2 preserves each fiber of the fibration by planes

$$\mathbf{R}^2 \to \mathbf{H} \times \mathbf{R} \stackrel{\mathrm{Im}(w)}{\to} (0, \infty) ,$$

and acts by a discrete lattice of translations on them. Thus the action of G' is free and discontinuous, and the quotient $G' \backslash \mathbf{H} \times \mathbf{R}$ is a torus bundle over $(0, \infty)$, which is naturally diffeomorphic to $\mathbf{T}^2 \times (0, \infty)$, where $\mathbf{T}^2$ is the quotient of $\mathbf{R}^2$ by the translations $(x, z) \mapsto (x + a_i, z + b_i)$, $i = 1, 2$. The relations

$$g_0 \circ g_1 \circ g_0^{-1} = g_1^a \circ g_2^b \quad \text{and} \quad g_0 \circ g_2 \circ g_0^{-1} = g_1^c \circ g_2^d$$

show that the group G' is normal in G. The element g_0 acts on the quotient $G' \backslash \mathbf{H} \times \mathbf{R} = \mathbf{T}^2 \times (0, \infty)$ by a linear automorphism on the first factor, and multiplication by α on the second. Thus the action of G on $\mathbf{H} \times \mathbf{R}$ is free, discontinuous and cocompact; one of its fundamental domains is $\mathbf{T}^2 \times [\alpha, 1]$. We will denote by

- $\overline{\mathcal{F}}$ the foliation of $\mathbf{H} \times \mathbf{R}$ defined by $dz = 0$; it is invariant by the group G.

- $(\mathbf{T}^2 \times (0, \infty), \mathcal{F})$ the quotient of $(\mathbf{H} \times \mathbf{R}, \overline{\mathcal{F}})$ by G'.
- $(M, \mathcal{F}_q)$ the quotient of $(\mathbf{H} \times \mathbf{R}, \overline{\mathcal{F}})$ by G.

Let $ds^2 = \frac{|dw|}{\text{Im}(w)}$, be a Riemannian metric on the leaves of $\overline{\mathcal{F}}$: it is invariant under the group G. We denote by ds^2 the induced metric on the leaves of $\mathcal{F}$ and $\mathcal{F}_q$.

Let N be a compact smooth manifold and $\Phi : N \to N$ be a diffeomorphism. We denote $(N_\Phi, \mathcal{F}_\Phi)$ the quotient of the product $\mathbf{T}^2 \times (0, \infty) \times N$, together with the product foliation $\mathcal{F} \times \{\cdot\}$, by the map (g_0, Φ). This is a compact foliated manifold fibering over $(M, \mathcal{F}_q)$ with fiber N. The Riemannian metric on the leaves of $\mathcal{F}_\Phi$ induced by g is still denoted by g. Recall that a subset of N_Φ is $\mathcal{F}_\Phi$-saturated if it is a union of leaves.

Lemma 2.1. *To a Φ-invariant closed subset F of N, we associate the $\mathcal{F}_\Phi$-saturated closed subset of N_Φ defined as the quotient of $\mathbf{T}^2 \times (0, \infty) \times F$ by (g_0, Φ). This is a bijection between the Φ-invariant closed subsets of N and the $\mathcal{F}_\Phi$-saturated subsets of N_Φ. In particular, if all the orbits of Φ are dense, all the leaves of $\mathcal{F}_\Phi$ are also dense.*

Proof. Let $\mathcal{M}$ be a $\mathcal{F}_\Phi$-saturated closed subset of N_Φ, and $\widetilde{\mathcal{M}}$ its pullback to $\mathbf{T}^2 \times (0, \infty) \times N$. It is a closed $\mathcal{F} \times \{\cdot\}$-invariant subset. Since $\mathcal{F}$ intersects a set $\mathbf{T}^2 \times \{\cdot\}$ of $\mathbf{T}^2 \times (0, \infty)$ on a linear irrational foliation of the 2-torus, every leaf of $\mathcal{F}$ is dense in $\mathbf{T}^2 \times (0, \infty)$. Thus $\widetilde{\mathcal{M}}$ must be of the form $\mathbf{T}^2 \times (0, \infty) \times F$, where F is a closed subset. The set F must be Φ-invariant for $\mathcal{M}$ to be $\mathcal{F}_\Phi$-saturated. □

The form $f\, dz$ is g_0-invariant on $\mathbf{T}^2 \times (0, \infty)$ if and only if $f \circ g_0 = \alpha f$. For instance, the form

$$\omega = \text{Im}(w)\, dz$$

is g_0-invariant and vanishes on $\mathcal{F}$. Denote by vol the volume form corresponding to ds^2 on the leaves, and by $V = \text{vol} \wedge \omega$ the volume form on $\mathbf{T}^2 \times (0, \infty)$: both of them are g_0-invariant. The following result is a consequence of Petite's work;[11] however we have preferred to give an independent proof of it in our particular situation.

Lemma 2.2. *The harmonic measures on $\mathcal{F}_\Phi$ are the images of measures of the form $V \otimes \mu$ on $\mathbf{H} \times \mathbf{R} \times N$, where μ is a Φ-invariant measure.*

Proof. Let us prove that if μ is a Φ-invariant measure, $V \otimes \mu$ defines a harmonic measure for $\mathcal{F}_\Phi$. First it is invariant by the group H generated by $G' \times \{\mathrm{id}\}$ and by the transformation (g_0, Φ). Consider the coordinates (w, z, p) of $\mathbf{H} \times (0, \infty) \times N$. We have

$$V \otimes \mu = \mathrm{Im}(w) \, \mathrm{vol}_g \otimes (dz \otimes d\mu) \,.$$

Because $\mathrm{Im}(w)$ is harmonic on the leaves of $\overline{\mathcal{F}}$, $V \otimes \mu$ is harmonic, by Garnett's desintegration lemma.[9] Moreover, it is invariant by the group H, so it defines a harmonic measure on the quotient $H \backslash \mathbf{H} \times \mathbf{R} \times N$.

The converse is harder. Let m be a harmonic measure on $\mathcal{F}_\Phi$. It can be pushed to a harmonic measure m' on the foliation $\mathcal{F}_q$ of $g_0 \backslash \mathbf{T}^2 \times (0, \infty)$. By[4,9], since $\mathcal{F}_q$ does not carry any invariant measure, and since all leaves are dense, we know that $\mathcal{F}_q$ has a unique[b] harmonic measure. This measure is necessarily the volume form V.

Let $\tilde{m}$ be the Radon measure on $\mathbf{H} \times (0, \infty) \times N$, obtained as the "pullback" of the measure m. By Garnett's decomposition lemma,[9]

$$\tilde{m} = \varphi \, \mathrm{vol} \wedge d\nu(z, p) \,,$$

where ν is a Radon measure on $\mathbf{R} \times N$, and $\varphi \in L^1_{\mathrm{loc}}(\mathrm{vol} \otimes \nu)$ is a non negative measurable function, harmonic on ν-a.e. leaf of $\overline{\mathcal{F}} \times \{\cdot\}$. Moreover, we can suppose that $\varphi(\sqrt{-1}, z, p) \equiv 1$. Let I be a compact interval in $\mathbf{R}$, and $f_I : \mathbf{R} \times (0, \infty) \to (0, \infty)$ the function

$$f_I(w) = \int_{I \times N} \varphi(w, z, p) \, d\nu(z, p) \,.$$

It is harmonic. Because $\tilde{m}$ is mapped to V by the projection $(w, z, p) \mapsto (w, z)$, we have $f_I = \mathrm{cst} \cdot \mathrm{Im}(w)$. But the harmonic function $\mathrm{Im}(w)$ is extremal in the cone of positive harmonic function on the upper half-plane, so that for ν a.e. (z, p)

$$\varphi(w, z, p) = \mathrm{Im}(w) \,.$$

Recall that ν is mapped by $(z, p) \mapsto z$ to the Lebesgue measure dz. By Fubini's theorem, there is a family of probability measures μ_z on N, depending measurably on z, such that

$$\nu = \int \mu_z \, dz \,.$$

The map $(x, z) \mapsto \mu_z$ is invariant by the translations $(x, z) \mapsto (x + a_i, z + b_i)$, $i = 1, 2$. It induces a measurable map from $\mathbf{T}^2$ to the space of probability measures on N which is constant on the irrational foliation defined by

---

[b]In[9], Garnett only stated this statement for the stable folation of the geodesic flow of a compact surface of curvature -1; however, her proof also works for the foliation $\mathcal{F}_q$.

$dx_+ = 0$. Because this foliation is ergodic with respect to the Lebesgue measure, $\mu_z = \mu$ is constant for almost every point z with respect to Lebesgue measure. The measure μ is Φ-invariant since $\tilde{m}$ is invariant by (g_0, Φ). □

Recall that Furstenberg[7] has shown that there is an analytic diffeomorphism Φ of the 2-torus $\mathbf{T}^2$, whose orbits are dense, but with uncountably many invariant measures. Lemmas 2.1 and 2.2 show that the foliation $\mathcal{F}_\Phi$ of the 5-manifold N_Φ satisfies the conclusions of the abstract.

It is possible to make a slightly different construction in the holomorphic world. We follow the construction of p. 174[1]. Let $M = (m_{ij}) \in SL(3, \mathbf{Z})$ be a matrix with one real eigenvalue $\alpha > 1$, and two complex conjugate eigenvalues β and $\overline{\beta}$ with modulus $|\beta| = |\overline{\beta}| < 1$. Let (a_1, a_2, a_3) be a real eigenvector of M corresponding to α and (b_1, b_2, b_3) an eigenvector of M corresponding to β. Since (a_1, a_2, a_3), (b_1, b_2, b_3) and $(\overline{b_1}, \overline{b_2}, \overline{b_3})$ are independent over $\mathbf{C}$, the vectors (a_1, b_1), (a_2, b_2) and (a_3, b_3) are independent over $\mathbf{R}$.

Let N be a compact complex manifold and Ψ be a complex automorphism of N. Let G be the group of automorphisms of the product $\mathbf{C} \times \mathbf{H} \times N$ generated by the maps

$$g_0(w, z, p) = (\alpha w, \beta z, \Psi(p))$$

$$g_i(w, z, p) = (w + a_i, z + b_i, p), \quad i = 1, 2, 3 .$$

These transformations preserve the holomorphic foliation $\mathcal{F}$ defined by $\{w = \text{cst}, p = \text{cst}\}$. As before, the group G acts freely, properly and discontinuously on $\mathbf{C} \times \mathbf{H} \times N$. The quotient is denoted by N_Ψ, and the quotient of $\mathcal{F}$ by $\mathcal{F}_\Psi$. Lemmas 2.1 and 2.2 also work for the foliation $\mathcal{F}_\Psi$.

It is not known wether there exists an automorphism of a compact complex manifold with every orbit dense, but with several invariant measures. However, if we choose Ψ to be a hyperbolic linear automorphism of a torus, it is possible to find a minimal closed Ψ-invariant subset supporting several Ψ-invariant measures. The foliation $\mathcal{F}_\Psi$ has an $\mathcal{F}_\Psi$-saturated subset carrying several harmonic measures.

3. Questions

- Let $\mathcal{F}$ be a smooth foliation of a compact manifold, and ds_i^2, $i = 1, 2$ be Riemannian metrics on the leaves. Is it true[c] that the convex compact

---

[c]There is an example[5] of a smooth codimension 1 foliation of a compact 3-manifold with the following properties:

subset of harmonic measures for ds_1^2 and ds_2^2 respectively are image of each other by an affine automorphism of the space of finite measures? This is true for transversally conformal foliations,[4] in which case the compact subsets of harmonic measures are simplex of the dimension given by the number of minimal closed $\mathcal{F}$-saturated sets.
- Does there exist a non uniquely-ergodic foliation by surfaces of a 4-manifold, which does not carry a transverse invariant measure?
- Is there a holomorphic foliation by curves of a complex projective/Kähler manifold which is not uniquely ergodic?
- Is there a biholomorphism of a compact complex manifold with every orbit dense and more than one invariant measure?

Acknowledgement

This work was inspired by a conversation with Étienne Ghys and Samuel Petite. I warmly thank both of them.

References

1. W. Barth, C. Peters, A. Van de Ven, *Compact complex surface* Springer-Verlag, Berlin 1984.
2. A. Candel, Adv. Math. **176** (2003), no. 2, 187-247.
3. A. Candel, L. Conlon, *Foliations. II.* Graduate Studies in Mathematics, 60. AMS, Providence, RI, 2003.
4. B. Deroin, V. Kleptsyn, GAFA vol. 17 (2007) **4** 1043-1105.
5. B. Deroin, C. Vernicos, In preparation.
6. J. E. Fornaess, N. Sibony, arXiv:0606744.
7. H. Furstenberg, Amer. J. Math. **83** (1961), 573-601.
8. H. Furstenberg, Ann. of Math. (2) **77** (1963) 477-515.
9. L. Garnett, J. Funct. Anal. **51** (1983) no. 3, 285-311.
10. Y. Guivarc'h, A. Raugi, Z. Wahrsch. Verw. Gebiete **69** (1985), no. 2, 187-242.
11. S. Petite, Erg. Th. Dyn. Syst. **26** (2006), no. 4, 1159-1176.
12. J. Plante, Ann. of Math. **102** (1975), 327-361.
13. M. Ratner, Ann. of Math. (2) **134** (1991), no. 3, 545-607.
14. D. Ruelle, D. Sullivan, Topology **14**, 319-327.

- it has dense leaves
- it does not carry any transverse invariant measure
- there is a Riemannian metric g on the leaves which is smooth on the leaves and continuous transversally, such that there are several Δ_g-harmonic measures.

This contrast with the fact that if the metric is transversally Hölder, then there is a unique harmonic measure.[4] Hence one has to suppose that the metrics ds_i^2 are at least transversally Hölder for this question to have a positive answer.

ON THE UNIFORM SIMPLICITY OF DIFFEOMORPHISM GROUPS

Takashi Tsuboi

*Graduate School of Mathematical Sciences, The University of Tokyo
Komaba Meguro, Tokyo 153-8914, Japan
E-mail: tsuboi@ms.u-tokyo.ac.jp*

We show the uniform simplicity of the identity component $\text{Diff}^r(M^n)_0$ of the group of C^r diffeomorphisms $\text{Diff}^r(M^n)$ ($1 \leq r \leq \infty$, $r \neq n+1$) of the compact connected n-dimensional manifold M^n with handle decomposition without handles of the middle index $n/2$. More precisely, for any elements f and g of such $\text{Diff}^r(M^n)_0 \setminus \{\text{id}\}$, f can be written as a product of at most $16n + 28$ conjugates of g or g^{-1}, which we denote by $f \in (C_g)^{16n+28}$. We have better estimates for several manifolds. For the n-dimensional sphere S^n, for any elements f and g of $\text{Diff}^r(S^n)_0 \setminus \{\text{id}\}$ ($1 \leq r \leq \infty$, $r \neq n+1$), $f \in (C_g)^{12}$, and for a compact connected 3-manifold M^3, for any elements f and g of $\text{Diff}^r(M^3)_0 \setminus \{\text{id}\}$ ($1 \leq r \leq \infty$, $r \neq 4$), $f \in (C_g)^{44}$.

Keywords: Diffeomorphism group, uniformly perfect group, uniformly simple group, commutator subgroup

Mathematics Subject Classification 1991: Primary 57R52, 57R50; Secondary 37C05

1. Introduction

In 1947, Ulam and von Neumann[22] announced the following theorem.

Theorem 1.1 (Ulam-von Neumann[22]). *The group of orientation preserving homeomorphisms of the 2-dimensional sphere S^2 is a simple group. Moreover there is a positive integer N such that for any orientation preserving homeomorphisms f and g of S^2, f can be written as a product of N conjugates of g if g is not the identity.*

In 1958, Anderson[1] showed the following theorem.

Theorem 1.2 (Anderson[1]). *Let $\text{Homeo}(S^n)_0$ denote the identity component of the group of homeomorphisms of the n-dimensional sphere S^n. For*

$n = 1, 2, 3$ and for elements f and $g \in \mathrm{Homeo}(S^n)_0 \setminus \{\mathrm{id}\}$, f can be written as a product of at most 6 conjugates of g or g^{-1}.

In 1960, Fisher[5] showed that for a compact connected manifold M^n of dim $n \leq 3$, $\mathrm{Homeo}(M^n)_0$ is a simple group.

Here, a group G is said to be *simple* if G contains no nontrivial proper normal subgroups. Equivalently, G is simple if, for $f \in G$ and $g \in G \setminus \{e\}$, f can be written as a product of conjugates of g or g^{-1}.

In 1970, Epstein[2,4] showed that for certain groups such as the group of C^r diffeomorphisms ($r \leq \infty$) where we can apply the fragmentation technique, the perfectness implies the simplicity.

Here a group G is said to be *perfect* if the abelianization of G is a trivial group. Equivalently, G is perfect if any element of G can be written as a product of commutators.

For a manifold M^n, let $\mathrm{Diff}^r(M^n)$ denote the group of C^r diffeomorphisms of M^n, and $\mathrm{Diff}^r_c(M^n)$, the group of C^r diffeomorphisms of M^n with compact support ($1 \leq r \leq \infty$). Here the *support* $\mathrm{supp}(f)$ of a diffeomorphism f of M^n is defined to be the *closure* of $\{x \in M^n \mid f(x) \neq x\}$. Let $\mathrm{Diff}^r(M^n)_0$ and $\mathrm{Diff}^r_c(M^n)_0$ denote the identity components of $\mathrm{Diff}^r(M^n)$ and $\mathrm{Diff}^r_c(M^n)$ with respect to the C^r topology, respectively.[2]

Herman-Mather-Thurston[2,7,10,11,15] showed the perfectness of the identity component $\mathrm{Diff}^r_c(M^n)_0$ of the group of C^r diffeomorphisms ($1 \leq r \leq \infty$, $r \neq n+1$) of an n-dimensional manifold M^n with compact support, which implies the simplicity of the group when M^n is connected.

For $g \in G$, let C_g denote the union of the conjugate classes of g and of g^{-1}. Then G is simple if $G = \bigcup_{k=1}^{\infty} (C_g)^k$ for any element $g \in G \setminus \{e\}$. For a simple group G, we can define an interesting distance function on the set $\{C_g \mid g \in G \setminus \{e\}\}$ by

$$d(C_f, C_g) = \log \min\{k \mid C_f \subset (C_g)^k \text{ and } C_g \subset (C_f)^k\}.$$

Definition 1.1. We say that G is uniformly simple if there is a positive integer N such that, for $f \in G$ and $g \in G \setminus \{e\}$, f can be written as a product of at most N conjugates of g or g^{-1}: $G = \bigcup_{k=1}^{N} (C_g)^k$.

In other words, G is uniformly simple if the distance function d on $\{C_g \mid g \in G \setminus \{e\}\}$ is bounded.

There are simple groups which are not uniformly simple. For example, the direct limit A_∞ of the alternate groups A_n, the identity component of the group of volume preserving diffeomorphisms with compact support of $\mathbf{R}^n$ ($n \geq 3$), etc.

If an infinite group is uniformly simple, then it is uniformly perfect. Here a group G is said to be *uniformly perfect* if there is a positive integer N such that any element $f \in G$ can be written as a product of at most N commutators. By using the results of Herman-Mather-Thurston,[2,7,10,11,15] we showed[21] the uniform perfectness of $\mathrm{Diff}^r(M^n)_0$ ($1 \leq r \leq \infty$, $r \neq n+1$) for the compact n-dimensional manifold M^n with handle decomposition without handles of the middle index $n/2$.

We show in this paper, the uniform simplicity of the identity component $\mathrm{Diff}^r(M^n)_0$ ($1 \leq r \leq \infty$, $r \neq n+1$) of the group of diffeomorphisms of the compact connected n-dimensional manifold M^n with handle decomposition without handles of the middle index $n/2$. This uniform simplicity (in particular, the estimates on the number of conjugates) follows from certain improvement of the proof in[21] of the uniform perfectness of $\mathrm{Diff}^r(M^n)_0$ (see also Remark 3.3).

Our results in this paper are as follows.

Theorem 1.3. *For the n-dimensional sphere S^n ($n \geq 1$), for any elements f and g of $\mathrm{Diff}^r(S^n)_0 \setminus \{\mathrm{id}\}$ ($1 \leq r \leq \infty$, $r \neq n+1$), f can be written as a product of at most 12 conjugates of g or g^{-1}.*

For a handle decomposition, let c be the order of the set of indices which appears as the indices of handles in the handle decomposition. In the following theorems, for a manifold M^n, $c(M^n)$ denotes the minimum of such numbers c among the handle decompositions of M^n without the middle index $n/2$ (if n is even). Of course, $c(M^n) \leq n+1$.

Theorem 1.4. *Let M^{2m} be a compact connected $(2m)$-dimensional manifold with handle decomposition without handles of index m, then for any elements f and g of $\mathrm{Diff}^r(M^{2m})_0 \setminus \{\mathrm{id}\}$ ($1 \leq r \leq \infty$, $r \neq 2m+1$), f can be written as a product of at most $16c(M^{2m})+8$ conjugates of g or g^{-1}.*

Theorem 1.5. *Let M^{2m+1} be a compact connected $(2m+1)$-dimensional manifold, then for any elements f and g of $\mathrm{Diff}^r(M^{2m+1})_0 \setminus \{\mathrm{id}\}$ ($1 \leq r \leq \infty$, $r \neq 2m+2$), f can be written as a product of at most $16c(M^{2m+1})+12$ conjugates of g or g^{-1}.*

Since $c(M^n) \leq n+1$, we have the following corollary.

Corollary 1.1. *Let M^n be a compact connected n-dimensional manifold with handle decomposition without handles of index $n/2$. For any elements f and g of $\mathrm{Diff}^r(M^n)_0 \setminus \{\mathrm{id}\}$ ($1 \leq r \leq \infty$, $r \neq n+1$), f can be written as a product of at most $16n + 28$ conjugates of g or g^{-1}.*

In many cases, we have a better estimate on the number of conjugates. In particular, for a compact connected 3-dimensional manifolds M^3, we have the following.

Corollary 1.2. *Let M^3 be a compact connected 3-dimensional manifold. For any elements f and g of $\mathrm{Diff}^r(M^3)_0 \setminus \{\mathrm{id}\}$ ($1 \leq r \leq \infty$, $r \neq 4$), f can be written as a product of at most 44 conjugates of g or g^{-1}.*

In Section 2, we review the results of our previous paper[21] and give the necessary improvement. In Section 3, we give the proofs of theorems. There we also remark that for the n-dimensional sphere S^n, any element $f \in \mathrm{Diff}(S^n)_0$ can be written as a product of 3 commutators, and for a compact $(2m+1)$-dimensional manifold M^{2m+1}, any element $f \in \mathrm{Diff}(M^{2m+1})_0$ can be written as a product of 5 commutators.

2. Uniform perfectness of diffeomorphism groups

In Theorem 4.1[21], we showed the following.

Theorem 2.1 (Tsuboi[21]). *Let M^n be the interior of a compact n-dimensional manifold with handle decomposition with handles of indices not greater than $(n-1)/2$, then any element of $\mathrm{Diff}_c^r(M^n)_0$ ($1 \leq r \leq \infty$, $r \neq n+1$) can be written as a product of two commutators.*

To discuss the uniform simplicity, we use an improvement of this theorem. In the proof of this theorem, we used a nice Morse function on M^n to find a k-dimensional complex K^k differentiably embedded in M^n ($k \leq (n-1)/2$) which is a deformation retract of M^n, and an isotopy $\{H_t\}_{t \in [0,1]}$ ($H_0 = \mathrm{id}$) with a neighborhood V of K^k such that $(H_1)^j(V)$ ($j \in \mathbb{Z}$) are disjoint. We will use the Morse function on M^n and the associated handle decomposition to show the following theorem.

Theorem 2.2. *Let M^n be the interior of a compact n-dimensional manifold with handle decomposition with handles of indices not greater than $(n-1)/2$. Let c be the order of the set of indices appearing in the handle decomposition. Then any element of $\mathrm{Diff}_c^r(M^n)_0$ ($1 \leq r \leq \infty$, $r \neq n+1$) can be written as a product of two commutators. Moreover, if M^n is connected, any element of $\mathrm{Diff}_c^r(M^n)_0$ can be written as a product of $4c+1$ commutators with support in balls.*

To prove Theorem 2.2, we review the Morse functions and handle decompositions. Before the beginning of the proof of Theorem 2.2, let f denote a Morse function and we fix notations as in[21].

Let $f : M^n \longrightarrow \mathbf{R}$ be a Morse function on a compact connected n-dimensional manifold M^n such that $f(M^n) = [0, n]$, the set of critical points of index k is contained in $f^{-1}(k)$ ($k = 0, \ldots, n$) and $f^{-1}(0)$ and $f^{-1}(n)$ are one point sets.

Put $W_k = f^{-1}([0, k + 1/2])$, and then this W_k is a compact manifold with boundary $\partial W_k = f^{-1}(k+1/2)$. Let c_k be the number of critical points of index k. Then the manifold W_k is diffeomorphic to the manifold obtained from W_{k-1} by attaching c_k handles of index k ($k = 0, \ldots, n$). This means the following.

Let $D^k \times D^{n-k}$ be the product of the k-dimensional disk D^k and the $(n-k)$-dimensional disk D^{n-k}. Let $\varphi_i : (\partial D^k) \times D^{n-k} \longrightarrow \partial W_{k-1}$ ($i = 1, \ldots, c_k$) be diffeomorphisms with disjoint images. Let

$$W'_k = W_{k-1} \cup_{\bigsqcup_{i=1}^{c_k} \varphi_i} \bigsqcup_{i=1}^{c_k} (D^k \times D^{n-k})_i$$

be the space obtained from the disjoint union $W_{k-1} \sqcup \bigsqcup_{i=1}^{c_k} (D^k \times D^{n-k})_i$ by identifying

$$x \in ((\partial D^k) \times D^{n-k})_i \subset (D^k \times D^{n-k})_i$$

with $\varphi_i(x) \in \partial W_{k-1} \subset W_{k-1}$.

In this paper, we consider that W'_k is a submanifold with corner of W_k and $W_k \setminus W'_k$ is diffeomorphic to $\partial W_k \times (-\infty, k + 1/2]$ (which is shown by using the flowlines of the gradient flow Ψ_t). The handles $(D^k \times D^{n-k})_i$ ($i = 1, \ldots, c_k$) of index k are contained in the interior of W_k. Then we have the sequence

$$D^n \cong W_0 \subset W'_1 \subset W_1 \subset \cdots \subset W'_k \subset W_k \subset \cdots \subset W'_n = W_n = M^n.$$

By choosing a Riemannian metric on the manifold M^n, the Morse function f defines the gradient vector field and the gradient flow Ψ_t. The fixed points of the gradient flow Ψ_t are precisely the critical points of f. The core disk and the co-core disk of a handle of a handle decomposition of M^n correspond to the local stable manifold and the local unstable manifold of the corresponding fixed point p of the gradient flow Ψ_t, respectively.[13,14] Let e_i^k and e'^{n-k}_i denote the global stable manifold and the global unstable manifold, respectively, for the fixed point p of Ψ_t which is a critical point of index k of f. Then e_i^k and e'^{n-k}_i are diffeomorphic to $\mathbf{R}^k$ and $\mathbf{R}^{n-k}$, respectively. Then we know that the global stable manifolds and the global unstable manifolds of fixed points of Ψ_t form the cell decomposition $\bigcup_{k=0}^{n} \bigcup_{i=1}^{c_k} e_i^k$ and the dual cell decomposition $\bigcup_{k=0}^{n} \bigcup_{i=1}^{c_k} e'^{n-k}_i$ of M^n, respectively.[13] The dual cell decomposition is the cell decomposition for the

Morse function $n-f$. Consider the k-skeleton $X^{(k)}$ of the cell decomposition and the $(n-k-1)$-skeleton $X'^{(n-k-1)}$ of the dual cell decomposition:

$$X^{(k)} = \bigcup_{j \leq k} \bigcup_{i=1}^{c_j} e_i^j \quad \text{and} \quad X'^{(n-k-1)} = \bigcup_{j \geq k+1} \bigcup_{i=1}^{c_j} e_i'^{n-j}.$$

$X^{(k)}$ and $X'^{(n-k-1)}$ are compact sets. The boundary ∂W_k of W_k is transverse to the gradient flow Ψ_t, and hence $M \setminus (X^{(k)} \cup X'^{(n-k-1)})$ is diffeomorphic to $\partial W_k \times \mathbf{R}$ by the map

$$\partial W_k \times \mathbf{R} \ni (x,t) \longmapsto \Psi_t(x) \in M \setminus (X^{(k)} \cup X'^{(n-k-1)}).$$

Moreover $\Psi_t(\partial W_k)$ converges to $X^{(k)}$ as $t \longrightarrow -\infty$ and to $X'^{(n-k-1)}$ as $t \longrightarrow \infty$. Hence, $M \setminus X'^{(n-k-1)}$ is diffeomorphic to the interior $\text{int}(W_k)$ of W_k and $X^{(k)}$ is a deformation retract of both W_k and $M \setminus X'^{(n-k-1)}$:

$$X^{(k)} \subset \text{int}(W_k) \subset W_k \subset M \setminus X'^{(n-k-1)}.$$

Hence we call $X^{(k)}$ the core complex of W_k.

The core disks $(D^k \times \{0\})_i$ is in the stable manifold for the gradient flow Ψ_t of the critical point $(\{0\} \times \{0\})_i$ of index k. We may consider the flow Ψ_t on the handle $(D^k \times D^{n-k})_i$ of index k is in the form of a direct product of linear flows. Then the stable manifold e_i^k is written as

$$e_i^k = \bigcup_{t \in (-\infty, 0]} \Psi_t((D^k \times \{0\})_i) \quad \text{or} \quad e_i^k = \bigcap_{\tau \in (-\infty, 0]} \bigcup_{t \in (-\infty, \tau]} \Psi_t((D^k \times D^{n-k})_i).$$

Using the gradient flow Ψ_t, for any neighborhood V of $X^{(k)}$ and for any compact subset A in $\text{int}(W_k)$, we can construct an isotopy $\{G_t : \text{int}(W_k) \longrightarrow \text{int}(W_k)\}_{t \in [0,1]}$ with compact support such that $G_0 = \text{id}_{\text{int}(W_k)}$, $G_t|X^{(k)} = \text{id}_{X^{(k)}}$ $(t \in [0,1])$ and $G_1(A) \subset V$. A similar statement is true for $X^{(k)} \subset M \setminus X'^{(n-k-1)}$.

We prove the following lemma which is the core complex version of Lemma 4.3[21].

Lemma 2.1. *Let M^n be a compact n-dimensional manifold. Let $X^{(k)}$ be the k skeleton of the cell decomposition associated with a Morse function on M^n. Let L^ℓ be a compact set which is a union of finitely many images of $\mathbf{R}^s$ $(s < \ell)$ under differentiable maps. If $k + \ell + 1 \leq n$ then there is an isotopy $\{F_t : M^n \longrightarrow M^n\}_{t \in [0,1]}$ $(F_0 = \text{id})$ such that $F_1(X^{(k)}) \cap L^\ell = \emptyset$.*

Proof. We construct the isotopy F_t, skeleton by skeleton. Assume that for $u \leq k-1$, there is an isotopy $\{F_t^u\}_{t \in [0,1]}$ $(F_0^u = \text{id})$ such that $F_1^u(X^{(u)}) \cap L^\ell = \emptyset$. Then there is a neighborhood U_u of $X^{(u)}$ such that $F_1^u(U_u) \cap L^\ell = \emptyset$.

Let $u + 1 \leq k$. Since the number of $(u + 1)$-dimensional cells of $X^{(k)}$ is c_{u+1}, there is a negative real number τ_{u+1} such that, for the $(u + 1)$-dimensional cells e_i^{u+1} ($i = 1, \ldots, c_k$) of $X^{(k)}$, $\Psi_{\tau_{u+1}}((\partial D^{u+1} \times D^{n-u-1})_i) \subset U_u$. Since there are only finitely many handles of index $u + 1$, we can take τ_{u+1} uniformly on i.

We define F_t^{u+1} with support in $\bigcup_{i=1}^{c_{u+1}} \Psi_{\tau_{u+1}}((D^{u+1} \times D^{n-u-1})_i)$. Note that

$$\bigcup_{i=1}^{c_{u+1}} \Psi_{\tau_{u+1}}((D^{u+1} \times D^{n-u-1})_i) \quad (\subset W'_{u+1})$$

is a union of disjoint closed balls in M^n. Since $\Psi_{\tau_{u+1}}((\partial D^{u+1} \times D^{n-u-1})_i) \subset U_u$, there is a disk $(D'^{u+1} \times \{0\})_i \subset (\text{int}(D^{u+1}) \times \{0\})_i$ such that

$$\Psi_{\tau_{u+1}}(((D^{u+1} \setminus \text{int}(D'^{u+1})) \times D^{n-u-1})_i) \subset U_u.$$

Hence

$$X^{(u+1)} \cap L^\ell \subset \bigcup_{i=1}^{c_{u+1}} \Psi_{\tau_{u+1}}((\text{int}(D'^{u+1}) \times \{0\})_i).$$

We have the projection

$$p = \text{proj}_2 \circ \Psi_{-\tau_{u+1}} : \Psi_{\tau_{u+1}}((D^{u+1} \times D^{n-u-1})_i) \longrightarrow D^{n-u-1}.$$

Since $p(\Psi_{\tau_{u+1}}((D^{u+1} \times D^{n-u-1})_i) \cap L^\ell)$ is a finite union of images of $\boldsymbol{R}^s$ ($s \leq \ell \leq n-k-1 \leq n-u-2$) under differentiable maps, it is a measure zero subset of D^{n-u-1}, and since L^ℓ is compact, it is a nowhere dense subset of D^{n-u-1}. Take a point q close to 0 in the complement of

$$p(\Psi_{\tau_{u+1}}((D^{u+1} \times D^{n-u-1})_i) \cap L^\ell).$$

Let $\{F_t^{u+1}\}_{t \in [0,1]}$ ($F_0^{u+1} = \text{id}$) be the isotopy with support in $\bigcup_{i=1}^{c_{u+1}} \Psi_{\tau_{u+1}}((D^{u+1} \times D^{n-u-1})_i)$ such that

$$F_t^{u+1}(\Psi_{\tau_{u+1}}(x, 0)) = \Psi_{\tau_{u+1}}(x, t\mu(x))$$

for $\Psi_{\tau_{u+1}}(x, 0) \in (D'^{u+1} \times D^{n-u-1})_i$, where $\mu : \text{int}(D'^{u+1}) \longrightarrow [0, 1]$ is a C^∞ function with compact support such that $\mu(x) = 1$ for $x \in D''^{u+1} \subset \text{int}(D'^{u+1})$ such that

$$\Psi_{\tau_{u+1}}(((D^{u+1} \setminus \text{int}(D''^{u+1})) \times D^{n-u-1})_i) \subset U_u$$

and

$$X^{(u+1)} \cap L^\ell \cap \Psi_{\tau_{u+1}}((D^{u+1} \times D^{n-u-1})_i) \subset \Psi_{\tau_{u+1}}((D''^{u+1} \times \{0\})_i).$$

Thus we obtain an isotopy $\{F_t^{u+1}\}_{t \in [0,1]}$ such that $F_1^{u+1}(X^{(u+1)}) \cap L^\ell = \emptyset$.

Then we define F_t to be the composition of $F_t^k, \ldots, F_t^0$. □

Remark 2.1. Note that the support of the isotopy $\{F_t^u\}_{t \in [0,1]}$ is contained in a disjoint union of balls, hence it is contained in a larger embedded ball V_u. Note also that we can choose F_1^u which is a commutator with support in the ball. It is because we can take a ball $V_u' \subset \overline{V_u'} \subset V_u$ which contains the support of the isotopy $\{F_t^u\}_{t \in [0,1]}$, and choose an element $\alpha \in \text{Diff}_c^r(V_u)$ such that $\alpha(V_u') \cap V_u' = \emptyset$ and $\alpha(V_u') \cap X^{(k)} = \emptyset$, Then $F_1^u \alpha (F_1^u)^{-1} \alpha^{-1}$ coincides with F_1^u on $X^{(k)}$.

Proof of Theorem 2.2. By applying Lemma 2.1 to the core complex $X^{(k)}$ of M^n with respect to $X^{(k)}$ itself, there is an isotopy $\{F_t\}_{t \in [0,1]}$ ($F_0 = \text{id}$) such that $F_1(X^{(k)}) \cap X^{(k)} = \emptyset$. Then there is a neighborhood W of $X^{(k)}$ such that $W \cap F_1(W) = \emptyset$. By using the gradient flow, we can construct an isotopy $\{G_t\}_{t \in [0,1]}$ ($G_0 = \text{id}$) such that $G_1(F_1(\overline{W})) \subset W$. Then for $g = G_1 \circ F_1$ and $U = W \setminus G_1(F_1(\overline{W}))$, $g^j(U)$ ($j \in \mathbf{Z}$) are disjoint (see Lemma 4.5[21]).

Note here that $F_1 = F_1^k \circ \cdots \circ F_1^0$ is a product of c commutators with support in balls by Remark 2.1, where $F_t^u = \text{id}$ if there are no handles of index u.

On the other hand, G_1 is defined by using the gradient flow. However, G_1 can also be written as a product of isotopies with support in neighborhoods of

$$(D^u \times D^{n-u})_i \cup \bigcup_{t \in [0,\infty)} \Psi_t((D^u \times \partial D^{n-u})_i)$$

which shrink these sets to the core disks $(D^u \times \{0\})_i$, where $i = 1, \ldots, c_u$; $u = 0, \ldots, k$. These neighborhoods are balls and the product G_1^u of these isotopies for the handles of the same index u is with support in a disjoint union of balls. Hence it is also supported in a larger embedded ball. By an argument similar to that in Remark 2.1, G_1^u can be replaced by a commutator with support in the ball without changing $G_1^u|(F_1(\overline{W}))$. Hence $G_1 = G_1^0 \circ \cdots \circ G_1^k$ is also a product of c commutators with support in balls.

Now any element $f \in \text{Diff}_c^r(M^n)_0$ ($1 \leq r \leq \infty, r \neq n+1$) is conjugate to an element with support in U by an isotopy constructed from the gradient flow Ψ_t. We may assume that the support of f is contained in U.

By the results of Herman-Mather-Thurston,[2,7,10,11,15] f can be written as a product of commutators such that the support of each commutator is contained in an embedded ball.

Hence we can write $f = [a_1, b_1] \cdots [a_k, b_k]$, where the supports of a_i and b_i are contained in a ball V_i in U. We put

$$H = \prod_{i=1}^{k} g^{k-i}([a_1, b_1] \cdots [a_i, b_i]) g^{i-k},$$

where $g = G_1 \circ F_1$. Then H is an element of $\text{Diff}_c^r(M^n)_0$ and

$$H^{-1}gHg^{-1} = ([a_1, b_1] \cdots [a_k, b_k])^{-1} \prod_{i=0}^{k-1} g^{k-i}[a_{i+1}, b_{i+1}] g^{i-k}$$

$$= f^{-1} \prod_{i=0}^{k-1} g^{k-i}[a_{i+1}, b_{i+1}] g^{i-k}$$

$$= f^{-1} \Big[\prod_{i=0}^{k-1} g^{k-i} a_{i+1} g^{i-k}, \prod_{i=0}^{k-1} g^{k-i} b_{i+1} g^{i-k} \Big].$$

By putting $A = \prod_{i=0}^{k-1} g^{k-i} a_{i+1} g^{i-k}$ and $B = \prod_{i=0}^{k-1} g^{k-i} b_{i+1} g^{i-k}$, f can be written as a product of two commutators: $f = [A, B][g, H^{-1}]$.

Now, note that the supports of A and B are contained in a disjoint union $\bigcup_{i=1}^{k} g(V_i)$ of balls $g(V_i)$. Thus the supports of A and B are contained in a larger embedded ball.

Since F_1 and G_1 can be written as products of c commutators with support in balls, $g = G_1 \circ F_1$ can be written as a product of $2c$ commutators with support in balls and $[g, H^{-1}] = g(H^{-1} g^{-1} H)$ can be written as a product of $4c$ commutators with support in balls. Thus f can be written as a product of $4c + 1$ commutators with support in balls. □

Remark 2.2. In many cases, we can construct F_1 such that $(F_1)^j(W)$ ($j \in \mathbf{Z}$) are disjoint. In this case, we use F_1 and W in the place of g and U, and f is written as a product of $2c + 1$ commutators with support in balls. In particular, for a 3-dimensional handle body H^3, this is the case, where $c = 2$. Hence any element of $\text{Diff}_c^r(H^3)_0$ ($1 \leq r \leq \infty$, $r \neq 4$) can be written as a product of 5 commutators with support in balls.

3. Uniform simplicity of the diffeomorphism groups

First we review how the perfectness of $\text{Diff}_c^r(\mathbf{R}^n)_0$ implies the simplicity of $\text{Diff}_c^r(M^n)_0$ for a connected manifold M. That is, we have the following lemma which is now well known.

Lemma 3.1. *Let M^n be a connected n-dimensional manifold. Let g be a nontrivial element of $\mathrm{Diff}^r_c(M^n)_0$. Assume that $f \in \mathrm{Diff}^r_c(M^n)_0$ is written as a product of commutators $[a_i, b_i]$ ($i = 1, \ldots, k$): $f = [a_1, b_1] \cdots [a_k, b_k]$, where a_i and b_i are with support in an embedded ball $U_i \subset \overline{U}_i \subset M^n$. Then f can be written as a product of $4k$ conjugates of g and g^{-1}.*

Proof. Since g is a nontrivial element of $\mathrm{Diff}^r_c(M^n)_0$, there is an open ball $U \subset \overline{U} \subset M^n$ such that $g(U) \cap U = \emptyset$. Then any commutator $[a, b]$ in $\mathrm{Diff}^r_c(U)_0$ can be written as a product of 4 conjugates of g or g^{-1}. For, if $a, b \in \mathrm{Diff}^r_c(U)_0$, then by putting $c = g^{-1}ag$, we have $cb = bc$ and

$$aba^{-1}b^{-1} = gcg^{-1}bgc^{-1}g^{-1}b^{-1}$$
$$= gcg^{-1}c^{-1}cbgc^{-1}b^{-1}bg^{-1}b^{-1}$$
$$= g(cg^{-1}c^{-1})(bcgc^{-1}b^{-1})(bg^{-1}b^{-1}).$$

Now for $f = [a_1, b_1] \cdots [a_k, b_k]$, there are balls U_i such that $\mathrm{supp}(a_i)$, $\mathrm{supp}(b_i) \subset U_i$. By the ball theorem, there is a diffeomorphism $h_i \in \mathrm{Diff}^r(M^n)_0$ such that $h_i(U_i) = U$. Since $h_i[a_i, b_i]h_i^{-1}$ is with support in U, it can be written as a product of 4 conjugates of g or g^{-1}: $h_i[a_i, b_i]h_i^{-1} \in (C_g)^4$. Hence $[a_i, b_i] \in (C_g)^4$ and $f = [a_1, b_1] \cdots [a_k, b_k] \in (C_g)^{4k}$. □

Before proving Theorem 1.3, we give a remark which makes a better estimate on the number of commutators than our previous one (Theorem 5.2[21]).

Remark 3.1. In Theorem 5.2[21], we showed that any element $f \in \mathrm{Diff}(S^n)_0$ can be written as a product of 4 commutators. However, we can in fact write $f \in \mathrm{Diff}(S^n)_0$ as a product of 3 commutators with support in embedded balls. The reason is as follows: By Theorem 5.1[21], for $f \in \mathrm{Diff}^r(S^n)_0$, we have the decomposition $f = g \circ h$, where $g \in \mathrm{Diff}^r_c(S^n \setminus Q^0)_0$ and $h \in \mathrm{Diff}^r_c(S^n \setminus P^0)_0$ for some points P^0 and $Q^0 \in S^n$. We have a closed ball $\overline{V}$ containing the support of the isotopy of g and take a diffeomorphism $\alpha \in \mathrm{Diff}^r_c(S^n \setminus Q^0)_0$, such that $\alpha(V) \cap V = \emptyset$ and $P^0 \notin \alpha(V)$. Then

$$f = (g\alpha g^{-1}\alpha^{-1}) \circ (\alpha g^{-1}\alpha^{-1}h)$$

and

$$\mathrm{supp}(\alpha g^{-1}\alpha^{-1}h) \subset \alpha(V) \cup \mathrm{supp}(h) \not\ni P^0.$$

Thus $g\alpha g^{-1}\alpha^{-1} \in \mathrm{Diff}^r_c(S^n \setminus Q^0)_0$ and $\alpha g^{-1}\alpha^{-1}h \in \mathrm{Diff}^r_c(S^n \setminus P^0)_0$. Here $\alpha g^{-1}\alpha^{-1}h$ can be written as a product of 2 commutators by Theorem 2.1 (Theorem 4.1[21]). Since $S^n \setminus Q^0$ and $S^n \setminus P^0$ are diffeomorphic to $\boldsymbol{R}^n$ and any commutator of $\text{Diff}_c^r(\mathbf{R}^n)_0$ is with support in a ball, f can be written as a product of 3 commutators with support in embedded balls.

Proof of Theorem 1.3. By Remark 3.1, any element $f \in \text{Diff}^r(S^n)_0$ can be written as a product of 3 commutators with support in embedded balls. By Lemma 3.1, f is written as a product of $4 \cdot 3 = 12$ conjugates of γ or γ^{-1} for any nontrivial element $\gamma \in \text{Diff}^r(S^n)_0$. □

By using Theorem 2.2, and Theorem 5.2[21], the proof of Theorem 1.4 is straightforward.

Proof of Theorem 1.4. Let M^{2m} be a compact connected $(2m)$-dimensional manifold with handle decomposition without handles of index m. For M^{2m}, from the handle decomposition, we obtain P^{m-1} and $Q^{m-1} \subset M^{2m}$ such that $P^{m-1} \subset M^{2m} \setminus Q^{m-1}$ and $Q^{m-1} \subset M^{2m} \setminus P^{m-1}$ are deformation retracts. By Theorem 5.2[21], any element f of $\text{Diff}^r(M^{2m})_0$ can be decomposed as $f = g \circ h$, where $g \in \text{Diff}_c^r(M^{2m} \setminus k(Q^{m-1}))_0$ and $h \in \text{Diff}_c^r(M^{2m} \setminus P^{m-1})_0$. Then by Theorem 2.2, g and h can be written as products of $4c(M^{2m} \setminus k(Q^{m-1})) + 1$ and $4c(M^{2m} \setminus P^{m-1}) + 1$ commutators with support in balls if $1 \leq r \leq \infty$, $r \neq 2m+1$, respectively. Since

$$c(M^{2m} \setminus k(Q^{m-1})) + c(M^{2m} \setminus P^{m-1}) = c(M^{2m}),$$

f can be written as $4c(M^{2m}) + 2$ commutators with support in balls. By Lemma 3.1, for any nontrivial element $\gamma \in \text{Diff}^r(M^{2m})_0$, f can be written as a product of $16c(M^{2m}) + 8$ conjugates of γ or γ^{-1}. □

Before proving Theorem 1.5, we give a better estimate on the number of commutators than our previous one (Theorem 6.1[21]).

Remark 3.2. In Theorem 6.1[21], we showed that for a compact $(2m+1)$-dimensional manifold M^{2m+1}, any element $f \in \text{Diff}^r(M^{2m+1})_0$ ($1 \leq r \leq \infty$, $r \neq 2m+2$) can be written as a product of 6 commutators. We can in fact write $f \in \text{Diff}^r(M^{2m+1})_0$ as a product of 5 commutators. The reason is just as follows: For a compact connected $(2m+1)$-dimensional manifold M^{2m+1}, we obtain P^m and $Q^m \subset M^{2m+1}$ from the handle decomposition such that $P^m \subset M^{2m+1} \setminus Q^m$ and $Q^m \subset M^{2m+1} \setminus P^m$ are deformation retracts. By Theorem 6.2[21], any element f of $\text{Diff}^r(M^{2m+1})_0$ can be decomposed as $f = a \circ g \circ h$, where a is with support in a disjoint union of balls, $g \in \text{Diff}_c^r(M^{2m+1} \setminus k(Q^m))_0$ and $h \in \text{Diff}_c^r(M^{2m+1} \setminus k'(P^m))_0$. By an argument similar to that in Remark 2.1 or 3.1, the diffeomorphism a can be replaced by a commutator with support in the ball by changing g. Since g and h can be written as products of two commutators by Theorem 2.1 (Theorem 4.1[21]), f can be written as products of 5 commutators.

Proof of Theorem 1.5. By Remark 3.2, any element f of $\mathrm{Diff}^r(M^{2m+1})_0$ is decomposed as $f = a \circ g \circ h$, where a is a commutator with support in the ball, $g \in \mathrm{Diff}^r_c(M^{2m+1} \setminus k(Q^m))_0$ and $h \in \mathrm{Diff}^r_c(M^{2m+1} \setminus k'(P^m))_0$. Then by Theorem 2.2, g and h can be written as products of $4c(M^{2m+1} \setminus k(Q^m)) + 1$ and $4c(M^{2m+1} \setminus k'(P^m)) + 1$ commutators with support in balls if $1 \leq r \leq \infty$, $r \neq 2m+2$, respectively. Since

$$c(M^{2m} \setminus k(Q^m)) + c(M^{2m+1} \setminus k'(P^m)) = c(M^{2m}),$$

f can be written as $4c(M^{2m+1}) + 3$ commutators with support in balls. By Lemma 3.1, for any nontrivial element $\gamma \in \mathrm{Diff}^r(M^{2m+1})_0$, f can be written as a product of $16c(M^{2m+1}) + 12$ conjugates of γ or γ^{-1}. □

Proof of Corollary 1.2. By Remark 2.2, for a 3-dimensional open handle body H^3, any element of $\mathrm{Diff}^r_c(H^3)$ ($1 \leq r \leq \infty$, $r \neq 4$) can be written as a product of 5 commutators with support in balls. Now any element $f \in \mathrm{Diff}^r(M^3)_0$, can be decomposed as $f = a \circ g \circ h$ as in the proof of Theorem 1.5. Since g and h can be written as products of 5 commutators with support in balls, f can be written as 11 commutators with support in balls. By Lemma 3.1, for any nontrivial element $\gamma \in \mathrm{Diff}^r(M^3)_0$, f can be written as a product of 44 conjugates of γ or γ^{-1}. □

Remark 3.3. The uniform simplicity of the groups we treated also follows from a proposition of Burago-Ivanov-Polterovich (Proposition 1.15[3]), our previous remark (Remark 6.6[21]) and Lemma 3.1. We note here that the fragmentation norm[3] of an element of $\mathrm{Diff}^r(S^n)_0$ is at most 2, that of an element of $\mathrm{Diff}^r(M^{2m})_0$ for M^{2m} with handle decomposition without handles of index m is at most $2c(M^{2m}) + 2$, that of an element of $\mathrm{Diff}^r(M^{2m+1})_0$ is at most $2c(M^{2m+1}) + 3$. The reason is that for $g = G_1 \circ F_1$ which we used in the proof of Theorem 2.2,

$$\begin{aligned} G_1 \circ F_1 &= G_1^0 \circ \cdots \circ G_1^k \circ F_1^k \circ \cdots \circ F_1^0 \\ &= (G_1^0 \circ F_1^0) \circ (F_1^0)^{-1} \circ (G_1^1 \circ F_1^1) \circ (F_1^0) \\ &\quad \circ (F_1^1 \circ F_1^0)^{-1} \circ (G_1^2 \circ F_1^2) \circ (F_1^1 \circ F_1^0) \\ &\quad \circ \cdots \circ (F_1^{k-1} \circ \cdots \circ F_1^0)^{-1} \circ (G_1^k \circ F_1^k) \circ (F_1^{k-1} \circ \cdots \circ F_1^0) \end{aligned}$$

and $G_1^u \circ F_1^u$ ($0 \leq u \leq k$) is with support in a union of disjoint balls, hence is with support in a larger ball. Hence $g = G_1 \circ F_1$ can be written as a product of c diffeomorphisms with support in embedded balls.

Acknowledgement

The author is partially supported by Grant-in-Aid for Scientific Research 20244003, Grant-in-Aid for Exploratory Research 18654008, Japan Society for Promotion of Science, and by the Global COE Program at Graduate School of Mathematical Sciences, the University of Tokyo.

References

1. R. D. Anderson, Amer. J. of Math. **80** (1958), 955–963.
2. A. Banyaga, *The structure of classical diffeomorphism groups, Mathematics and its Applications*, vol. 400, Kluwer Academic Publishers Group, Dordrecht (1997) xii+197 pp.
3. D.Burago, S.Ivanov and L.Polterovich, Advanced Studies in Pure Math. **52**, Groups of Diffeomorphisms (2008), pp. 221–250.
4. D. B. A. Epstein, Compositio Math. **22** (1970), 165–173.
5. G. M. Fisher, Transactions Amer. Math. Soc. **97** (1960) 193–212.
6. S. Haller and J. Teichmann, Annals of Global Analysis and Geometry **23**, 53-63 (2003).
7. M. Herman, Publ. Math. I. H. E. S. **49** (1979), 5–234.
8. D. Kotschick, Advanced Studies in Pure Math. **52**, Groups of Diffeomorphisms (2008), pp. 401–413.
9. J. Mather, Topology **10** (1971), 297–298.
10. J. Mather, Comm. Math. Helv. **48** (1973) 195–233.
11. J. Mather, Comm. Math. Helv. **49** (1974), 512–528, **50** (1975), 33–40, and **60** (1985), 122–124.
12. S. Matsumoto and S. Morita, Trans. Amer. Math. Soc. **94** (1985), 539–544.
13. J. Milnor, *Morse theory*, Annals of Mathematics Studies, No. 51, Princeton University Press, Princeton, N.J. (1963), vi+153 pp.
14. J. Milnor, *Lectures on the h-cobordism theorem*, Princeton University Press, Princeton, N.J. (1965), v+116 pp.
15. W. Thurston, Bull. Amer. Math. Soc. **80** (1974), 304–307.
16. T. Tsuboi, Ann. Inst. Fourier **31** (2) (1981) 1–59.
17. T. Tsuboi. Advanced Studies in Pure Math. **5**, Foliations (1985), pp. 37-120.
18. T. Tsuboi, Annals of Math. **130** (1989), 227–271.
19. T. Tsuboi, Proceedings of Foliations: Geometry and Dynamics, Warsaw 2000, World Scientific, Singapore (2002), pp. 421–440.
20. T. Tsuboi, Foliations 2005, Lodz, World Scientific, Singapore (2006), pp. 411–430.
21. T. Tsuboi, Advanced Studies in Pure Math. **52**, Groups of Diffeomorphisms (2008), pp. 505–524.
22. S. M. Ulam and J. von Neumann, Bull. Amer. Math. Soc. **53** (1947), 508.
23. H. Whitney, Ann. of Math., **45** (1944), 247–293.

ON BENNEQUIN'S ISOTOPY LEMMA AND THURSTON'S INEQUALITY

Y. Mitsumatsu

Department of Mathematics, Chuo University
Bunkyo-ku, Tokyo, 112-8551, Japan
E-mail: yoshi@math.chuo-u.ac.jp

In this exposition we survey convergences of contact structures to foliations from the point of view of Thurston's and Thurston–Bennequin's inequalities featuring the case when the structures are associated with spinnable structures and their Dehn fillings. Also we review previous works by the author and others[6] and[12] as the main motivation for the above.

1. Thurston's Inequality

Let us first recall Thurston's inequality for foliations of codimension 1 on 3-manifolds and Thurston–Bennequin's inequality for contact structures, which we can regard as topological expressions of '(pseudo-)convexity'. Unless otherwise specified, throughout the article all 3-manifolds are oriented and compact, foliations and contact structures are both tangentially and transversely oriented in a coherent way, and so are embedded surfaces as well.

Let $\mathcal{F}$ be a foliation of codimension one on a closed oriented 3-manifold M. Assuming that $\mathcal{F}$ has no Reeb components, Thurston showed the following inequality for arbitrary embedded closed surface Σ of genus $g > 0$.

Thurston's Absolute Inequality (cf.[16])

$$|\langle e(T\mathcal{F}), [\Sigma]\rangle| \leq |\chi(\Sigma)| = 2g - 2.$$

Here $e(\cdot)$ and $\chi(\cdot)$ denote the Euler class and the Euler characteristic.

As a principle the following relative inequality is more refined, which also Thurston proved under the same assumption. Let Σ be any Seifert surface whose oriented boundary $L = \partial \Sigma$ is a positively transverse link to $\mathcal{F}$. As Σ is homotopically equivalent to a 1-complex, there exists a trivialization X of $T\mathcal{F}|_\Sigma$. Let L^X denote the shift of L along $X|_L$. Consider the linking number $\text{lk}(L, L^X)$ between L and L^X, which is also regarded as the relative Euler number $-\langle e(T\mathcal{F}), [\Sigma, L] \rangle$ under the boundary condition that $T\mathcal{F}|_L$ is trivialized by $T\mathcal{F} \cap T\Sigma$ along L.

Thurston's Relative Inequality (cf.[16]) $\quad \text{lk}(L, L^X) \leq -\chi(\Sigma)$.

These inequalities have their complete analogues in contact topology. Simply replacing $T\mathcal{F}$ with an oriented contact plane field ξ in Thurston's inequalities, we obtain so called *Thurston-Bennequin's* absolute and relative inequalities for oriented contact structures. In the contact case, because the relative inequality is definitely stronger than the absolute one, we usually refer to the relative one simply as Thurston-Bennequin's inequality.

Not only because these inequalities look alike in the foliation and contact cases, through convergences of contact structures to foliations (or perturbations of foliations to contact structures, *vice versa*) which are studied under the name of 'confoliations' in,[3] their similarities are drawing our attentions.

Now let us take an isotopic family $\{\xi_t\}$ (or more generally an isomorphic sequence $\{\xi_n\}$) of contact structures which converges to a foliation $\mathcal{F}$ as plane fields in C^0-topology. Then as an oriented $\mathbb{R}^2$-bundles, the ξ_t's and $T\mathcal{F}$ are clearly isomorphic, if the absolute inequality holds for one, then for the other the corresponding absolute one holds as well.

On the other hand, concerning the relative inequalities the situation is more delicate. If Thurston-Bennequin's relative inequality holds for ξ_n's, namely ξ_n's are tight, then Thurston's relative inequality holds for $\mathcal{F}$. But the converse is not true in general. If an oriented link is positively transverse to $\mathcal{F}$, so is it to ξ_n's for large enough n as well. However a transverse link to a ξ_n needs not to be transverse to $\mathcal{F}$ in general. For example, there exists an isotopic family of overtwisted contact structures which converges to the standard contact structure ξ_0 on S^3 (cf.[8,9]).

Example 1.1. A *bad convergence* like the above happens each time when we take a spinnable structure (=open book decomposition) of M^3 with monodromy φ which supports a tight contact structure ξ_φ as its Thurston–Winkelnkemper construction ([6,17]). In such a case, the contact structure can be isotopically deformed and converges to a spinnable foliation $\mathcal{F}_\varphi$ ([6,9]). ξ_φ and $\mathcal{F}_\varphi$ satisfy the four inequalities. Now we can arrange the Lutz twist

$\xi_{\varphi LT}$ of ξ_φ and its isotopic family so as to converge to $\mathcal{F}_\varphi$ by the same construction as in.[8,9]

One of the aim of this article is to show that we can avoid such bad convergences for contact sructures and foliations associated with a spinnable structure or its Dehn filling. Here, the Dehn filling implies the one performed on the complement of the Reeb component (*i.e.*, the neighbourhood of the binder). In this article, we call this just as the Dehn filling of the spinnable structure.

In Bennequin's pioneering work[1] it is shown that Thurston–Bennequin's relative inequality implies the tightness of a contact structure. Conversely the elimination lemma due to Eliashberg[2] and Giroux[4] tells that the tightness implies both the absolute and Thurston–Bennequin's relative inequality. Therefore the relative one implies the absolute one, but the implication is not direct and to show it we should once pass through the tightness. This indirectness appears more seriously on the foliation side, namely, Thurston's relative inequality does not imply the absolute one.

The known examples of foliations for which the absolute inequality fails but the relative one holds are very limited. They are all not more than a variant of $\zeta = \{S^2 \times \{*\}\}$ on $S^2 \times S^1$. It is pausible that there is essentially no more such foliations on prime 3-manifolds.

The following diagram summarizes the situations explained above.

Diagram 1.2.

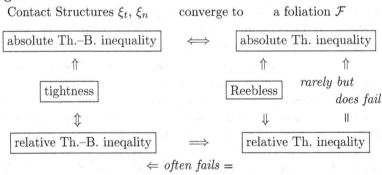

In the final section we expalin the bottom arrow form right to left holds for foliations and contact structures associated with spinnable structures and their Dehn fillings. We close this section by raising the following problem.

Problem 1.3. Specify the class of foliations for which the right side vertical arrow from bottom to top fails.

2. Thurston's Absolute Inequality for Spinnable Foliations

Here we review the results in[6] very briefly in a limited situation. Let $F \cong \Sigma_g^1$ be a compact oriented surface of genus g with one bondary component. The mapping class group $\mathcal{M}_g^1$ of $F = \Sigma_g^1$ relative to the boundary admits a generating system $\{\tau_0, ..., \tau_{2g}\}$ which are right handed Dehn-twists along the simple closed curves $C_0, ..., C_{2g}$ indicated in the following figure. This system is called the *Dehn-Lickorish-Humphries (DHL* for short) *generators*.

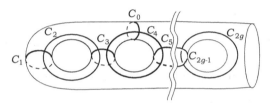 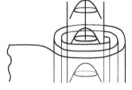

Dehn-Lickorish-Humphries generators   *Spinnable Foliation*

Let M_φ be the 3-manifold which admits a spinnable structure with page F and monodromy $\varphi \in \mathcal{M}_g^1$. Naturally associated with this spinnable structure is the spinnable foliation $\mathcal{F}_\varphi$. For a detailed description, see.[6] Also a contact structure ξ_φ is canonically associated by the Thurston-Winkelnkemper construction ([17]) to the monodromy φ. Moreover we can place $\mathcal{F}_\varphi$ and ξ_φ in such a way that by the vector field X obtained as the intersection $X = \xi_\varphi \wedge T\mathcal{F}_\varphi$ of the two 2-plane fields generates a flow with which ξ_φ is isotoped and converging to $T\mathcal{F}_\varphi$.

Let us assume that in a presentation of φ with DHL-generators, neither of τ_2 and τ_4 appears. Then the monodromy φ is presented as

$$\varphi = \tau_0^{j_0} \tau_1^{j_1} \tau_3^{j_3} \cdot \prod_{k=4}^{l} \tau_{i_k}^{j_k}, \quad (i_k \in \{5, 6, \ldots, 2g\}, k = 1, \ldots, l).$$

Let us also assume $j_0 j_1 j_3 \neq 0$ because otherwise it is shown as Theorem a in[6] that $e(T\mathcal{F}) = 0 \in H^2(M_\varphi; \mathbb{Q})$, so that Thurston's absolute inequality vacantly holds.

Theorem b, c. *The condition*
$$\frac{1}{j_0} + \frac{1}{j_1} + \frac{1}{j_3} = 0$$
is equivalent to $e(T\mathcal{F}) \neq 0 \in H^2(M_\varphi; \mathbb{Q})$. *In this case necessarily Thurston's absolute inequality fails.*

The above condition implies at least one of j_0, j_1, j_3 is negative. If every exponent is non-negative, a theorem of Loi-Piergallini[7] asserts that $(M_\varphi,)$ is Stein-fillable and the corresponding foliation and contact structure satisfy the four inequalities. The above theorem says that the condition implies the strong 'non-convexity'.

3. Dehn Filling

Concerning the specific monodromies studied in the previous section, it turned out in the work of[12] that if we perform a Dehn filling along the Reeb component of M_φ and take an appropriate foliation, in most cases Thurston's absolute inequality is recovered. In this section let us recall this phenomenon.

Let ℓ_0 be a longitude of the boundary of the complement of the interior of the Reeb component of $\mathcal{F}_\varphi$. Namely, ℓ_0 is the oriented boundary of a page in terms of the spinnable structure. Also let m_0 denote the meridian, which bounds a disk of the Reeb component in M_φ. Now remove the Reeb component from $(M_\varphi, \mathcal{F}_\varphi)$ and paste back the Reeb component with new meridian $m = -p\ell_0 + qm_0$ for a primitive integer vector (p, q). The resultant manifold M_r is called the *Dehn filling* with coefficient $r = -q/p$. (This might be different from the conventional one. The reason for this convention on the coefficient is to make the following description simpler.) The resultant foliation is denoted by $\mathcal{F}_r$. We also obtain a contact structure ξ_r on M_r by more or less the same construction as Thurston-Winkelnkemper's with least twisting on the solid torus.

If $r = 0$ we obtain M_φ and nothing is changed. If $r = \infty$, the boundaries of the pages become meridians and M_r is just a mapping torus of $\hat{\varphi} : \hat{F} \to \hat{F}$ which is the natural extension of φ to the filling up $\hat{F}$ of F by a disk. For integral Dehn fillings, the following proposition is easy to see.

Proposition 3.1. *For any integer $r = n \in \mathbb{Z}$, the triple $(M_r, \mathcal{F}_r, \xi_r)$ is isomorphic to $(M_{\varphi(n)}, \mathcal{F}_{\varphi(n)}, \xi_{\varphi(n)})$ which is associated with the monodoromy $\varphi(n) = \varphi \circ \tau^n_{\partial' F}$, where $\partial'F \subset \text{int}F$ denotes a simple closed curve parallel to ∂F.*

For the sake of simplicity, we assume that the holonomy φ is a composition of only τ_0's, τ_1's, and τ_3's, namely, $\varphi = \tau_0^{j_0}\tau_1^{j_1}\tau_3^{j_3}$. We also keep on assuming the same condition as in the previous section

$$\frac{1}{j_0} + \frac{1}{j_1} + \frac{1}{j_3} = 0$$

which assures $e(T\mathcal{F}_r) \neq 0 \in H^2(M_r; \mathbb{Q})$ for any coefficient r.

The following result is based on a theorem of Sela ([15]) which grew out of Gabai's fundamental work on sutured manifold decomposition.

Theorem 3.2 ([12]). *For all $r \in \mathbb{Q} \cup \{\infty\}$ but finitely many exceptions, $\mathcal{F}_r$ satisfies Thurston's absolute inequality.*

At least in this situation, even under the presence of Reeb component, Thurston's absolute inequality tends to hold.

The following proposition is also well known and not difficult to show.

Proposition 3.3. *For any monodoromy φ, $\varphi(n)$ is written as a product of only right-handed Dehn twists for $n \gg 0 \in \mathbb{N}$. Consequently, for such n, the structures are Stein fillable and satisfy the four inequalities.*

The set $T_\varphi = \{r; \xi_r \text{ is tight}\}$ of coefficients includes the above integers. By the work of Honda-Kazez-Matić,[5] we also know that $T_\varphi \cap \mathbb{Z}$ is included in $\{n \in \mathbb{Z}; \varphi(n) \text{ is right-veering }\}$.

Now our interests on these examples amount to the following question.

Problem 3.4. Are ξ_r fillable or tight for non-integral $r \gg 0$?

4. Bennequin's Isotopy Lemma

In the final section we explain that Bennequin's isotopy lemma which was extended to the case of spinnable structures in[10] has a slight generalization, namely to the case of Dehn fillings.

Generalized Bennequin's Isotopy Lemma *For Dehn fillings associated with any spinnable structure of any coefficient r, $\mathcal{F}_r$ and ξ_r can be placed so that the following statement holds:*

> *Any positively transverse link to ξ_r can be isotoped through such links to one which is also positively transverse to $\mathcal{F}_r$.*

Consequently, the bottom horizontal arrow from right to left in Diagram 1.2, which is labeled as 'often fails', holds for Dehn fillings of spinnable structures.

In the above statement 'placed' implies the following: We can fix the placement of $\mathcal{F}_r$. Then, for a natural realization of ξ_r, we can find an isotopy by which ξ_r is reset well with respect to $\mathcal{F}_r$.

This lemma holds even for a spinnable structure with a page F which has more than one boundary components. In such a case, the coefficient r should be regarded as a multi-index respecting each boundary component. However, in what follows, only for the simplicity, the arguments go as if F has only one boundary component.

A proof of Bennequin's isotopy lemma in the case of spinnable structures is very roughly outlined in.[10] Of course the basic ideas are all due to Bennequin ([1]). In this section we give a fairy quick explain the proof of the above lemma, focusing our attntion on the point where the proof is modified from the case of spinnable structures to that of Dehn fillings. The detailed proof will appear in the forthcoming paper.[11]

The first step of the isotopy makes the link avoid the Reeb component. This is not at all difficult in Bennequin's original case, which can be regarded as the spinnable structure with monodromy $\varphi = \text{Id}_{D^2}$, but it is not the same even in the case of spinnable structure. This is overlooked in.[10] This can be fixed but requires a compicated explanation. So it is detailed in.[11] Between spinnable structures and Dehn fillings, no change is needed for this step.

In the second step of isotopy, on the pages, the link is devided into short pieces each of which are always positive or always negative with respect to the pages. Here instead of considering foliations, we take care of only pages, because away from the boundary they are the same.

The vector field of the intersection of the foliation and contact structure (see the description in Section 2) has positive divergence. The pieces which are negative with respect to the foliation flows out to the boundary by this vector field. We have to take care of self-intersections. This is achieved by further breaking the pieces into shorter ones. This step has no change from.[10]

Now we come to the important step. On the boundary of the solid torus, a small piece of the link is positively transverse to the contact structure but negative with respect to the pages. The piece is easily isotoped relative to the end points in the solid torus 'to the other side' of the boundary so that the piece is positively transverse with respect to the pages. This process is easier in the case of spinnable structures, as seen in the following.

In the case of a Dehn filling M_r, before realizing the above isotopy, first we change the original spinnable structure M_φ into some $M_{\varphi(n)}$, because M_r is considered to be also a Dehn filling of $M_{\varphi(n)}$ for any $n \in \mathbb{Z}$. It is easy to find such n that the new meridian is positive with respect to the pages. The new meridian m for $M_{\varphi(n)}$ is $m_0 - n\ell_0$.

Then, on the boundary of the solid torus, to describe our Dehn filling, we still have a freedom for the choice of the longitude ℓ up to $\mathbb{Z} \cdot m$.

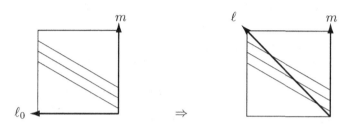

Thin lines indicate the boundary of pages.

We can take ℓ so that on the $(-\ell)$-m plane the boundary of the pages has a positive slope less than 1. The contact structure on the boundary draws a line field on the boundary torus which is close to the boundary of the pages. So we put a contact sturcture on the solid torus which is presented as $\ker[d\theta + r^2 d\omega]$, where (r, ω, θ) $(r \leq R)$ is the cylindrical coordinate for the Reeb component as a solid torus. An appropriate choice of radius R makes it fit to the contact structure on the pages.

From the situation we see that on the $(-\ell)$-m plane the slope of the contact structure is positive and steeper than that of the pages. The piece of the link is positive with respect to the contact structure but negative with respect to the pages. Therefore its direction is 'down-left' and pinched in a narrow sector between the pages and the contact structures.

Then we can push the piece in the same way as in the case of spinnable structures. The result of this isotopy is, as drawn in the figure, transverse to $\mathcal{F}_r$. This completes all the process of isotopies.

Acknowledgements

The author is supported by Grant-in-Aid for Scientific Research 18340020.

References

1. D. Bennequin, *Astérisque* **107-108** (1983), 83–161.

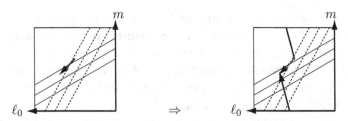

Dotted thin lines indicate the intersection with contact structre.
Thick short arrow is a piece of link.
Thick arrow in the right figure is the result of isotopy.

2. Y. Eliashberg, *Ann. Inst. Fourier* **42** (1991), 165–192.
3. Y. Eliashberg and W. Thurston, *Confoliations*, A.M.S. University Lecture Series, **13** (1998).
4. E. Giroux, *Comment. Math. Helvetici* **66** (1991), 637–677.
5. K. Honda, W. Kazez and G. Matić, *Invent. Math.* **169** (2007), 427–449.
6. H. Kodama, Y. Mitsumatsu, S. Miyoshi and A. Mori, *On Thurston's inequality for spinnable foliations*, preprint, submitted to the Proceedings of PAULFEST.
7. A. Loi and R. Piergallini, *Invent. Math.* **143**(2001), 325-348.
8. Y. Mitsumatsu, Foliations and contact structures on 3-manifolds, *Foliations: geometry and dynamics (Warsaw, 2000)*, 75–125, World Sci. Publ., River Edge, NJ, 2002.
9. Y. Mitsumatsu, Convergence of contact structures to foliations, *Foliations 2005*, 353–371, World Sci. Publ., Hackensack, NJ, 2006.
10. Y. Mitsumatsu and A. Mori, On Benequin's isotopy lemma, Appendix to,[9] *Foliations 2005*, 365-371, World Sci. Publ., Hackensack, NJ, 2006.
11. Y. Mitsumatsu and A. Mori, *A generalization of Bennequin's Isotopy Lemma*, in preparation.
12. S. Miyoshi and A. Mori: Reeb components and Thurston's ineauality, to appear in the proceedings of PAULFEST.
13. Sh. Morita, *Math. Proc. Camb. Phil. Soc.* **105** (1989), 79–101.
14. Sh. Morita, *J. Diff. Geom.* **47** (1997), 560–599.
15. Z. Sela: *Israel J. Math.* **69** (1990), 371–378.
16. W. Thurston: *Norm on the homology of 3-manifolds*, Memoirs of the AMS, **339** (1986), 99–130.
17. W. Thurston and E. Winkelnkemper, *Proc. Amer. Math. Soc.* **52** (1975), 345–347.

ON THE JULIA SETS OF COMPLEX CODIMENSION-ONE TRANSVERSALLY HOLOMORPHIC FOLIATIONS

Taro Asuke[*]

*Graduate School of Mathematical Sciences, University of Tokyo
3-8-1 Komaba, Meguro, Tokyo 153-8914, Japan
E-mail: asuke@ms.u-tokyo.ac.jp*

The Fatou-Julia decomposition for foliation is previously introduced by Ghys, Gomez-Mont and Saludes.[8] On the other hand, another Fatou-Julia decomposition can be considered.[3] We will briefly explain how the difference is, and discuss properties of the Julia sets. Some examples are also given. The details including proofs of results will appear elsewhere.

Keywords: Transversally holomorphic foliations, Fatou sets, Julia sets

1. Introduction

The Fatou-Julia decomposition is one of the most basic concepts in studying complex dynamical systems. It is expected that the Fatou-Julia decomposition also exists for transversally holomorphic foliations of complex codimension one. Such a decomposition is firstly introduced by Ghys, Gomez-Mont and Saludes.[8] The decomposition is introduced by using vector fields invariant under the holonomy and related to deformations of foliations. On the other hand, the Fatou set is usually defined in terms of normal families. One can indeed ask if holonomies form a normal family, and define another Fatou-Julia decomposition.[3] Two decompositions are in a simple relation, namely, the Fatou set in the sense of Ghys, Gomez-Mont and Saludes is always contained in the Fatou set in our sense. When studying characteristic classes, the Julia sets are usually preferable to be small. From this viewpoint, the decomposition based on normal families is useful. On the other hand, relation to deformations is less obvious. Our definition of Julia sets is similar to those of mapping iterations so that one can expect they have common properties, for example, the Julia set is expected to

---

[*]The author is partially supported by Grant-in Aid for Scientific research (No. 19684001).

have hyperbolic (expanding or contracting) holonomies. It is true under certain assumptions. In this article, we will explain the definition as well as some properties of Julia sets from this point of view. We will present some examples which illustrates properties of Julia sets.

This article is based on a talk which the author gave at "VIII International Colloquium on Differential Geometry" held in Santiago de Compostela during 7–11 of July, 2008. The author will express his gratitude to the organizers for their warm hospitality. He will also grateful to the referee for valuable comments. Especially, Proposition 2.1 and its proof is suggested by him. The details including proofs of results will appear elsewhere.[3] Some additional account can be found in another review of related results.[4]

2. Definition of the Fatou and Julia sets

Let M be a closed manifold and let $\mathcal{F}$ be a transversally holomorphic foliation of M. We assume that the complex codimension of $\mathcal{F}$ is equal to one. Let T be a complete transversal for $\mathcal{F}$, namely, every leaf of $\mathcal{F}$ meets T at least once, and the holonomy pseudogroup associated with T consists of biholomorphic local diffeomorphisms.

Before defining the Fatou set of $\mathcal{F}$, we will define the Fatou set of (Γ, T) as follows. Since M is compact, we may assume that the number of connected components of T is finite and that each component is an open disc in $\mathbb{C}$. Moreover, we can find another complete transversal T' by slightly shrinking the complete transversal T. Let (Γ', T') be the holonomy pseudogroup obtained by restricting Γ to T'. (Γ', T') is called a reduction of (Γ, T).

Definition 2.1. Let (Γ, T) and (Γ', T') be as above.

(1) A connected open subset U of T' is called a *Fatou neighborhood* if the following conditions are satisfied.

   (a) The germ of any element of Γ' defined on a neighborhood of a point in U extends to an element of Γ defined on the whole U.
   (b) Let
   $$\Gamma_U = \left\{ \gamma \in \Gamma \;\middle|\; \begin{array}{l} \gamma \text{ is defined on } U \text{ and the extension of} \\ \text{the germ of an element of } \Gamma' \text{ as in (a)} \end{array} \right\},$$
   then Γ_U is a normal family.

(2) The union of Fatou neighborhoods is called the *Fatou set* of (Γ', T') and denoted by $F(\Gamma')$. The complement of the Fatou set is called the *Julia set* of (Γ', T') and denoted by $J(\Gamma')$.
(3) The *Fatou set* of (Γ, T) is the Γ-orbit of $F(\Gamma')$, namely, $F(\Gamma) = \Gamma(F(\Gamma'))$. The *Julia set* of (Γ, T) is the complement of $F(\Gamma)$ and denoted by $J(\Gamma)$.

It is known that (Γ, T) is compactly generated.[10] The definition of the Fatou and Julia sets actually works for compactly generated pseudogroups.

Remark 2.1.

(1) Γ_U is a normal family if any subfamily $\mathscr{F}$ of Γ_U admits a subfamily which uniformly converges on compact sets. This implies that asymptotic behavior of holonomies on U is tame.
(2) We can choose T so that T can be embedded as a bounded subset of $\mathbb{C}$ because M is compact. Then, Γ_U is always a normal family by virtue of Montel's theorem. On the other hand, it is necessary to fix a domain of definition in order to speak of normal families. This leads to the first condition in Definition 2.1.

The notion of Fatou set is closely related with complete pseudogroups[9] as follows. A pseudogroup (Γ, T) is *complete* if for each pair of points x and y of T, there are open neighborhoods U of x and V of y with the following properties, namely, if $\gamma_{x'}$ is the germ of an element of Γ of which the source contains $x' \in U$ and the target contains $y' \in V$ then there is an element $\gamma \in \Gamma$ of which the germ at x' is equal to $\gamma_{x'}$. Such a pair (U, V) is called a *completeness pair*.

Complete pseudogroups appear for example as the holonomy pseudogroups of Riemannian foliations of closed manifolds.[1,9] Theorem 2.1 below suggests that these notions are of the same nature. It is straightforward that Γ is complete on the Fatou set in a uniform way. Actually the completeness of Γ on $F(\Gamma)$ is used in the study of the structure of the Fatou sets. Moreover, the following proposition holds. The proposition together with the proof is suggested by the referee, to whom the author is grateful.

Proposition 2.1. *Let F be the union of $x \in T'$ such that there is a completeness pair (U_y, V_y) for any $y \in T'$. Then $F = F(\Gamma')$.*

Proof. Since (Γ, T) is compactly generated, we may assume that T' is a finite union of open balls in $\mathbb{C}$. Hence it suffices to verify that the condition (1)-(a) in Definition 2.1 holds precisely on F. Suppose that $x \in F$. Since the closure $\overline{T'}$ in T is compact, we can find a finite set of completeness pairs $\{(U_i, V_i)\}_{1 \leq i \leq n}$ for T such that $\overline{T'}$ is covered by $\{V_i\}$. Then $U = U_1 \cap \cdots \cap U_n$ is a Fatou neighborhood. The inclusion $F(\Gamma') \subset F$ is clear. □

Remark 2.2.

(1) If (Γ, T) is not compactly generated, then completeness is not preserved under equivalences of pseudogroups.[1] It is also the case for the Fatou-Julia decomposition. See Example 4.5.
(2) The closure of complete pseudogroups of local isometries can be defined.[9] Hence, by Theorem 2.1 below, the closure of $\Gamma|_{F(\Gamma)}$ can be defined. Let Γ_1 be the closure in this sense. On the other hand, another closure of $\Gamma|_{F(\Gamma)}$ can be defined as follows. Let U be a Fatou neighborhood and let $\overline{\Gamma_U}$ be the closure of Γ_U in the space of local biholomorphic diffeomorphisms with respect to the compact-open topology. We denote by Γ_2 the pseudogroup generated by $\overline{\Gamma_U}$, where U runs through Fatou neighborhoods. Then Γ_2 can be regarded as a closure of $\Gamma|_{F(\Gamma)}$. It can be shown that Γ_1 and Γ_2 coincide.

$F(\Gamma)$ and $J(\Gamma)$ are Γ-invariant. Hence the saturation of $F(\Gamma)$ in M makes a sense.

Definition 2.2. The Fatou set of $\mathcal{F}$ is the saturation of $F(\Gamma)$ in M and denoted by $F(\mathcal{F})$. The Julia set of $\mathcal{F}$ is the saturation of $J(\Gamma)$ in M and denoted by $J(\mathcal{F})$. The connected components of $F(\Gamma)$ and $J(\Gamma)$ are called the Fatou components and the Julia components, respectively.

$F(\Gamma)$ and $F(\mathcal{F})$ are open. Of course, $J(\mathcal{F})$ is the complement of $F(\mathcal{F})$.

The following lemma justifies the above definition.

Lemma 2.1.

(1) $F(\Gamma)$ is independent of the choice of (Γ', T').
(2) $F(\mathcal{F})$ is independent of the choice of (Γ, T).

Let $x \in F(\mathcal{F})$ and let l be a leaf path originated from x. The definition says that if l becomes long, then the holonomy along l will converge to a mapping after choosing a subsequence. Hence the dynamics in the Fatou set will be mild. Indeed, the following property of the Fatou set is fundamental.

Theorem 2.1. $\mathcal{F}$ is transversally Hermitian[13] when restricted to $F(\mathcal{F})$.

This is shown as follows. First an invariant metric which is locally Lipschitz continuous can be explicitly constructed. Once this is done, it turns out that holonomies satisfy a first order differential equation which is Lipschitz continuous. This allows to apply classical results and arguments of H. Cartan on local Lie groups,[5] and it follows that $\mathcal{F}$ restricted to $F(\mathcal{F})$ is transversally Hermitian.

Theorem 2.1 is quite relevant in the study of $F(\mathcal{F})$ and also of $J(\mathcal{F})$. For example, one can introduce the notion of critical exponent and conformal measures for Γ analogous to the case of mapping iterations. The construction strongly depends on the above theorem.

The Fatou and Julia sets of foliations are firstly introduced by Ghys, Gomez-Mont and Saludes.[8] The decomposition is defined by using holonomy invariant vector fields. Namely, if there is a section of the complex normal bundle of certain regularity which does not vanish at $x \in M$, then x belongs to the Fatou set. We refer the original article for the details and only remark the following fact.

Proposition 2.2. Let $F_{GGS}(\mathcal{F})$ be the Fatou set in the sense of Ghys, Gomez-Mont and Saludes, then $F_{GGS}(\mathcal{F})$ is contained in $F(\mathcal{F})$. The inclusion can be strict.

Roughly speaking, the reason is as follows. If there is a holonomy invariant vector field, then it induces a vector field on T' invariant under Γ'. If this vector field does not vanish at $p \in T'$, then it does not vanish also at $\gamma(p)$ for any $\gamma \in \Gamma'$. One can integrate the vector field near $\gamma(p)$, $\gamma \in \Gamma'$, at the same time and obtain a holonomy defined on a neighborhood of p. The resulting holonomy does not necessarily belong to Γ', but it belongs to Γ. In view of Theorem 3.1 below, it is preferable that the Julia set is small. On the other hand, the decomposition by Ghys, Gomez-Mont and Saludes is directly related to deformations of foliations. Our Julia set is too large to admit such vector fields.

The difference between $F(\mathcal{F})$ and $F_{GGS}(\mathcal{F})$ is not quite large. Actually $F(\mathcal{F})$ has a similar structure to that of $F_{GGS}(\mathcal{F})$. The difference largely occurs in the following situations:

(1) There is a holonomy of finite order. If such a holonomy exists, the corresponding leaf belongs to $F(\mathcal{F})$ but not to $F_{GGS}(\mathcal{F})$.
(2) It can be shown that the foliation naturally induces a Riemannian foliation $\widetilde{\mathcal{F}}$ in the unit normal bundle on $F(\mathcal{F})$. If there is a leaf, say, $\widetilde{L}$ of

$\widetilde{\mathcal{F}}$ which meets a fiber of the unit normal bundle more than once, the leaves of $\mathcal{F}$ contained in the projection of $\widetilde{L}$ fail to belong to $F_{GGS}(\mathcal{F})$.

The both cases are easily realized as suspensions.

3. Some properties of Julia sets

The Julia set is expected to play a role of minimal sets of real codimension-one foliations. For example, we have the following weak version of Duminy's theorem.[7,11]

Theorem 3.1.

(1) The Godbillon-Vey class vanishes if $J(\mathcal{F})$ is empty.
(2) The imaginary part of the Bott class vanishes if $J(\mathcal{F})$ is empty.

This is a straightforward consequence of Theorem 2.1 and the definition of characteristic classes. A precise version which concerns the Godbillon measure in the sense of Heitsch-Hurder[11] and the residue of the imaginary part of the Bott class[2] can be also shown.

Since the definition of $J(\mathcal{F})$ is similar to that of the Julia sets for mapping iterations, one can expect they have common or similar properties. For example, it can be shown that $J(\mathcal{F})$ contains at most a finite number of closed leaves. One of the significant properties of the Julia sets for mapping iterations and the limit sets of Kleinian groups is that they contain many hyperbolic fixed points except elementary cases. We do not know if it is also the case for $J(\mathcal{F})$, however, $J(\mathcal{F})$ can be characterized as follows. In order to make the idea clearer, we give the statement in terms of foliations but in an ambiguous way. The precise statement is made in terms of pseudogroups.

Theorem 3.2. *A point $z \in M$ belongs to $J(\mathcal{F})$ if and only if there are two sequences, $\{z_n\}$ in M which converges to z and $\{\gamma_n\}$ of holonomies, such that each γ_n is associated to a path originated from z_n and $\lim_{n \to \infty} |\gamma'_n|_{z_n} = +\infty$. Here the case where $z_n = z$ for all n is allowed.*

Remark 3.1. Theorem 3.2 suggests that $J(\mathcal{F})$ contains a hyperbolic holonomy (i.e., the absolute value of the differential is not equal to one) associated to a loop under some mild condition. Indeed, some results are known. First, an analogy of conical limit sets in the context of complex dynamical systems (of mapping iterations) can be also introduced for $J(\mathcal{F})$, and it can be shown that $J(\mathcal{F})$ contains a hyperbolic holonomy if the conical limit set is non-empty. Second, according to a recent result of Deroin and Kleptsyn,[6] there exists a hyperbolic holonomy under the absence of transversal invariant measures. Leaves with hyperbolic holonomies are always contained in $J(\mathcal{F})$ so that $J(\mathcal{F})$ also contains a hyperbolic holonomy if $\mathcal{F}$ does not admit any transversal invariant measure. It seems quite difficult for $\mathcal{F}$ to admit an invariant measure unless $J(\mathcal{F})$ is empty or $\mathcal{F}$ contains closed leaves. It is also known that the number of closed leaves contained in $J(\mathcal{F})$ is finite. Finally, it is known that the support of the Godbillon measure contains leaves of exponential growth if it is non-empty.[12] A precise version of Theorem 3.1 implies that $J(\mathcal{F})$ contains leaves of exponential growth if the Godbillon-Vey class of $\mathcal{F}$ is non-trivial.

Some notions concerning the Julia sets of mapping iterations and the limit sets of Kleinian groups will be also valid for $J(\mathcal{F})$. For example, if $F(\Gamma)$ is non-empty then we can consider the critical exponent of Γ as follows. For a Kleinian group, the critical exponent is defined by looking at the convergence of Poincaré series. A direct analogue of the Poincaré series for Γ will be $\sum_{\gamma \in \Gamma_x} |\gamma'|_x^s$, where Γ_x denotes the subset of Γ which consists of elements defined near x, and γ' denotes the usual differential viewed as a function which is defined by fixing an embedding of T into $\mathbb{C}$. However, it can be shown that the sum does not converge for any s even if $x \in F(\Gamma)$. This can be avoided as follows. There is an invariant metric on $F(\Gamma)$. If we write this metric as $g^2 |dz|^2$ on T, then $|\gamma'|_x g(\gamma(x)) = g(x)$. Hence the sum can be replaced with $\sum_{\gamma \in \Gamma_x} \frac{1}{g(\gamma(x))^s}$. To be precise, we should consider the integral $\int_{T'} g^{-s} dm_g$, where dm_g denotes the volume form with respect to g on T' ($1/g$ is considered to be 0 on $J(\Gamma)$). In this way we can speak of critical exponent of Γ. We refer to the original article[3] for details with some further studies.

4. Examples

It is known that the number of closed leaves in $J(\mathcal{F})$ is finite as mentioned in Remark 3.1. However, the number can be arbitrarily large if the manifold M is not fixed.

Example 4.1. Let $[z_0 : z_1 : z_2]$ be the homogeneous coordinates of $\mathbb{C}P^2$ and let $\mathbb{C}^2 = \{[z_0 : z_1 : z_2] \in \mathbb{C}P^2 \,|\, z_i \neq 0\}$, and let (u_1, u_2) the inhomoge-

neous coordinates on $\mathbb{C}^2$. Let X be a vector field on $\mathbb{C}^2$ given by
$$X = \lambda_1 u_1 \frac{\partial}{\partial u_1} + \lambda_2 u_2 \frac{\partial}{\partial u_2}.$$
We assume that $\lambda_1 \lambda_2 \neq 0$, $\lambda_1 \neq \lambda_2$ and $\lambda_1/\lambda_2 \notin \mathbb{R}$, then X induces a singular foliation $\mathcal{F}$ of $\mathbb{C}P^2$ with three singularities $p_1 = [0:0:1]$, $p_2 = [0:1:0]$ and $p_3 = [1:0:0]$. Let $L_i = \{[z_0:z_1:z_2] \in \mathbb{C}P^2 \,|\, z_i = 0\}$, D_i be a small round ball centered at p_i and $S_i \approx S^3$ be its boundary. The condition $\lambda_1/\lambda_2 \notin \mathbb{R}$ implies that $\mathcal{F}$ is transversal to S_i. Let $M = \mathbb{C}P^2 \setminus (D_1 \cup D_2 \cup D_3)$ and let M_3 be its double, then M_3 naturally inherits a transversally holomorphic foliation $\mathcal{F}_3$ induced from $\mathcal{F}$. The foliation $\mathcal{F}_3$ has three compact leaves $\mathcal{L}_0$, $\mathcal{L}_1$ and $\mathcal{L}_2$, namely, the leaves induced from L_0, L_1 and L_2. Then, one can show that $F(\mathcal{F}_3) = M_3 \setminus (\mathcal{L}_0 \cup \mathcal{L}_1 \cup \mathcal{L}_2)$. The number of the Julia components can be arbitrarily large as follows. Let M' be a copy of M and let $\partial M' = S_1' \cup S_2' \cup S_3'$. Let M_1 be the manifold with boundary obtained by gluing M with M' along S_1 and S_1', and S_2 and S_2', then $\partial M_1 = S_3 \cup S_3'$. Let $\mathcal{F}_4$ be the natural foliation of the double M_4 of M_1, then $J(\mathcal{F}_4)$ consists of 4 connected components. In general, let $N_1, \cdots, N_{r-2}$ be copies of M_1 and let M_r be the manifold obtained by gluing them. Let $\mathcal{F}_r$ be the naturally induced foliation of M_r, then $J(\mathcal{F}_r)$ consists of r connected components. We remark that this construction can be also described by using blowing-ups.

Example 4.2. Let Γ be a Kleinian group and let $\mathbb{C}P^1 = \Omega(\Gamma) \sqcup \Lambda(\Gamma)$ be the decomposition into the domain of discontinuity and the limit set. Let $\mathcal{F}$ be a suspension of this action, then $F(\mathcal{F})$ is the suspension of $\Omega(\Gamma)$ and the $J(\mathcal{F})$ is the suspension of $\Lambda(\Gamma)$.

Example 4.3. There is a transversally Hermitian foliation $\mathcal{F}$ such that $J_{GGS}(\mathcal{F})$ is the whole manifold (Example 8.6[8]), where $J_{GGS}(\mathcal{F})$ denotes the Julia set in the sense of Ghys, Gomez-Mont and Saludes. On the other hand, $J(\mathcal{F})$ is empty. In particular, $F(\mathcal{F}) \supsetneq F_{GGS}(\mathcal{F})$.

Example 4.4. There is a foliation $\mathcal{F}$ of a connected manifold such that the interior of $J_{GGS}(\mathcal{F})$ is non-empty without being the whole manifold (Example 8.9[8]). It is obtained by modifying a certain Fatou component into a Julia set (in the sense of Ghys, Gomez-Mont and Saludes). However, it is easily seen that this modification does not change the Fatou component into a Julia set in our sense and that the interior of $J(\mathcal{F})$ is empty.

It seems unknown whether there exist foliations whose Julia set has non-empty interior. On the other hand, the Julia set of foliations can be the whole manifold. Such an example can be obtained by taking suspensions as in Examples 4.2.

We do not know if there is a reasonable extension of the Fatou-Julia decomposition to not necessarily closed manifolds (or non-compactly generated pseudogroups). Indeed, it is relevant to choose (Γ', T') so that the closure of T' in T is compact in defining the Fatou set.

Example 4.5. Let $D(r)$ be the disc in $\mathbb{C}$ of radius r and let $\mathcal{F}$ be the foliation of $M = (-1,1) \times D(1)$ with leaves $(-1,1) \times \{z\}$. If M itself is regarded as a foliation atlas, then the Fatou set should be the whole M. On the other hand, let $i \in \mathbb{Z}$ and define a foliation atlas as follows. For $i > 0$, let $\{V_j^{(i)}\}_{j=1,2,\ldots}$ be an open covering of $D(1)$ by discs of radius 2^{-i}. Let $W_j^{(i)} = \left(-1 + 1/2^{-i+1}, -1 + 1/2^{-i-1}\right) \times V_j^{(i)}$ and $T_j^{(i)} = \{-1+1/2^{-i}\} \times V_j^{(i)}$. Giving an order to $\{W_j^{(i)}\}$, let $\{W_j^{(i)}\} = \{W_1', W_2', \cdots\}$ and $\{T_j^{(i)}\} = \{T_1', T_2', \cdots\}$. Set then $U_0 = (-1/2, 1/2) \times D(1)$, $T_0 = \{0\} \times D(1)$, and $U_i = W'_{|i|}$, $T_i = T'_{|i|}$ for $i \neq 0$. Let (Γ', T') be a pseudogroup obtained by shrinking T'. Note that the closure of T' in T is non-compact. Applying the definition to (Γ, T) and (Γ', T'), the Fatou set is empty.

Note that this construction can be done in a foliation chart. The above example also shows that if we directly apply Definition 2.1 to a foliation of an open set, then the Fatou-Julia decomposition will depend on the choice of the realization of the holonomy pseudogroup.

References

1. J. A. Álvarez López and X. M. Masa, *Topology Appl.* **155**, 544–604 (2008).
2. T. Asuke, *Topology* **43**, 289–317 (2004).
3. T. Asuke, A Fatou-Julia decomposition of transversally holomorphic foliations, submitted, (2008).
4. T. Asuke, On the Fatou-Julia decomposition of transversally holomorphic foliations of complex codimension one, submitted, (2008).
5. H. Cartan, Sur les Groupes de Transformations Analytiques, Hermann, Paris (1935).
6. B. Deroin and V. Kleptsyn, *Geom. Funct. Anal.* **17**, 1043–1105 (2007).
7. G. Duminy, L'invariant de Godbillon-Vey d'un feuilletage se localise dans les feuilles ressort, preprint, (1982).
8. E. Ghys, X. Gómez-Mont and J. Saludes, *Essays on Geometry and Related Topics: Memoires dediés à André Haefliger*, eds. E. Ghys, P. de la Harpe, V. F. R. Jones, V. Sergiescu and T. Tsuboi, Monographie de l'Enseignement Mathématique, Vol. 38 2001 pp. 287–319.
9. A. Haefliger, *A fête of topology*, Academic Press, Boston, MA (1988), pp. 3–32.

10. A. Haefliger, *Foliations: geometry and dynamics (Warsaw, 2000)*, World Sci. Publ., River Edge, NJ (2002), pp. 174–197.
11. J. Heitsch and S. Hurder, *J. Diff. Geom.* **20**, 291–309 (1984).
12. S. Hurder, *J. Diff. Geom.* **23**, 347–365 (1986).
13. P. Molino, *Riemannian foliations*, Progress in Mathematics, no. 73 (Birkhäuser, Boston, 1988). Translated by G. Cairns.

SINGULAR RIEMANNIAN FOLIATIONS ON SPACES WITHOUT CONJUGATE POINTS

Alexander Lytchak

Mathematisches Institut, Universität Bonn
Beringstr. 1, 53115 Bonn, Germany
E-mail: lytchak@math.uni-bonn.de

We describe the topological structure of cocompact singular Riemannian foliations on Riemannian manifolds without conjugate points. We prove that such foliations are regular and developable and have regular closures. We deduce that in some cases such foliations do not exist.

Keywords: Negative curvature, focal points, geodesic flow

1. Introduction

Riemannian manifolds of non-negative curvature often admit large groups of isometries. Moreover, there are many famous examples of Riemannian foliations on such spaces, like the Hopf fibrations and of singular Riemannian foliations, such as isoparametric foliations. Singular Riemannian foliations on non-negatively curved manifolds tend to be homogeneous and seem to be rather rigid objects. On the other hand, (singular) Riemannian foliations on such spaces are often related to other rigidity questions (cf.[1-8]).

If one changes the sign of the curvature then the situation seems to be completely different on the first glance. For instance, in a simply connected negatively curved manifold there are infinite-dimensional families of Riemannian submersions to the real line and there seem to be no hope of getting any kind of control of such objects. However, for *compact* manifold of non-positive curvature the situation seems again be very similar to the "rigid" non-negatively curved world. The first indication is the famous result of Bochner[9] that describes connected isometry groups of such spaces. In particular, the isometry group of a compact negatively curved manifold turns out to be finite. Indeed, this is the case for any Riemannian metric on such manifolds, since they have positive minimal volume.[10] In[11] it is shown that the existence of a Riemannian flow on a compact manifold forces its minimal volume to be zero, thus Riemannian flows do not exist on compact negatively curved manifolds. Finally, A. Zeghib proved in Theorem F[12] that on a compact negatively curved manifold there are no (regular) Riemannian foliations at all.

Remark 1.1. Previously, the non-existence of regular Riemannian foliations on compact negatively curved manifolds was claimed in[13] and, in special cases in[14,15]. However, these proofs are not correct, cf.[16] and the discussion in pp. 1435-1436[12].

Here, we generalize the non-existence theorem to singular Riemannian foliations, a broad generalization of regular Riemannian foliations and isometric group actions. We prove:

Theorem 1.1. *Singular Riemannian foliations do not exist on compact negatively curved manifolds.*

Remark 1.2. In[17] the non-existence result was proved under the assumption that the singular Riemannian foliations has horizontal sections, i.e., that the horizontal distribution in the regular part is integrable.

In fact, in analogy with[17] we prove in a broader context that a singular Riemannian foliation on a compact negatively curved manifold cannot have singular leaves, i.e., it must be a regular Riemannian foliation. Then we apply[12]. Our main result result used in Theorem 1.1 describes the topology of singular Riemannian foliations in the following more general situation.

Theorem 1.2. *Let M be a complete Riemannian manifold without conjugate points and let $\mathcal{F}$ be a singular Riemannian foliation on M such that the space of leaves $M/\mathcal{F}$ has bounded diameter with respect to the quotient pseudo-metric. Then $\mathcal{F}$ is a regular foliations and has a regular closure $\bar{\mathcal{F}}$. The quotient $M/\bar{\mathcal{F}}$ is a good Riemannian orbifold without conjugate points. The leaves of the lift $\tilde{\mathcal{F}}$ of $\mathcal{F}$ to the universal covering $\tilde{M}$ of M are closed and contractible. They are given by a Riemannian submersion $p: \tilde{M} \to B$ to a contractible manifold B.*

In the case of a simply connected total space M we deduce from the last part of Theorem 1.2:

Corollary 1.1. *Let M be a complete, simply connected Riemannian manifold without conjugate points. Then there are no non-trivial singular Riemannian foliations $\mathcal{F}$ on M with a bounded quotient $M/\mathcal{F}$.*

Remark 1.3. In the case $M = \mathbb{R}^n$ the last result was recently shown in[18] using different methods.

The proof of Theorem 1.2 is divided into a geometric and a topological part. In the geometric part, similar to[17], we analyze the structure of $\bar{\mathcal{F}}$ and prove that regular leaves of $\bar{\mathcal{F}}$ do not have focal points (this already implies the first two claims in our theorem). The idea of the proof is that focal points of regular leaves correspond either to crossings of singular leaves or to conjugate points in the quotient. Now, the Poincare recurrence theorem for the quasi-geodesic flow on the quotient $M/\bar{\mathcal{F}}$ (cf. Theorem 1.6[19]; here we use the compactness of the quotient) tells us that the existence of a single focal point would imply the existence of a horizontal geodesic with arbitrary many focal points. (This is a modified form of the statement that on a compact Riemannian manifold with uniformly bounded number of conjugate points along all geodesics, there are no conjugate points at all). However, the absence of conjugate points on M implies that each leaf has at most $\dim(M)$ focal points along any horizontal geodesic. This contradiction finishes the geometric part of the proof.

The remaining part of the proof is finished by using the following purely topological observation.

Proposition 1.1. *Let M be an aspherical manifold with a complete Riemannian metric. Let $\mathcal{F}$ be a Riemannian foliation on M with dense leaves. Then the leaves of the lift $\tilde{\mathcal{F}}$ of $\mathcal{F}$ to the universal covering $\tilde{M}$ are closed and contractible. The lifted foliation $\tilde{\mathcal{F}}$ is given by a Riemannian submersion $p : \tilde{M} \to B$ onto a contractible homogeneous manifold B.*

2. Preliminaries

A *transnormal system* $\mathcal{F}$ on a Riemannian manifold M is a decomposition of M into smooth, injectively immersed, connected submanifolds, called leaves, such that geodesics emanating perpendicularly to one leaf stay perpendicularly to all leaves. A transnormal system $\mathcal{F}$ is called a *singular Riemannian foliation* if there are smooth vector fields X_i on M such that for each point $p \in M$ the tangent space $T_pL(p)$ of the leaf $L(p)$ through p is given as the span of the vectors $X_i(p) \in T_pM$. We refer to[8,19,20] for more on singular Riemannian foliations. Examples of singular Riemannian foliations are (regular) Riemannian foliations and the orbit decomposition of an isometric group action.

If M is complete then leaves of a transnormal system $\mathcal{F}$ are equidistant and the distance between leaves define a natural pseudo-metric on the space of leaves. This pseudo-metric space is bounded if and only if some finite tubular neighborhood of a leaf coincides with the whole space. If $\mathcal{F}$ is *closed*, i.e., if leaves of $\mathcal{F}$ are closed then the quotient $B = M/\mathcal{F}$ is a complete, locally compact, geodesic metric space, that is compact if and only if it is bounded. Moreover, B is an Alexandrov space with curvature locally bounded below. If it is compact, its Hausdorff measure is finite.

Let $\mathcal{F}$ be a singular Riemannian foliation on the Riemannian manifold M. The *dimension of* $\mathcal{F}$, $\dim(\mathcal{F})$, is the maximal dimension of its leaves. For $s \leq \dim(\mathcal{F})$ denote by Σ_s the subset of all points $x \in M$ with $\dim(L(x)) = s$. Then Σ_s is an embedded submanifold of M and the restriction of $\mathcal{F}$ to Σ_s is a Riemannian foliation. For a point $x \in M$, we denote by Σ^x the connected component of Σ_s through x, where $s = \dim(L(x))$. We call the decomposition of M into the manifolds Σ^x the *canonical stratification* of M. The subset $\Sigma_{\dim(\mathcal{F})}$ is open, dense and connected in M. It is the *regular stratum* M. It will be denoted by M_0 and will also be called the set or regular points of M. All other strata Σ^x are called *singular strata*.

Let $\mathcal{F}$ be a singular Riemannian foliation on a complete Riemannian manifold M. Then the decomposition $\bar{\mathcal{F}}$ of M into closures of leaves of $\mathcal{F}$ is a transnormal system, that we will call the closure of $\mathcal{F}$. The restriction of $\bar{\mathcal{F}}$ to each stratum Σ of M (with respect to $\mathcal{F}$) is a singular Riemannian foliation.

For a transnormal system $\mathcal{F}$ on M, we will call a point $x \in M$ regular if its leaf is regular, i.e., if it has the maximal dimension. The closure of a singular leaf of a singular Riemannian manifold $\mathcal{F}$ on a complete Riemannian manifold M is a singular leaf of $\bar{\mathcal{F}}$. In particular, if $\bar{\mathcal{F}}$ does not have singular leaves then $\mathcal{F}$ is a (regular) Riemannian foliation.

Let $M, \mathcal{F}, \bar{\mathcal{F}}$ be as above. Then M gets a canonical stratification with respect to $\bar{\mathcal{F}}$ that is finer than the canonical stratification with respect to $\mathcal{F}$, such that the restriction of $\bar{\mathcal{F}}$ to each stratum is a Riemannian foliation. The main stratum M_0 is again open and dense. This defines a canonical stratification of the quotient $B = M/\bar{\mathcal{F}}$ into smooth Riemannian orbifolds. The main stratum M_0 is projected to the main stratum B_0 of B that is open and dense in B. If B is compact, the orbifold B_0 has finite volume.

Horizontal geodesics of the transnormal system $\bar{\mathcal{F}}$ are projected to concatenations of geodesics in B. Each horizontal geodesic in the regular part M_0 is projected to an orbifold-geodesic in B_0. Let γ_1 and γ_2 be horizontal geodesics whose projections η_1 and η_2 to B coincide initially. Then η_1 and η_2 coincide on the whole real line (cf.[19,21] for the case of singular Riemannian foliation and[18,22] for the case of closed transnormal systems). Therefore the geodesic flow on M restricted to the space of horizontal vectors projects to a "quasi-geodesic" flow on the "unit tangent bundle" of B. Note, finally, that for each regular leaf L of $\bar{\mathcal{F}}$ and each horizontal geodesic γ starting on L, each intersection point of γ with a singular leaf is a focal point of L along γ.

We finish this section with an easy application of the Poincaré's recurrence theorem:

Lemma 2.1. *Let B_0 be a (non-necessarily) complete Riemannian orbifold with finite volume. Let V be a non-empty open subset of the unit tangent bundle U_0 of B_0. Assume that the geodesic flow $\phi_t(v)$ is defined for all $v \in V$ and all $t > 0$. Let a positive real number T be given. Then there is a non-empty open subset $V_0 \subset V$ and $\bar{T} > T$ such that $\phi_{\bar{T}}(V_0) \subset V$, and such that $\phi_t(v)$ is defined for all $v \in V_0$ and all $t \in [-T, 0]$.*

3. Geometric arguments

Using the preparation from the last section, we can now easily prove the geometric part of Theorem 1.2.

Let $M, \mathcal{F}$ be as in Theorem 1.2. Consider the closure $\bar{\mathcal{F}}$ of $\mathcal{F}$. Let B denote the compact quotient $B = M/\bar{\mathcal{F}}$ with the projection $q : M \to B$. Let B_0 be the regular part of B, i.e., the set of all regular leaves of $\bar{\mathcal{F}}$ in M.

We are going to prove that all regular leaves of $\bar{\mathcal{F}}$ have no focal points in M. Assume the contrary. Denote by M_0 the regular part of M (with respect to $\bar{\mathcal{F}}$; the original singular foliation $\mathcal{F}$ will not be used in this section). Let $\mathcal{H}$ be the horizontal distribution on M_0. Let $\mathcal{H}^1$ be the space of unit vectors in $\mathcal{H}$, with the foot point projection $p : \mathcal{H}^1 \to M$. For $h \in \mathcal{H}^1$ let $\gamma^h : [0, \infty) \to M$ denote the horizontal geodesic starting in the direction of h. By $L(h)$ we denote the leaf of $\bar{\mathcal{F}}$ through the foot point $p(h) \in M$. By $f(h)$ we will denote the $L(h)$-index of γ^h, i.e., the number of $L(h)$-focal points along γ^h. By Λ^h we denote the Lagrangian space of normal Jacobi fields along γ^h that consists of $L(h)$-Jacobi fields (cf.[23]). As in[23], we denote for an interval $I \subset (0, \infty)$ by $\mathrm{ind}_{\Lambda^h}(I)$ the number of $L(h)$-focal points along γ^h in $\gamma^h(I)$.

Since there are no conjugate points in the manifold M, the function f is bounded by $\dim(M)$ on $\mathcal{H}^1$ (Corollary 1.2[23]). Let m be the maximum of the function f, that is positive by our assumption. Choose some $h_0 \in \mathcal{H}^1$ with $f(h_0) = m$. Choose some $T > 0$ such that all (precisely m, when counted with multiplicity) $L(h_0)$-focal points along γ^{h_0} come before T, i.e., $\mathrm{ind}_{\Lambda^{h_0}}((0, T)) = m$. By continuity of indices and maximality of m, we find an open neighborhood V of h_0 in $\mathcal{H}^1$, with $\mathrm{ind}_{\Lambda^h}((0,T)) = m$, for all $h \in V$.

Since each intersection of γ^h with a singular leaf happens in a focal point, for all $h \in V$, the geodesic $\gamma^h : [T,\infty) \to M$ does not intersect singular leaves. Thus, $\gamma^h([T,\infty))$ is contained in M_0 and, for its projection $\eta^h = q \circ \gamma^h$, we have $\eta^h([T,\infty) \subset B_0$. Due to Lemma 2.1, we find an open subset V_0 of V and some $\bar{T} > T$ such that for all $h \in V_0$ we have $\gamma_h[0,\infty) \subset M_0$ and $(\gamma^h)'(\bar{T}) \in V_0$.

Choose now some $h \in V_0$. Since γ^h is contained in M_0, the projection $\eta^h = q \circ \gamma^h$ is an orbifold-geodesic in B_0. Moreover, $L(h)$-focal points along γ^h correspond to conjugate points along η^h. For the Jacobi equation along η^h (in terms of,[23] this is the transversal Jacobi equation introduced in[8]), we have the following picture. The point $\eta^h(\bar{T})$ has at least one conjugate point along η^h in the interval $(\bar{T}, \bar{T} + T)$ (in fact, there are precisely m such points counted with multiplicities). Therefore, $\eta^h(0)$ has at least one conjugate point along η^h in the interval $(\bar{T}, \bar{T} + T)$ (Corollary 1.3[23]). Since $\bar{T} > T$, by assumption, we get an $L(h)$-focal point $\gamma^h(t)$ along γ^h for some $t > T$, in contradiction to $\mathrm{ind}_{\Lambda^h}(0,\infty) = \mathrm{ind}_{\Lambda^h}(0,T)$.

Thus, we have proved, that all regular leaves of $\bar{\mathcal{F}}$ have no focal points. Hence $\bar{\mathcal{F}}$ has no singular leaves. Therefore, $\mathcal{F}$ and $\bar{\mathcal{F}}$ are regular Riemannian foliation. Moreover, since focal points of leaves of a closed regular Riemannian foliation correspond to conjugate points in the quotient orbifold, we deduce that the quotient $B = M/\bar{\mathcal{F}}$ has no conjugate points.

4. Topological arguments

First, we are going to prove Proposition 1.1. Thus let M be an aspherical manifold with a complete Riemannian metric. Let $\mathcal{F}$ be a Riemannian foliation on M with dense leaves. Let $\tilde{M}$ be the universal covering of M. Denote by Γ the group of deck transformations of $\tilde{M}$. Let $\tilde{\mathcal{F}}$ be the lift of $\mathcal{F}$ to $\tilde{M}$ and denote by $\mathcal{F}_1$ the closure of $\tilde{\mathcal{F}}$. Since $\tilde{F}$ is invariant under the action of Γ, so is its closure $\mathcal{F}_1$. Thus $\mathcal{F}_1$ induces a singular Riemannian foliation $\mathcal{F}_2$ on M whose leaves contain the leaves of $\mathcal{F}$. Since the leaves of $\mathcal{F}$ are dense so must be the leaves of $\mathcal{F}_2$. In particular, $\mathcal{F}_2$ and, therefore, $\mathcal{F}_1$ must be regular Riemannian foliations. Consider the Riemannian orbifold $B = \tilde{M}/\mathcal{F}_1$. Since $\mathcal{F}_2$ has dense leaves, the natural isometric action of Γ on B must have dense orbits. In particular, B must be a homogeneous Riemannian manifold.

Thus, the projection $p : \tilde{M} \to B$ is a Riemannian submersion. From the long exact sequence of the fibration p (and the contractibility of $\tilde{M}$) we deduce that B must be simply connected. Since B is homogeneous, its homotopy and homology groups are finitely generated. From the long exact sequence of p we deduce that the homotopy groups of the fibers L of p (these are leaves of $\mathcal{F}_1$) are abelian and finitely generated. Hence, the homology groups of L are finitely generated as well. Now, we can apply the spectral sequence for the fiber bundle p, as in p. 599^4, and deduce that the homology groups of L and B must vanish in positive degrees. We conclude that L and B are contractible.

It remains to prove that the leaves of $\tilde{F}$ are closed, i.e., that $\tilde{\mathcal{F}}$ and $\mathcal{F}_1$ coincide. Assume the contrary and take a non-closed leaf L. Then its closure $\bar{L}$ is a leaf of $\mathcal{F}_1$, hence it is contractible. Thus the restriction of $\tilde{\mathcal{F}}$ to $\bar{L}$ is a Riemannian foliation with dense leaves on a complete, contractible manifold $\bar{L}$. But this is impossible.[24] This finishes the proof of Proposition 1.1.

Now we can finish the proof of Theorem 1.2. We already now, that the closure $\bar{\mathcal{F}}$ is a regular Riemannian foliation on M. Moreover, the leaves of $\bar{\mathcal{F}}$ have no focal points, and $M/\bar{\mathcal{F}}$ is a Riemannian orbifold without conjugate points. Now, the proof of Theorem 2^{25} reveals that the lift $\mathcal{F}_1$ of $\bar{\mathcal{F}}$ to the universal covering $\tilde{M}$ is a simple foliation. Moreover, the quotient $\hat{B} = \tilde{M}/\mathcal{F}_1$ is a Riemannian manifold without conjugate points. From the long exact sequence we deduce that $\hat{B}$ is simply connected. Therefore, it is diffeomorphic to $\mathbb{R}^n$. Each leaf L of $\mathcal{F}_1$ has no focal points. Therefore, its normal exponential map is a diffeomorphism. Thus the distance function $d_x : L \to \mathbb{R}$ to each point $x \in M \setminus L$ is a Morse function on L with only one critical point. Therefore, L is diffeomorphic to a Euclidean space as well.

In particular, the leaves of $\bar{\mathcal{F}}$ are aspherical. From Proposition 1.1 we deduce that the lift $\tilde{\mathcal{F}}$ of $\mathcal{F}$ to $\tilde{M}$ has closed and contractible leaves. In particular, all leaves of $\tilde{\mathcal{F}}$ have trivial fundamental group and therefore no holonomy. Therefore, $\tilde{\mathcal{F}}$ is a simple foliation. From the long exact sequence we deduce that the quotient $B_1 = \tilde{M}/\tilde{\mathcal{F}}$ is a contractible manifold.

Acknowledgments

The author was supported in part by the SFB 611 *Singuläre Phänomene und Skalierung in mathematischen Modellen*.

References

1. D. Gromoll and K. Grove, *Ann. Sci. École Norm. Sup* **20**, 227 (1987).
2. D. Gromoll and G. Walschap, *J. Differential Geom.* **28**, 143 (1988).
3. G. Thorbergsson, *Ann. of Math.* **133**, 429 (1991).
4. L. Guijarro and P. Petersen, *Ann. Sci. École Norm. Sup.* **30**, 595 (1997).
5. D. Gromoll and G. Walschap, *J. Differential Geom.* **57**, 233 (2001).

6. D. Gromoll and G. Walschap, *Asian J. Math.* **1**, 716 (2001).
7. B. Wilking, *Invent. Math.* **144**, 281 (2001).
8. B. Wilking, *Geom. Funct. Anal.* **17**, 1297 (2007).
9. S. Bochner, *Bull. Amer. Math. Soc.* **52**, 776 (1946).
10. M. Gromov, *Inst. Hautes Études Sci. Publ. Math.* **56**, 213 (1982).
11. Y. Carrière, *Math. Z.* **186**, 393 (1984).
12. A. Zeghib, *Amer. J. Math.* **117**, 1431 (1995).
13. P. Walczak, *Ann. Global Anal. Geom.* **9**, 83 (1991).
14. H. Kim and G. Walschap, *Indiana Univ. Math. J.* **41**, 37 (1992).
15. G. Walschap, *Amer. J. Math.* **115**, 1189 (1993).
16. P. Walczak, *Ann. Global Anal. Geom.* **9**, p. 325 (1991).
17. D. Töben, *Math. Z.* **255**, 427 (2007).
18. C. Boltner, On the structure of equidistant foliations of $\mathbb{R}^n$, PhD thesis, Augsburg2007. arXiv:math.DG/0712.0245.
19. A. Lytchak and G. Thorbergsson, arXiv:math.DG/0709.2607, (2007).
20. P. Molino, *Riemannian foliations* (Birkhäuser Boston, Inc., Boston, MA, 1988).
21. M. Alexandrino and D. Töben, *Proc. Amer. Math. Soc.* **136**, 3271 (2008).
22. A. Lytchak, Allgemeine Theorie der Submetrien und verwandte mathematische Probleme, PhD thesis, Bonn2001.
23. A. Lytchak, arXiv:math.DG/0708.2651; to appear in Differential Geom. Appl, (2007).
24. A. Haefliger, *A fête of topology*, (Academic Press, Boston, MA, 1988) pp. 3–32.
25. J. Hebda, *Indiana Univ. Math. J.* **35**, 321 (1986).

VARIATIONAL FORMULAE FOR THE TOTAL MEAN CURVATURES OF A CODIMENSION-ONE DISTRIBUTION

V. Rovenski

Mathematical Department, University of Haifa, Haifa, 31905, Israel
E-mail: rovenski@math.haifa.ac.il
http://math.haifa.ac.il/ROVENSKI/rovenski.html

P. Walczak

Katedra Geometrii, Uniwersytet Łódzki, ul. Banacha 22, Łódź, Poland
E-mail: pawelwal@math.uni.lodz.pl
http://math.uni.lodz.pl/~pawelwal/

We develop variational formulae for the total m-th mean curvatures of codimension-one distributions on a compact Riemannian manifold M^{n+1}. Based on the method of our recent work Rovenski–Walczak (2007) we show that the total generalized mean curvatures over $M(c)$ of constant curvature c do not depend on the choice of k orthonormal vector fields, that for $k = 1$ was proved by Brito–Langevin–Rosenberg (1981). We calculate the variations of the total generalized mean curvatures for k orthonormal vector fields on a compact Riemannian manifold M.

Keywords: Distribution, vector field, Riemannian metric, mean curvatures, variation, Newtonian transformation, integral formula

0. Introduction

Let $\mathcal{F}$ be a codimension-one foliation with a unit normal N on a compact Riemannian manifold M^{n+1}, and $A_N = -(\nabla N)^\perp$ the *Weingarten operator* of the leaves of $\mathcal{F}$. The *elementary symmetric functions* $\sigma_m(A)$ of a $n \times n$ matrix A are defined by the equality $\sum_m \sigma_m(A)\, t^m = \det(I_n + tA)$, where I_n is the unit matrix. Brito, Langevin & Rosenberg[2] (generalizing result by Asimov[1]) have shown that for M^{n+1} of constant curvature c the integrals

$$I_m(N) = \int_M \sigma_m(A_N)\, d\mathrm{vol} \tag{1}$$

depend on n, m, c and $\mathrm{vol}(M)$ only, namely,

$$I_m(N) = \begin{cases} c^{m/2} \binom{n/2}{m/2} \mathrm{vol}(M), & n, m \text{ even,} \\ 0, & n \text{ or } m \text{ odd.} \end{cases} \qquad (2)$$

In this case, $I'_m(0) = 0$ for any variation $N(s)$. One may show that

$$\sigma_m(A_N) = \sigma_m(-\nabla N), \quad 1 \le m \le n. \qquad (3)$$

In [3,4] we generalized a result by Asimov [1] and Brito–Langevin–Rosenberg [2], see (2), dropping their restriction that M has constant sectional curvature.

For $\boldsymbol{\lambda} = (\lambda_1, \ldots, \lambda_m) \in \mathbb{Z}_+^m$ and $\mathbf{t} = (t_1, \ldots t_k) \in \mathbb{R}_+^m$ denote by $\mathbf{t}^{\boldsymbol{\lambda}} = \Pi_{i=1}^k t_i^{\lambda_i}$. Set $|\boldsymbol{\lambda}| = \sum_i \lambda_i$. The *mean curvatures* $\sigma_{\boldsymbol{\lambda}}(A_1, \ldots A_k)$ of a set of order n quadratic matrices $A_1, \ldots A_k$ are defined by the polynomial equality

$$\det(I_n + t_1 A_1 + \ldots + t_k A_k) = \sum_{|\boldsymbol{\lambda}| \le n} \sigma_{\boldsymbol{\lambda}}(A_1, \ldots A_k) \mathbf{t}^{\boldsymbol{\lambda}}. \qquad (4)$$

Theorem 0.1 ([3]). *Let N be a unit vector field on a compact locally symmetric space M^{n+1}. Then for any $m > 0$*

$$\int_M g_m(N)\, d\mathrm{vol} = 0, \qquad g_m(N) = \sum_{\|\boldsymbol{\lambda}\| = m} \sigma_{\boldsymbol{\lambda}}(B_1(N), \ldots B_m(N)), \qquad (5)$$

where $\boldsymbol{\lambda} = (\lambda_1, \ldots, \lambda_m)$ is a sequence of nonnegative integers, $\|\boldsymbol{\lambda}\| = \lambda_1 + 2\lambda_2 + \ldots + m\lambda_m$ and $B_{2k}(N) = \frac{(-1)^k}{(2k)!} R_N^k$, $B_{2k+1}(N) = \frac{(-1)^k}{(2k+1)!} R_N^k \nabla N$.

The formula (5) for few initial values of m, $m = 1, 2, 3$, reads as [3]

$$\int_M \sigma_1(\nabla N)\, d\mathrm{vol} = 0, \qquad \int_M \sigma_2(\nabla N) - \frac{1}{2} \mathrm{Ric}(N, N)\, d\mathrm{vol} = 0,$$

$$\int_M \sigma_3(\nabla N) - \frac{1}{2} \mathrm{Ric}(N, N) \mathrm{Tr}(\nabla N) + \frac{1}{3} \mathrm{Tr}(R_N \nabla N)\, d\mathrm{vol} = 0. \qquad (6)$$

Reilly[5] developed variational formulae for the functional (1) of a "single" hypersurface $M \subset \mathbb{R}^N$.

A **problem** we are interested in is *to develop variational formulae for the functionals $I_m(N)$ ($m > 0$), N being a unit vector field on a compact M.* We do not assume that the orthogonal distribution $N^\perp$ is integrable.

Extending our methods [3,4] (see Theorem 0.1) we generalize (2).

Theorem 0.2. *Let $N_1, \ldots N_k$ ($k > 1$) be an orthonormal system of vector fields on a compact Riemannian manifold M^{n+1} with the condition $R_{N_i} = c I_n$ ($i \le k$) (for example, M has a constant curvature c). Then the integrals*

$$I_{\boldsymbol{\lambda}}(N_1, \ldots N_k) = \int_M \sigma_{\boldsymbol{\lambda}}(\nabla N_1, \ldots \nabla N_k)\, d\mathrm{vol} \qquad (7)$$

do not depend on $N_1, \ldots N_k$: they depend on n, k, $\boldsymbol{\lambda} = (\lambda_1, \ldots \lambda_k)$, c and vol(M) only. In particular, $I_\lambda(N_1, \ldots N_k) = 0$ when $c = 0$ or when either $|\boldsymbol{\lambda}|$ or n is odd.

Hence, the natural extension of the **problem** is *to develop variational formulae for the functionals (7) for any compact Riemannian manifold with k orthonormal vector fields $N_1, \ldots, N_k$*.

The proposed variational formulae for $I_\lambda(N_1, \ldots N_k)$ can be applied to k-webs and also to a *k-regular* unit vector field N on a Riemannian manifold, i.e., such a field N for which N itself and its $k-1$ derivatives, $\nabla_N N, \nabla_N(\nabla_N N), \ldots (\nabla_N)^{k-1} N$, are point-wise linearly independent.

We will present the 1-st variation of (1) in the form

$$I'_m(0) = \int_M \dot{\sigma}_m(\nabla N)\, d\text{vol} = \int_M \langle a_L(N), \xi \rangle\, d\text{vol},$$

where "dot" denotes differentiation with respect to s at $s = 0$. Therefore, the *Euler-Lagrange equation* for (1) has the form $a_L(N)^\perp = 0$, where $(\cdot)^\perp$ is the orthogonal to N component. Similarly, the *Euler-Lagrange equations* for (7) constitute a system $a_L^i(N_1, \ldots N_k)^{\perp_i} = 0$ ($1 \le i \le k$).

1. Main Results

We will call the vector field div $K = \text{Tr}(\nabla K) = \sum_i (\nabla_{e_i} K)e_i$, the *divergence of a linear operator* $K : TM \to TM$. Let R be the curvature tensor and $R_N = R(\cdot, N)N$ the Jacobi operator. The Ricci curvature in a direction N is $\text{Ric}(N, N) = \sum_i \langle R(e_i, N)N, e_i \rangle$, where $\{e_i\}$ is a local orthonormal basis of $N^\perp$. Denote by $\mathcal{R}(S)$ the vector valued bilinear form $(X, Y) \mapsto R(X, N)SY$, where S is a linear operator on TM.

Definition 1.1. The *Newtonian transformations* of a $n \times n$ matrix A are defined inductively by $T_0(A) = I_n$, $T_m(A) = \sigma_m(A) I_n - A T_{m-1}(A)$ for $1 \le m \le n$. Hence, the m-th Newtonian transformation (or tensor) is

$$T_m(A) = \sigma_m(A) I_n - \sigma_{m-1}(A) A + \ldots + (-1)^m A^m, \qquad 0 \le m \le n. \quad (8)$$

Theorem 1.1. *Let N be a unit vector field on a compact Riemannian manifold M^{n+1}. Then N is a critical point of the functional (1) for $m > 0$ if and only if the following Euler-Lagrange equation holds:*

$$\left[\text{div}\, T^t_{m-1}(\nabla N) \right]^{\perp_N} = 0. \quad (9)$$

Here $(\,\cdot\,)^t$ is a conjugated (transposed) operator. Clearly, $\operatorname{div} T_0^t(\nabla N) = \operatorname{div} I_n = 0$. For $m = 2$, the left-hand-side of (9) reads as

$$[\operatorname{div} T_1^t(\nabla N)]^{\perp N} = [\nabla \sigma_1(\nabla N) - \operatorname{div}(\nabla N^t)]^{\perp N}, \quad \text{and so on.}$$

Denote by $\widetilde{\operatorname{Ric}}(N) = \sum_i R(N, e_i)\, e_i$ the vector field dual **via** $\langle \cdot \rangle$ to the linear form $X \mapsto \sum_i \langle R(e_i, N) X, e_i \rangle$ for any $X \in TM$. Let $\mathcal{I}_{2,N}$ be the bilinear form on the space of vector fields orthogonal to N given by

$$\mathcal{I}_{2,N}(X, Y) = \operatorname{Ric}(X, Y) - \operatorname{Ric}(N, N)\langle X, Y \rangle.$$

With this notation, we have the following.

Corollary 1.1. *A unit vector field N on a compact Riemannian manifold M is a critical point for the functional $I_2(N)$, see (1), if and only if*

$$\widetilde{\operatorname{Ric}}(N)^{\perp} = 0, \tag{10}$$

where $(\cdot)^{\perp}$ is the orthogonal to N component. It is a point of local minimum if the form $\mathcal{I}_{2,N}$ is positive definite on the space of sections of $N^{\perp}$.

Remark 1.1. For a map between compact Riemannian spaces $f : M^m \to N$, the *energy* is defined to be $\mathcal{E}(f) = \frac{1}{2} \int_M \sum_{a=1}^m \langle df(e_a), df(e_a) \rangle \, d\operatorname{vol}$ (see [6], etc). A unit vector field on a Riemannian manifold M can be considered as a map between M and $T_1 M$, the unit tangent bundle equipped with the Sasaki metric. Then, the *energy of a unit vector field N* on a compact Riemannian manifold M^n can be expressed by the formula

$$\mathcal{E}(N) = \frac{1}{2} \operatorname{vol}(M) + \frac{1}{2} \int_M |\nabla N|^2 \, d\operatorname{vol}. \tag{11}$$

The integral in (11), up to some constants, is the *total bending* $\mathcal{B}(N)$ of N:

$$\mathcal{B}(N) = c_n \int_M |\nabla N|^2 \, d\operatorname{vol}, \qquad c_n = 1/((n-1)\operatorname{vol}(S^n)). \tag{12}$$

Recall that $\operatorname{vol}(S^{n-1}) = \frac{2\pi^{n/2}}{\Gamma(n/2)}$. The problem of minimizing $\mathcal{E}(N)$ or $\mathcal{B}(N)$ has been studied by several authors, see [7,8]. Let N be a unit vector field on $M \setminus S$, M^{n+1} being a compact Riemannian manifold and S a finite subset of M. If $n \geq 2$, then [8]

$$\mathcal{E}(N) \geq \frac{1}{2n-2} \int_M \operatorname{Ric}(N, N) \, d\operatorname{vol} + \frac{n+1}{2} \operatorname{vol}(M).$$

Hence the minimal value of $I_2(N)$ is useful for estimation from below of the energy of a vector field on M.

Corollary 1.2. *A a unit vector field N on a compact Riemannian manifold M is a critical point of the functional $I_3(N)$ if and only if*

$$[(\operatorname{div} N)\widetilde{\operatorname{Ric}}(N) + \frac{1}{2}\nabla \operatorname{Ric}(N,N) + \operatorname{Tr}(\mathcal{R}(\nabla N))]^{\perp} = 0. \qquad (13)$$

Remark 1.2. The last term in (13) vanishes when $N^{\perp}$ is integrable and that the Euler-Lagrange equation is satisfied for all N's on M of constant sectional curvature. Moreover, the first two terms in there vanish when M is Einstein.

Next we compute the first variation of the total mean curvatures for a finite orthonormal system of vector fields on M.

Theorem 1.2. *An orthonormal set $\{N_1, \ldots N_k\}$ of vector fields on a compact Riemannian manifold M^{n+1} is a critical point of the functional (7) for $\boldsymbol{\lambda} = (\lambda_1, \ldots \lambda_k)$ if and only if the following Euler-Lagrange equations hold:*

$$\sum_{1 \le r \le |\boldsymbol{\lambda}|} (-1)^{r-1} \sum_{i_2,\ldots i_r \le k} [(\widetilde{A}^t \nabla \sigma_{\boldsymbol{\lambda}_{i_1,i_2\ldots i_r}}(\vec{A}) + \sigma_{\boldsymbol{\lambda}_{i_1,i_2\ldots i_r}}(\vec{A}))\operatorname{div}\widetilde{A}^t]^{\perp i_1} = 0,$$

where $i_1 = 1, \ldots k$, $\vec{A} = (\nabla N_1, \ldots \nabla N_k)$, $\widetilde{A} = \nabla N_{i_2} \cdot \ldots \nabla N_{i_r}$, $(\cdot)^t$ is a conjugated (transposed) operator and $[\cdot]^{\perp j}$ is N_j-orthogonal component. Here we put $\boldsymbol{\lambda}_i = \boldsymbol{\lambda} - e_i = (\lambda_1, \ldots \lambda_i - 1, \ldots \lambda_k)$, $\boldsymbol{\lambda}_{i,j} = \boldsymbol{\lambda} - e_i - e_j$, etc.

2. Proofs

2.1. *Algebraic preliminaries*

Lemma 2.1. *Let $A(s) = \cos s\, A_1 + \sin s\, A_2$. Then for any $m > 0$,*

$$\sigma_m(A(s)) = \sum_{i+j=m} \sigma_{(i,j)}(A_1, A_2)(\cos s)^i (\sin s)^j.$$

Proof. By definition, we have

$$\sum_m \sigma_m(A(s))\, t^m = \det|I_n + tA(s)| = \det|I_n + (t\cos s)A_1 + (t\sin s)A_2|$$
$$= \sum_m \left(\sum_{i+j=m} \sigma_{(i,j)}(A_1, A_2)(\cos s)^i (\sin s)^j\right) t^m.$$

Comparing coefficients of two polynomials of t we get the claim. $\square$

Lemma 2.2. *Let $f_m(x,y) = \sum_{j=0}^m f_{m-j,j}\, x^{m-j} y^j$ be a homogeneous polynomial of degree $m > 0$. Then*

$$f_m(\cos s, \sin s) = \frac{1}{2^m}\sum_{j=0}^{[m/2]}[\alpha_j \cos((m-2j)s) + \beta_j \sin((m-2j)s)], \qquad (14)$$

where α_j, β_j depend linearly on f_m. The equality $f_m(\cos s, \sin s) \equiv f_0 \;(\forall s)$ for even m transforms into the linear system for the coefficients $f_{m-b,b}$,

$$\begin{aligned}\sum_{b=0,2,\ldots}^{[m/2]}(-1)^{b/2}\bar{C}(j,m-b,b)f_{m-b,b} &= \delta_{m/2}^j f_0, \\ \sum_{b=1,3,\ldots}^{[m/2]}(-1)^{(b-1)/2}\bar{C}(j,m-b,b)f_{m-b,b} &= 0, \quad j=0,\ldots[m/2].\end{aligned} \quad (15)$$

Here δ_i^j is the Kronecker's delta, and

$$C(j,a,b) = \sum_{0\le l\le j}(-1)^l \binom{a}{j-l}\binom{b}{l}, \quad \bar{C}(j,a,b) = C(j,a,b) + C(a+b-j,a,b). \quad (16)$$

In particular, $C(j,0,b) = (-1)^j\binom{b}{j}$ and $C(j,a,0) = \binom{a}{j}$.

Proof. By definition $f_m = \sum_{a+b=m} f_{a,b}(\cos s)^a(\sin s)^b$. Using Binomial Theorem and the formulae $\cos s = (e^{is} + e^{-is})/2$, $\sin s = (e^{is} - e^{-is})/(2i)$ where $i^2 = -1$, we obtain

$$(\cos s)^a(\sin s)^b = i^{-b}\frac{e^{i(a+b)s}}{2^{a+b}}(1+e^{-2is})^a(1-e^{-2is})^b =$$
$$i^{-b}\frac{e^{i(a+b)s}}{2^{a+b}}\sum_{j=0}^{a+b}e^{-2ijs}\sum_{k+l=j}(-1)^l\binom{a}{k}\binom{b}{l} = i^{-b}\frac{1}{2^{a+b}}\sum_{j=0}^{a+b}C(j,a,b)\,e^{i(a+b-2j)s}.$$

Hence we arrive at the Fourier polynomial expansion

$$(\cos s)^a(\sin s)^b = \begin{cases} \frac{(-1)^{b/2}}{2^{a+b}}\sum_{j=0}^{a+b}C(j,a,b)\cos((a+b-2j)s), & b \text{ even} \\ \frac{(-1)^{(b-1)/2}}{2^{a+b}}\sum_{j=0}^{a+b}C(j,a,b)\sin((a+b-2j)s), & b \text{ odd}. \end{cases} \quad (17)$$

Notice that for $a=0$ or $b=0$ (17) reduces to the well known formulae, Taylor expansion of the powers of cos and sin. Next we get

$$\sum_{j=0}^{m}[\tilde{\alpha}_j \cos((m-2j)s) + \tilde{\beta}_j \sin((m-2j)s)] =$$
$$\sum_{j=0}^{[m/2]}[(\tilde{\alpha}_j + \tilde{\alpha}_{m-j})\cos((m-2j)s) + (\tilde{\beta}_j + \tilde{\beta}_{m-j})\sin((m-2j)s)].$$

Finally, using notation (16), we obtain the system (14), where

$$\alpha_j = \sum_{b=0,2,\ldots}^{[m/2]}(-1)^{b/2}\bar{C}(j,m-b,b)f_{m-b,b},$$
$$\beta_j = \sum_{b=1,3,\ldots}^{[m/2]}(-1)^{(b-1)/2}\bar{C}(j,m-b,b)f_{m-b,b}.$$

The equality $f_m(\cos s, \sin s) \equiv f_0 \;(\forall s)$ for even m transforms into (15). □

2.2. Proof of Theorem 0.2

Let us shorten $C_i = \cos s_i$ and $S_i = \sin s_i$ for $i \leq k$, $|s_i| \leq \pi/2$. Consider a unit vector field $N(s) = C_1 N_1 + \sum_{i=2}^{k} S_1 \ldots S_{i-1} C_i N_i$ on M. Hence $\nabla N(s) = C_1 \nabla N_1 + \sum_{i=2}^{k} S_1 \ldots S_{i-1} C_i \nabla N_i$. By definition, we obtain

$$\sum_{0 \leq m \leq n} \sigma_m(\nabla N(s))\, t^m = \det |I_n + t \nabla N(s)| =$$
$$\det |I_n + t C_1 \nabla N_1 + t S_1 C_2 \nabla N_2 + \ldots + t S_1 \ldots S_{k-1} C_k \nabla N_k| =$$
$$\sum_{0 \leq m \leq n} \Big(\sum_{|\lambda|=m} \sigma_\lambda(\nabla N_1, \ldots \nabla N_k) C_1^{i_1} S_1^{m-i_1} C_2^{i_2} S_2^{m-i_1-i_2} \ldots C_{k-1}^{i_{k-1}} S_{k-1}^{i_k} \Big) t^m.$$

Comparing coefficients of the two polynomials of t, we get

$$\sigma_m(\nabla N(s)) = \sum_{|\lambda|=m} \sigma_\lambda(\nabla N_1, \ldots \nabla N_k) C_1^{i_1} S_1^{m-i_1} C_2^{i_2} S_2^{m-i_1-i_2} \ldots C_{k-1}^{i_{k-1}} S_{k-1}^{i_k}.$$

Notice that $\int_M \sigma_m(\nabla N(s))\, d\mathrm{vol} = C(c, n, m)$ is constant by conditions and (2). Applying Lemmas 2.1, 2.2, we obtain for all $i_1 < m$

$$\sum_{|\boldsymbol{\mu}|=m-i_1} \int_M [\sigma_{(i_1, \boldsymbol{\mu})}(\nabla N_1, \ldots \nabla N_k) C_2^{i_2} S_2^{m-i_1-i_2} \ldots C_{k-1}^{i_{k-1}} S_{k-1}^{i_k}]\, d\mathrm{vol} = \mathrm{const}.$$

Repeating this argument by induction, we obtain $I_\lambda(N_1, \ldots N_k) = \mathrm{const}$.

If $c = 0$ or if either $|\lambda|$ or n are odd, then we have $C(c, n, |\lambda|) = 0$; hence $I_\lambda(N_1, \ldots N_k) = 0$. This completes the proof of Theorem 0.2.

Example 2.1. To illustrate Theorem 0.2 for $k = 2$, consider orthonormal vector fields N_1, N_2 on a compact space form $M(c)$. If either m or n is odd then $I_{(i, m-i)}(\nabla N_1, \nabla N_2) = 0$. If both m and n are even then $X_i = I_{(i, m-i)}(\nabla N_1, \nabla N_2)$ $(i = 0, \ldots m)$ are solutions of the system, see (15),

$$\sum_{b=0,2,\ldots}^{m/2} (-1)^{b/2} \bar{C}(j, m-b, b) X_b = \delta_{m/2}^{j} c^{m/2} \binom{n/2}{m/2} \mathrm{vol}(M),$$
$$\sum_{b=1,3,\ldots}^{m/2} (-1)^{(b-1)/2} \bar{C}(j, m-b, b) X_b = 0 \quad (j = 0, \ldots m/2).$$

2.3. Variational formulae

Let N_s, $s \in (-\epsilon, \epsilon)$ be a smooth one-parameter family of unit vector fields on M, $N_0 = N$, a given unit vector field. Set $I(s) = I_2(N_s)$ for all s. One may calculate $\dot{\nabla} N = \nabla \xi$, where $\xi = dN_s/ds(0)$ is tangent to $N^\perp$.

Lemma 2.3. *For any linear operator $S : TM \to TM$ we have*

$$\mathrm{Tr}(S \nabla \xi) = \mathrm{div}(S\,\xi) - \langle \mathrm{div}\, S^t, \xi \rangle. \tag{18}$$

Proof. We calculate

$$\mathrm{Tr}(S\nabla\xi) = \sum_i \langle S\nabla_{e_i}\xi, e_i\rangle = \sum_i \left[\nabla_{e_i}\langle S\xi, e_i\rangle - \langle \nabla_{e_i}(S^t e_i), \xi\rangle\right]$$
$$= \mathrm{div}(S\,\xi) - \sum_i \left[\langle \nabla_{e_i}(S^t e_i) - S^t(\nabla_{e_i} e_i), \xi\rangle\right]$$
$$= \mathrm{div}(S\,\xi) - \langle \sum_i (\nabla_{e_i} S^t) e_i, \xi\rangle = \mathrm{div}(S\,\xi) - \langle \mathrm{div}\, S^t, \xi\rangle. \quad \square$$

Lemma 2.4. *If* $\boldsymbol{\lambda} = (\lambda_1, \lambda_2, \ldots \lambda_k)$ *with* $\lambda_j > 0$, $\forall j$, *and* $\vec{A}(s) = (A_1(s), \ldots A_k(s))$ ($s \geq 0$) *are smooth families of matrices, then*

$$\dot{\sigma}_{\boldsymbol{\lambda}}(\vec{A}(s)) = \sum_{i=1}^{k} \sigma_{(1,\boldsymbol{\lambda}_i)}(\dot{A}_i, \vec{A}). \tag{19}$$

The formula (19) reduces to the following one:

$$\dot{\sigma}_{\boldsymbol{\lambda}}(\vec{A}(s)) = \sum_{1 \leq m \leq |\boldsymbol{\lambda}|} (-1)^{m-1} \sum_{i_1,\ldots i_m \leq k} \sigma_{\boldsymbol{\lambda}_{i_1\ldots i_m}}(\vec{A})\,\mathrm{Tr}(\dot{A}_{i_1} A_{i_2} \ldots A_{i_m}). \tag{20}$$

Proof. Let $A_i(s) = A_i + s\dot{A}_i + o(s)$ and $\dot{A}_i = 0$ for $i \neq 1$. By (4),

$$\det(I_n + t_1(A_1+s\dot{A}_1)+\ldots+t_k A_k) = \det(I_n + st_1\dot{A}_1 + t_1 A_1 + \ldots + t_k A_k)$$
$$= \sum_{|\boldsymbol{\lambda}|\leq n} \sigma_{\boldsymbol{\lambda}}(\vec{A})\,\mathbf{t}^{\boldsymbol{\lambda}} + \sum_{\substack{|\boldsymbol{\lambda}|\leq n \\ \lambda_1>0}} \sigma_{(1,\boldsymbol{\lambda}_1)}(\dot{A}_1, \vec{A})\,\mathbf{t}^{\boldsymbol{\lambda}} s + o(s).$$

Hence, $\frac{d}{ds}\sigma_{\boldsymbol{\lambda}}(A_1 + s\dot{A}_1, A_2, \ldots A_k)_{|s=0} = \sigma_{(1,\boldsymbol{\lambda}_1)}(\dot{A}_1, \vec{A})$ for any $\boldsymbol{\lambda} = (\lambda_1, \lambda_2, \ldots \lambda_k)$ with $\lambda_1 > 0$. Using symmetries of these invariants, see Lemma 1 (identity II) in [3], we get (19). By Lemma 4 in [3] for $\boldsymbol{\lambda} = (\lambda_1, \lambda_2, \ldots \lambda_k)$, with $\lambda_j > 0$, $\forall j$, we have

$$\sigma_{(1,\boldsymbol{\lambda}_{i_1})}(\dot{A}_{i_1}, \vec{A}) = \sigma_{\boldsymbol{\lambda}_{i_1}}(\vec{A})\,\mathrm{Tr}(\dot{A}_{i_1}) - \sum_{i_2=1}^{k} \sigma_{(1,\boldsymbol{\lambda}_{i_1 i_2})}(\dot{A}_{i_1} A_{i_2}, \vec{A}).$$

Applying this formula again, we obtain

$$\sigma_{(1,\boldsymbol{\lambda}_{i_1 i_2})}(\dot{A}_{i_1} A_{i_2}, \vec{A}) = \sigma_{\boldsymbol{\lambda}_{i_1 i_2}}(\vec{A})\,\mathrm{Tr}(\dot{A}_{i_1} A_{i_2}) - \sum_{i_3=1}^{m} \sigma_{(1,\boldsymbol{\lambda}_{i_1 i_2 i_3})}(\dot{A}_{i_1} A_{i_2} A_{i_3}, \vec{A}).$$

Repeating this process, or by induction, we deduce (20). $\quad \square$

Example 2.2. If $A(s)$, $s \geq 0$, is a smooth family of matrices, then (20) reduces (using definition of Newtonian transformations) to the following:

$$\dot{\sigma}_m(A(s)) = \sum_{i=0}^{m-1} (-1)^i \sigma_{m-i-1}(A)\,\mathrm{Tr}(\dot{A} A^i) \stackrel{(8)}{=} \mathrm{Tr}\left(\dot{A}\, T_{m-1}(A)\right). \tag{21}$$

For small values of m, $m = 1, 2, 3$, (21) reads as $\dot{\sigma}_1(A(s)) = \mathrm{Tr}(\dot{A})$ and

$$\dot{\sigma}_2(A(s)) = \mathrm{Tr}(A)\,\mathrm{Tr}(\dot{A}) - \mathrm{Tr}(A\dot{A}),$$
$$\dot{\sigma}_3(A(s)) = \sigma_2(A)\,\mathrm{Tr}(\dot{A}) - \mathrm{Tr}(A)\,\mathrm{Tr}(A\dot{A}) + \mathrm{Tr}(A^2\dot{A}).$$

Proof of Theorem 1.1. Denote by $N_s = N + s\xi + o(s)$ a variation of the unit vector field N. We write down (18) with $S = T_{m-1}(\nabla N)$

$$\operatorname{Tr}(T_{m-1}(\nabla N)\nabla\xi) = \operatorname{div}(T_{m-1}(\nabla N)\xi) - \langle \operatorname{div} T^t_{m-1}(\nabla N), \xi \rangle.$$

Hence, integrating (21) and applying the Divergence Theorem, we obtain

$$I'_m(0) = \int_M \dot\sigma_m(\nabla N(s))|_{s=0}\, d\operatorname{vol} = -\int_M \langle \operatorname{div} T^t_{m-1}(\nabla N), \xi\rangle\, d\operatorname{vol}. \quad (22)$$

Since $\xi \perp N$ is arbitrary, the claim follows from (22). □

Proof of Theorem 1.2. Denote by $N_i(s) = N_i + s\xi_i + o(s)$ $(i \le k)$ variations of unit vector fields N_i. Then $\dot N_i = \xi_i$ and $(\nabla N_i)' = \nabla\xi_i$. In order to use (20), we apply (18) to find $\operatorname{Tr}(\widetilde A\nabla\xi_{i_1}) = \operatorname{div}(\widetilde A\,\xi_{i_1}) - \langle \operatorname{div} \widetilde A^t, \xi_{i_1}\rangle$. Also applying the well-known formula (where f, X are of a class C^1)

$$\operatorname{div}(fX) = f\operatorname{div}(X) + \langle \nabla f, X\rangle \quad (23)$$

to $f = \sigma_{\lambda_{i_1\ldots i_r}}(\vec A)$ and $X = \widetilde A\,\xi_{i_1}$, we arrive at

$$\sigma_{\lambda_{i_1\ldots i_r}}(\vec A)\operatorname{div}(\widetilde A\,\xi_{i_1}) = \operatorname{div}(\sigma_{\lambda_{i_1\ldots i_r}}(\vec A)\,\widetilde A\,\xi_{i_1}) - \langle \nabla\sigma_{\lambda_{i_1\ldots i_r}}(\vec A), \widetilde A\,\xi_{i_1}\rangle.$$

Collecting the terms, we obtain

$$\sigma_{\lambda_{i_1\ldots i_r}}(\vec A)\operatorname{Tr}(\widetilde A\nabla\xi_{i_1}) = \operatorname{div}(\sigma_{\lambda_{i_1\ldots i_r}}(\vec A)\,\widetilde A\,\xi_{i_1})$$
$$-\langle \sigma_{\lambda_{i_1\ldots i_r}}(\vec A)\operatorname{div}\widetilde A^t - \widetilde A^t\nabla\sigma_{\lambda_{i_1\ldots i_r}}(\vec A), \xi_{i_1}\rangle.$$

Integrating and applying the Divergence Theorem, we obtain

$$I'_\lambda(0) = \int_M \dot\sigma_\lambda(\vec A)\,d\operatorname{vol} =$$
$$-\int_M \sum_{1\le r\le|\lambda|}(-1)^{r-1}\sum_{i_1,\ldots,i_r}\langle \widetilde A^t\nabla\sigma_{\lambda_{i_1\ldots i_r}}(\vec A) + \sigma_{\lambda_{i_1\ldots i_r}}(\vec A)\operatorname{div}\widetilde A^t, \xi_{i_1}\rangle\, d\operatorname{vol}.$$

Since $\xi_i \perp N_i$ are arbitrary, the claim follows from the above formula. □

Proof of Corollary 1.1. Let $(-\epsilon, \epsilon) \ni s \to N_s$ be a smooth family of unit vector fields on M, $N_0 = N$, a given unit vector field. Set $I(s) = I_2(N_s)$ for all s. In view of (6) we have $I(s) = \int_M \operatorname{Ric}(N_s, N_s)\,d\operatorname{vol}$. Bi-linearity of the Ricci form implies that

$$2\,I'_2(0) = \int_M \operatorname{Ric}(N, \xi)\,d\operatorname{vol} = \int_M \langle \widetilde{\operatorname{Ric}}(N), \xi\rangle\,d\operatorname{vol},$$

$\xi = (dN_s/ds)(0)$ being the variation field for $s \to N_s$. From $\langle N_s, N_s\rangle = 1$ for all s, it follows that ξ is orthogonal to N. Since for such ξ one can define a variation N_s with $dN_s/ds(0) = \xi$, we get (10).

Assume now that N is a critical point for $I_2(N)$, hence $\widetilde{\operatorname{Ric}}(N)^\perp = 0$. Set $\xi = dN_s/ds(0)$ and $\eta = d^2N_s/ds^2(0)$. Then $\langle \xi, N\rangle = 0$ and $\langle \eta, N\rangle =$

$-\|\xi\|^2$. Therefore, $\eta = -\|\xi\|^2 N + \xi^\perp$ for some $\xi^\perp$ orthogonal to N. Since $\mathrm{Ric}(N, \xi^\perp) = 0$, we obtain

$$2I''(0) = \int_M (\mathrm{Ric}(\xi, \xi) + \mathrm{Ric}(N, \eta))\, d\mathrm{vol}$$
$$= \int_M (\mathrm{Ric}(\xi, \xi) - \mathrm{Ric}(N, N)\|\xi\|^2)\, d\mathrm{vol}. \qquad \Box$$

Remark 2.1. The solutions N of the Euler-Lagrange equation (10) are eigenvectors of the symmetric operator $\widetilde{\mathrm{Ric}} : TM \to TM$, they correspond to the critical values of the symmetric quadratic form $\mathrm{Ric}(N, N)$. The unit eigenvector X_1 with the minimal eigenvalue λ_1, i.e., $\widetilde{\mathrm{Ric}}(X_1) = \lambda_1 X_1$, corresponds to minimum of this form, $\min_{|X|=1} \mathrm{Ric}(X, X) = \mathrm{Ric}(X_1, X_1)$.

Proof of Corollary 1.2. As before, $\xi = dN_s/ds(0)$ is orthogonal to N. From (6) we know that

$$\int_M \sigma_3(\nabla N)\, d\mathrm{vol} = \tfrac{1}{2} \int_M \mathrm{Ric}(N, N)\, \mathrm{Tr}(\nabla N)\, d\mathrm{vol} - \tfrac{1}{3} \int_M \mathrm{Tr}(R_N \nabla N)\, d\mathrm{vol},$$

whenever $(M, \langle \cdot \rangle)$ is locally symmetric. We will use

$$\dot\sigma_1(R_N \nabla N) = \mathrm{Tr}((R_N \nabla N)')$$
$$= \mathrm{Tr}(R(\nabla N(\cdot), \xi)N + R(\nabla N(\cdot), N)\xi) + \mathrm{Tr}(R_N \nabla \xi).$$

Taking as before (see Corollary 1.1) a 1-parameter family of unit vector fields N_s and differentiating the function $s \mapsto I_3(s) = I_3(N_s)$ we obtain

$$I_3'(0) = \int_M \big(\mathrm{Tr}(\nabla N)\,\mathrm{Ric}(N, \xi) + \tfrac{1}{2}\mathrm{Ric}(N, N)\,\mathrm{div}\,\xi$$
$$- \tfrac{1}{3}\mathrm{Tr}\,R(\nabla \xi, N)N - \tfrac{1}{3}\mathrm{Tr}\,R(\nabla N(\cdot), \xi)N - \tfrac{1}{3}\mathrm{Tr}\,R(\nabla N(\cdot), N)\xi\big)\, d\mathrm{vol}.$$

Applying the formula (23) to $f = \mathrm{Ric}(N, N)$ and $X = \xi$, using the standard symmetries of the curvature tensor, the fact that our Riemannian manifold is locally symmetric ($\nabla R = 0$), and the Green formula ($\int_M \mathrm{div}\,X\,d\mathrm{vol} = 0$) we arrive at

$$I_3'(0) = \int_M \big\langle \mathrm{Tr}(\nabla N)\,\widetilde{\mathrm{Ric}}(N) + \tfrac{1}{2}\nabla \mathrm{Ric}(N, N) + \mathrm{Tr}(\mathcal{R}(\nabla N)),\ \xi \big\rangle\, d\mathrm{vol}, \quad (24)$$

The only difficulty in calculations leading to (24) is in writing the term $\mathrm{Tr}\langle R(\cdot, N)\xi, \nabla_{(\cdot)} N \rangle$ in the form

$$\mathrm{div}\,R(\xi, N)N + \langle \mathrm{Tr}(\cdot, N)\nabla N(\cdot),\ \xi \rangle + \langle \mathrm{Tr}\,R(\cdot, \nabla N(\cdot))N, \xi \rangle$$
$$+ \mathrm{Tr}((X, Y) \mapsto \langle (\nabla_X R)(\xi, N, N), Y \rangle$$

and observing that the last term vanishes due to the local symmetry of M.$\Box$

References

1. D. Asimov: Bull. Amer. Math. Soc. **84**, 131–133 (1978).
2. F. Brito, R. Langevin and H. Rosenberg, *J. Diff. Geom.* **16**, 19–50 (1981).
3. V. Rovenski and P. Walczak, Integral formulae on foliated symmetric spaces, *Preprint* **2007/13**, 1–27 (2007).
4. V. Rovenski and P. Walczak, *Proc. of the* 10-*th International Conference on DGA-2007, Olomouc, August 27–31, 2007,* (World Scientific, 2008), pp. 193–204.
5. R. Reilly, *J. Differential Geometry* **8**, 465–477 (1973).
6. J. Eells and L. Lemaire, *Bull. London Math. Soc.* **10**, 1–68 (1978).
7. F. Brito, *Differential Geometry and its Applications* **12**, 157–163 (2000).
8. F. Brito and P. Walczak, *Ann. Polon. Math.* **LXXXIII.3**, 269–274 (2000).

ON A WEITZENBÖCK-LIKE FORMULA FOR RIEMANNIAN FOLIATIONS

V. Slesar

Department of Mathematics, University of Craiova
Craiova 200585, Romania
E-mail: vlslesar@hotmail.com
www.ucv.ro

The transverse geometry of a Riemannian foliation has been intensively studied in the last period of time. In this paper we present the relations between the foliated structure of a Riemannian foliation and the classical Weitzenböck formula. We obtain a *transversal* Weitzenböck type formula which works in a more general setting than a recent Weitzenböck formula due to Y. A. Kordyukov.[1]

Keywords: Riemannian foliation, de Rham derivative Weitzenböck formula

1. Introduction

Let us consider the smooth, foliation $(M, \mathcal{F}, g)$ on a closed, n-dimensional manifold M, endowed with a Riemannian metric g. We start out by stating some basic facts concerning the *spectral sequence* and the *adiabatic limit* associated with the foliation. The dimension of the foliation will be denoted by p, the codimension by q and (Ω, d) will denote the de Rham complex on M. In the classical way[2] we get a bigrading for Ω, induced by the foliated structure and the bundle-like metric:

$$\Omega^{u,v} = C^\infty \left(\bigwedge^u T\mathcal{F}^{\perp *} \oplus \bigwedge^v T\mathcal{F}^* \right), \ u,v \in \mathbb{N}. \quad (1)$$

Then, the exterior derivative and the coderivative split into bihomogeneous differential components as follows:

$$d = d_{0,1} + d_{1,0} + d_{2,-1}, \ \delta = \delta_{0,-1} + \delta_{1,0} + \delta_{-2,1}, \quad (2)$$

where the indices correspond to the bigrading.

Using the above operators we can define the *transversal* Laplace operator:

Definition 1.1. The *transversal* Laplace operator is the Laplace type operator defined using $d_{1,0}$ and $\delta_{-1,0}$:

$$\Delta_\perp H := d_{1,0}\delta_{-1,0} + \delta_{-1,0}d_{1,0}. \tag{3}$$

The *transversal* Laplace operator has been intensively studied,[3] the interplay with *basic* de Rham cohomology being of particular interest.

The *adiabatic limit* of a foliation was introduced by E. Witten for a Riemannian bundle over the circle.[4] We decompose the metric $g = g_\perp \oplus g_\mathcal{F}$ with respect to the splitting $TM = T\mathcal{F}^\perp \oplus T\mathcal{F}$. Introducing a parameter $h > 0$, let us define the family of metrics

$$g_h = h^{-2}g_\perp \oplus g_\mathcal{F}.$$

The limit of the foliation $(M, \mathcal{F}, g_h)$ as $h \downarrow 0$ is known as the *adiabatic limit*, while the above parameter is known as the *adiabatic parameter*.

Considering the special case when the metric is bundle-like,[5] in this paper we present a Weitzenböck formula for the *transversal* Laplace operator when acting on arbitrary differential forms. The importance of this approach comes from the fact that recent work of J. A. Álvarez López and Y. A. Kordyukov[6,7] focussed on *differential spectral sequence* extend the interest area from *basic* forms to general differential forms.

2. Canonical differential operators defined on a Riemannian foliation

For the rest of this paper, we will consider a Riemannian foliation (i. e. endowed with a bundle-like metric).

We also consider local infinitesimal transformations $\{F_a\}$, $1 \leq a \leq q$, of $(M, \mathcal{F})$ orthogonal to the leaves, while $\{E_i\}$, $1 \leq i \leq p$, will be smooth local vector fields tangent to the leaves. Furthermore, assume that the system $\{F_a, E_i\}$ determines an orthonormal basis $\{f_a, e_i\}$ at any point where they are defined. Let us consider also the dual coframes $\{\theta^a, \omega^i\}$ for $\{F_a, E_i\}$, and $\{\alpha^a, \beta^i\}$ for $\{f_a, e_i\}$. We denote by $U^\mathcal{T}$ the transverse component and by $U^\mathcal{L}$ the leafwise component of a local tangent vector field U. For the sake of simplicity, the local vector fields $\{F_a\}$ will be called *basic* vector fields.[8] In what follows we consider arbitrary local tangent vector fields U and V. First of all, let us remind the Gray-O'Neill tensors fields A and T:[9]

$$T_U V := \nabla_{U^\mathcal{L}}^\mathcal{L} V^\mathcal{T} + \nabla_{U^\mathcal{L}}^\mathcal{T} V^\mathcal{L},$$

$$A_U V := \nabla_{U^\mathcal{T}}^\mathcal{L} V^\mathcal{T} + \nabla_{U^\mathcal{T}}^\mathcal{T} V^\mathcal{L},$$

where ∇ is the Levi-Civita connection associated to g. We canonically extend the above 0th order operators on differential forms. In accordance with the previous literature,[10] let us define the following metric connection:

$$\widetilde{\nabla}_U V = \pi_{\mathcal{F}} \nabla_U \pi_{\mathcal{F}} V + \pi_Q \nabla_U \pi_Q V.$$

As a consequence, the Levi Civita connection splits as follows:

$$\nabla_U = \widetilde{\nabla}_{U^{\mathcal{T}}} + \widetilde{\nabla}_{U^{\mathcal{L}}} + A_{U^{\mathcal{T}}} + T_{U^{\mathcal{L}}}. \qquad (4)$$

We take an orthonormal frame field $\{\mathcal{E}_i\}$ in the neighborhood of an arbitrary point $x \in M$ which induces an orthonormal basis $\{\epsilon_s\}$ for $T_x M$, with $1 \leq s \leq n$. If $\{\Theta^s\}$ and $\{\theta^s\}$ are the dual coframes for $\{\mathcal{E}_s\}$ and $\{\epsilon_s\}$ respectively, then the exterior derivative and its adjoint operator with respect to the canonical inner product on the Riemannian manifold can be written as follows:

$$d = \sum_s \Theta^s \wedge \nabla_{\mathcal{E}_s}, \quad \delta = -\sum_s i_{\mathcal{E}_s} \nabla_{\mathcal{E}_s}.$$

Using (4) and the orthonormal basis $\{f_a, e_i\}$ and $\{\alpha^a, \beta^i\}$, the above two operators split as follows:

$$d = \sum_a \alpha^a \wedge \widetilde{\nabla}_{f_a} + \sum_a \alpha^a \wedge A_{f_a} + \sum_i \beta^i \wedge \widetilde{\nabla}_{e_i} + \sum_i \beta^i \wedge T_{e_i}, \qquad (5)$$

$$\delta = -\sum_a i_{f_a} \widetilde{\nabla}_{f_a} - \sum_a i_{f_a} A_{f_a} - \sum_i i_{e_i} \widetilde{\nabla}_{e_i} - \sum_i i_{e_i} \wedge T_{e_i}.$$

Considering a foliated chart $\mathcal{U}$ on M, then[7]

$$\Omega^{u,v}(\mathcal{U}) = \Omega^u(\mathcal{U}/\mathcal{F}_\mathcal{U}) \wedge \Omega^{0,v}(\mathcal{U}) \equiv \Omega^u(\mathcal{U}/\mathcal{F}_\mathcal{U}) \otimes \Omega^{0,v}(\mathcal{U}).$$

Here $\Omega^u(\mathcal{U}/\mathcal{F}_\mathcal{U})$ denotes the space of *basic* forms of transversal degree u, defined on $\mathcal{U}$. Then, if we take $\alpha \in \Omega^u(\mathcal{U}/\mathcal{F}_\mathcal{U})$ and $\beta \in \Omega^{0,v}(\mathcal{U})$, just considering the induced bigrading, we get for the case of a Riemannian foliation:[11]

$$d_{1,0}(\alpha \wedge \beta) = \sum_a \alpha^a \wedge \widetilde{\nabla}_{f_a} \alpha \wedge \beta + \alpha \wedge (-1)^u \sum_a \alpha^a \wedge \widetilde{\nabla}_{f_a} \beta \qquad (6)$$

and for the co-differential operator:

$$\delta_{-1,0}(\alpha \wedge \beta) = -\sum_a i_{f_a} \widetilde{\nabla}_{f_a} \alpha \wedge \beta + i_{k^\natural} \alpha \wedge \beta - \sum_a i_{f_a} \alpha \wedge \widetilde{\nabla}_{f_a} \beta \qquad (7)$$

where $g(k^\natural, U) = k(U) := g(\sum_i T_{e_i} e_i, U)$, in other words k is the *mean curvature form*.

Remark 2.1. If β vanishes, then we get[2]

$$d_{1,0}\alpha = \sum_a \alpha^a \wedge \widetilde{\nabla}_{f_a}\alpha,$$

$$\delta_{-1,0}\alpha = -\sum_a i_{f_a}\widetilde{\nabla}_{f_a}\alpha + i_{k^\natural}\alpha.$$

3. Adiabatic limits and Riemannian foliations

The variation of Laplace operator, Levi-Civita connection and metric tensor field with respect to the *adiabatic parameter* has been studied in several previous papers.[6,12] In this section we recall the main results. Using the classical Koszul formula, we are able to express all the components of the Levi-Civita connection (determined by the transverse-tangent decomposition) as polynomials in h. We obtain:[12]

Proposition 3.1. *The following equalities relate the canonical Levi-Civita connections associated to the metrics g_h and g:*

$$\begin{aligned}
\nabla_{F_a}^{g_h,\mathcal{T}}\theta^b &= \nabla_{F_a}^{\mathcal{T}}\theta^b, & \nabla_{F_a}^{g_h,\mathcal{L}}\theta^a &= h^2\nabla_{F_a}^{\mathcal{L}}\theta^a, \\
\nabla_{E_i}^{g_h,\mathcal{T}}\omega^j &= \nabla_{E_i}^{\mathcal{T}}\omega^j, & \nabla_{E_i}^{g_h,\mathcal{L}}\omega^j &= \nabla_{E_i}^{\mathcal{L}}\omega^j, \\
\nabla_{F_a}^{g_h,\mathcal{T}}\omega^i &= \nabla_{F_a}^{\mathcal{T}}\omega^i, & \nabla_{F_a}^{g_h,\mathcal{L}}\omega^i &= \nabla_{F_a}^{\mathcal{T}}\omega^i, \\
\nabla_{E_i}^{g_h,\mathcal{T}}\theta^a &= h^2\nabla_{E_i}^{\mathcal{T}}\theta^a, & \nabla_{E_i}^{g_h,\mathcal{L}}\theta^a &= h^2\nabla_{E_i}^{\mathcal{L}}\theta^a
\end{aligned} \quad (8)$$

for any indices a, b, i and j, with $1 \leq a, b \leq q$ and $1 \leq i, j \leq p$, respectively.

In what follows let us now consider the classical Weitzenböck formula.[13] We take an orthonormal frame field $\{\mathcal{E}_s\}$ on the neighborhood of an arbitrary point $x \in M$ which induces an orthonormal basis $\{\epsilon_s\}$ for T_xM such that $\nabla_{\epsilon_s}\mathcal{E}_t = 0$, with $1 \leq s, t \leq n$. If $\{\Lambda^s\}$ (respectively $\{\lambda^s\}$) is the dual coframe of $\{\mathcal{E}_i\}$ (respectively $\{\epsilon_i\}$), considering that $d = \sum_s \Lambda^s \wedge \nabla_{\mathcal{E}_s}$ and $\delta = -\sum_s i_{\mathcal{E}_s}\nabla_{\mathcal{E}_s}$, we can express the Laplace operator:

$$\Delta = d\delta + \delta d = \nabla^*\nabla + K, \quad (9)$$

where $K =: \sum_{s<t} \lambda^s \cdot \lambda^t \cdot R_{\epsilon_s,\epsilon_t}$, and the dot stands for Clifford multiplication on differential forms.

We have the following canonical splittings of the tangent and cotangent bundles: $TM = T\mathcal{F}^\perp \oplus T\mathcal{F}$ and $TM^* = T\mathcal{F}^{\perp *} \oplus T\mathcal{F}^*$. The canonical transversal and leafwise projection operator will be denoted by $\mathrm{pr}^\mathcal{T}$

and $\mathrm{pr}^{\mathcal{L}}$ respectively. We can consider the *rescaling homomorphism* Θ_h : $(TM^*, g_h) \to (TM^*, g)$:

$$\Theta_h := h\,\mathrm{id}_{T\mathcal{F}^\perp{}^*} \oplus \mathrm{id}_{T\mathcal{F}^*}. \tag{10}$$

The induced *rescaling homomorphism* on differential forms or tensor fields will be denoted also by Θ_h. One can prove that the above *rescaling homomorphism* is in fact an isometry of Riemannian vector bundles.[14] This allow us to define the *rescaled* operators $\Delta_h := \Theta_h \Delta_{g_h} \Theta_h^{-1}$, $\nabla^h := \Theta_h \nabla^{g_h} \Theta_h^{-1}$ and $K^h := \Theta_h K_{g_h} \Theta_h^{-1}$.

Applying (9) for $\Theta_h^{-1}\omega$, where $\omega \in \Omega^r$, and by using that Θ_h is in fact an isometry of Riemannian vector bundles, we obtain the formula

$$\langle \Delta_h \omega, \omega \rangle = \langle \nabla^h \omega, \nabla^h \omega \rangle + \langle K^h \omega, \omega \rangle, \tag{11}$$

where the inner product is obtained by integrating on the closed Riemannian manifold M.[15]

We consider the covariant derivative ∇ induced on $\Omega^{u,v}$, with u and v satisfying $u + v = r$, $r \in \mathbb{N}$:

$$\nabla : \Omega^{u,v} \longrightarrow C^\infty(TM^*) \otimes C^\infty(\Lambda^r TM^*).$$

We refine the covariant derivative in the presence of the canonical projections operators $\mathrm{pr}^{\mathcal{T}}$ and $\mathrm{pr}^{\mathcal{L}}$—determined by the foliated structure, and the canonical projections $\pi_{u,v}$, $\pi_{u-1,v+1}$ and $\pi_{u+1,v-1}$-induced by the bigrading, defining the following six differential operators:

$$\nabla_{\mathcal{T},0,0} := (\mathrm{pr}^{\mathcal{T}} \otimes \pi_{u,v}) \circ \nabla, \qquad \nabla_{\mathcal{L},0,0} := (\mathrm{pr}^{\mathcal{L}} \otimes \pi_{u,v}) \circ \nabla,$$
$$\nabla_{\mathcal{T},-1,1} := (\mathrm{pr}^{\mathcal{T}} \otimes \pi_{u-1,v+1}) \circ \nabla, \; \nabla_{\mathcal{L},-1,1} := (\mathrm{pr}^{\mathcal{L}} \otimes \pi_{u-1,v+1}) \circ \nabla,$$
$$\nabla_{\mathcal{T},1,-1} := (\mathrm{pr}^{\mathcal{T}} \otimes \pi_{u+1,v-1}) \circ \nabla, \; \nabla_{\mathcal{L},1,-1} := (\mathrm{pr}^{\mathcal{L}} \otimes \pi_{u+1,v-1}) \circ \nabla.$$

The above operators can be naturally extended from $\Omega^{u,v}$ to Ω.

In the classical manner,[7] we choose a foliated chart $\mathcal{U}$ on M; then we obtain the following local description of the de Rham complex:

$$\Omega^{u,v}(\mathcal{U}) = \Omega^u(\mathcal{U}/\mathcal{F}_\mathcal{U}) \wedge \Omega^{0,v}(\mathcal{U}) \equiv \Omega^u(\mathcal{U}/\mathcal{F}_\mathcal{U}) \otimes \Omega^{0,v}(\mathcal{U}), \tag{12}$$

As a consequence, we consider $\alpha \in \Omega^u(\mathcal{U}/\mathcal{F}_\mathcal{U})$ and $\beta \in \Omega^{0,v}(\mathcal{U})$, and we evaluate the above operators acting locally on differential forms of the type $\alpha \wedge \beta$, the general formula being easy to obtain by linearity. Considering also how these operators change the bigrading, we write all the operators only using the Levi-Civita connection associated to g and the adiabatic parameter h:

$$\nabla^h_{\mathcal{T},0,0}(\alpha \wedge \beta) = h\left(\nabla^{g_h}_{\mathcal{T},0,0}\alpha \otimes \beta + \alpha \otimes \nabla^{g_h}_{\mathcal{T},0,0}\beta\right) = h\nabla_{\mathcal{T},0,0}(\alpha \wedge \beta),$$

$$\nabla^h_{\mathcal{L},0,0}(\alpha \wedge \beta) = \nabla^{g_h}_{\mathcal{L},0,0}(\alpha \wedge \beta) = \alpha \otimes \nabla_{\mathcal{L},0,0}\beta + h^2 \nabla_{\mathcal{L},0,0}\alpha \otimes \beta,$$

$$\nabla^h_{\mathcal{T},-1,1}(\alpha \wedge \beta) = \nabla^{g_h}_{\mathcal{T},-1,1}(\alpha \wedge \beta) = h^2 \nabla_{\mathcal{T},-1,1}\alpha \wedge \beta = h^2 \nabla_{\mathcal{T},-1,1}(\alpha \wedge \beta),$$

$$\nabla^h_{\mathcal{L},-1,1}(\alpha \wedge \beta) = h^{-1}\nabla^{g_h}_{\mathcal{L},-1,1}(\alpha \wedge \beta) = h\nabla_{\mathcal{L},-1,1}\alpha \wedge \beta = h\nabla_{\mathcal{L},-1,1}(\alpha \wedge \beta),$$

$$\nabla^h_{\mathcal{T},1,-1}(\alpha \wedge \beta) = h^2\nabla^{g_h}_{\mathcal{T},1,-1}(\alpha \wedge \beta) = h^2\,\alpha \wedge \nabla_{\mathcal{T},1,-1}\beta = h^2\nabla_{\mathcal{T},1,-1}(\alpha \wedge \beta),$$

$$\nabla^h_{\mathcal{L},1,-1}(\alpha \wedge \beta) = h\,\nabla^{g_h}_{\mathcal{L},1,-1}(\alpha \wedge \beta) = h\,\alpha \wedge \nabla_{\mathcal{L},1,-1}\beta = h\,\nabla_{\mathcal{L},1,-1}(\alpha \wedge \beta).$$

In the following we denote $\mathrm{id}_{\Omega^u(\mathcal{U}/\mathcal{F}_\mathcal{U})} \otimes \nabla_{\mathcal{L},0,0}$ by $\nabla^0_{\mathcal{L},0,0}$ and $\nabla_{\mathcal{L},0,0} \otimes \mathrm{id}_{\Omega^{0,\cdot}(\mathcal{U})}$ by $\nabla^2_{\mathcal{L},0,0}$.

In order to investigate the last term of (11), let us observe that the formulas (8) allow us to express the curvature components as polynomials in h:[12]

$$K^h = \sum_{l=0}^{4} h^l \cdot K^l$$

and in accordance with the bigrading, this means:

$$K^l = K^l_{-2,2} + K^l_{-1,1} + K^l_{0,0} + K^l_{1,-1} + K^l_{2,-2} \tag{13}$$

for $0 \leq l \leq 4$.

4. A transversal Weitzenböck formula

We remind some previous Weitzenböck formulas and Bochner techniques for Riemannian foliations. First of all, we refer to a vanishing result concerning the basic de Rham complex.[16] Let us consider Ω_b the de Rham complex of *basic* forms defined on the foliation and the *basic* cohomology groups H^r_b, for $0 \leq r \leq q$. We consider also

$$\tilde{\delta} := -\sum_a i_{f_a}\tilde{\nabla}_{f_a},$$

and define

$$\tilde{\Delta} := \tilde{\delta}d + d\tilde{\delta}.$$

If $\mathcal{U}$ is a foliated chart and N a transversal, then the Weitzenböck formula on N reads:

$$\Delta^N = \sum_a \tilde{\nabla}^*_{f_a}\tilde{\nabla}_{f_a} + K^N, \tag{14}$$

where Δ^N and K^N are the corresponding Laplace operator and canonical curvature expression for N. Using the fact that the action of $\tilde{\Delta}$ on Ω_b coincides with the action of Δ^N on the corresponding projection of Ω_b on N, the authors obtain the following vanishing result:[16]

Theorem 4.1. *Assume the transversal curvature operator is positive. Then $H_b^r = 0$, for $0 < r < q$.*

We state now another *transversal* Weitzenböck type formula which works in the setting of transversal bundle $\bigwedge T\mathcal{F}^{\perp *}$:[1]

Theorem 4.2. *We have the following formula*

$$\Delta_\perp = \sum_a \tilde{\nabla}^*_{f_a} \tilde{\nabla}_{f_a} + \sum_{a,b} \alpha^a \wedge i_{f_b}(R^\perp_{f_b, f_a} - 2 \cdot \tilde{\nabla}_{A_{f_b} f_a}) \quad (15)$$
$$+ \sum_a \alpha^a \wedge i_{\nabla_{f_a} k^\natural}.$$

Now, collecting the coefficients of h^2 in (11) and using the same technique,[12] we end up with the following general *transversal* Weitzenböck formula:

Theorem 4.3. *If $(M, \mathcal{F}, g)$ is a Riemannian foliations and ω is a smooth differential form of degree r defined on M, then the following equality holds:*

$$\langle \Delta_\perp \omega, \omega \rangle = 2 \langle \nabla^0_{\mathcal{L}, 0, 0} \omega, \nabla^2_{\mathcal{L}, 0, 0} \omega \rangle + \|\nabla_{\mathcal{T}, 0, 0} \omega\|^2 + \|\nabla_{\mathcal{L}, 1, -1} \omega\|^2 \quad (16)$$
$$+ \|\nabla_{\mathcal{L}, -1, 1} \omega\|^2 + \langle K^2_{0, 0} \omega, \omega \rangle.$$

The above formula works in a more general setting than a previous *transverse* Weitzenböck type formula[1] which works for transverse fiber bundle. In certain situations, a useful tool for studying the basic de Rham complex is the associated spectral sequence.[6] The spectral sequence terms do not contain only basic differential forms, so the relations (6), (7) and our *transversal* Weitzenböck type formula written for differential forms of arbitrary degree might help us to investigate the cohomology of a Riemannian foliation.

Remark 4.1. If we restrict our setting to the case of transverse fiber bundle and apply the above techniques, we obtain a *global* Weitzenböck formula which corresponds to the *pointwise* Weitzenböck formula (15), obtained by direct calculation.[1]

Acknowledgments

I would like to thank J.A. Álvarez López and Y.A. Kordyukov for helpful discussions. Also, I would like to thank the referee for observations and suggestions.

References

1. Y. A. Kordyukov, *Ann. Glob. Anal. Geom.* **34**, 195 (2008).
2. J. A. Álvarez López, *Ann. of Global Anal. and Geom.* **10**, 179 (1992).
3. Ph. Tondeur, *Geometry of Foliations*, (Birkhäuser-Verlag, 1997).
4. E. Witten, *Comm. Math. Phys.* **100**, 197 (1985).
5. B. Reinhart, *Ann. Math.* **69**, 119 (1959).
6. J. A. Álvarez López and Y. A. Kordyukov, *Geom. and Funct. Anal.* **10**, 977 (1992).
7. J. A. Álvarez López and Y. A. Kordyukov, *Compositio Mathematica* **125**, 129 (2001).
8. P. Molino, *Riemannian Foliations*, (Birkhäuser-Verlag, 1988).
9. B. O'Neill, *Michigan Math. J.* **13**, 459 (1966).
10. J. A. Álvarez López and Ph. Tondeur, *J. Funct. Anal.* **99**, 443 (1991).
11. V. Slesar, *Contemporary Geometry and Topology and Related Topics*, (Editors: D. Andrica and S. Moroianu, Cluj University Press, 2008).
12. V. Slesar, *Ann. Glob. Anal. Geom.* **32**, 87 (2007).
13. P. Petersen, *Riemannian Geometry*, (Springer-Verlag, 1998).
14. R. R. Mazzeo and R. B. Melrose, *J. Diff. Geom.* **31**, (1990).
15. M. Craioveanu, M. Puta, Th. M. Rassias, *Old and New Aspects in Spectral Geometry*, (Springer-Verlag, 2001).
16. M. Min-Oo, E. Ruh and Ph. Tondeur, *J. Reine Angew. Math.* **415**, 167 (1991).

DUALITY AND MINIMALITY FOR RIEMANNIAN FOLIATIONS ON OPEN MANIFOLDS

Xosé M. Masa

Departamento de Xeometría e Topoloxía Universidade de Santiago de Compostela
15782 Santiago de Compostela, Spain
E-mail: xose.masa@usc.es

We are interested in the duality properties of the cohomology of a singular Riemannian foliation. We present here the most simple case where there exists a general duality theorem.

Keywords: Singular Riemannian foliation, cohomological duality, minimality

1. The duality

Let $\mathcal{F}$ be a Riemannian foliation of dimension l and codimension k on a smooth manifold M. If M and $\mathcal{F}$ are orientable and M is closed, the de Rham spectral sequence of $\mathcal{F}$ is finite dimensional and satisfies the duality condition

$$E_2^{s,t}(\mathcal{F}) \cong E_2^{k-s,l-t}(\mathcal{F}).$$

Moreover the manifold admits a metric such that each leaf is minimal if and only if the basic cohomology of maximum degree, $E_2^{k,0}$, is non null.[1]

Without the hypothesis on M to be closed, this result can be extended. We assume, for the sake of simplicity, that manifolds and foliation are orientable.

Proposition 1.1. *Let* $\pi\colon M \to B$ *a locally trivial bundle with compact fibers,* $\mathcal{F}$ *a Riemannian foliation tangent to the fibers and compatible whith the bundle structure. Then the spectral sequece of the foliation satisfies*

$$E_2^{s,t}(\mathcal{F}) \cong E_{2,c}^{k-s,l-t}(\mathcal{F}),$$

where $E_{2,c}$ *denotes the spectral sequence with compact supports. If* B *admits a open finite covering of triviality for* π, *then the second term of the spectral sequence is finite-dimensional.*

Proof. We consider the sheaf $\mathcal{P}_{\mathcal{F}}^*$ of local basic forms of $\mathcal{F}$. One has
$$E_2^{s,t}(\mathcal{F}) \cong H^s H^t(M, \mathcal{P}_{\mathcal{F}}^*).$$
Let $\mathcal{E}_1^{*,t}$ denote the sheaf $\mathcal{H}^t(\pi, \mathcal{P}_{\mathcal{F}}^*)$ over B. Since it is fine, we have
$$E_2^{s,t}(\mathcal{F}) = \mathbb{H}^s(B, \mathcal{E}_1^{*,t}),$$
where $\mathbb{H}^*$ stands for the hypercohomology. So we have a spectral sequence
$$H^{s_1}(B, \mathcal{E}_2^{s_2,t}) \Rightarrow E_2^{s_1+s_2,t}(\mathcal{F}),$$
where $\mathcal{E}_2^{s,t}(U) = E_2^{s,t}(\mathcal{F}|_{\pi^{-1}(U)})$. Now, if v denotes the codimension of $\mathcal{F}$ in the fibers, the sheaves $\mathcal{E}_2^{s,t}$ and $\mathcal{E}_2^{v-s,l-t}$ are dual to each other. □

2. The minimality

Corollary 2.1. *Under the hypotheses of the above proposition, the foliation is minimal if and only if* $E_{2,c}^{k,0} \neq 0$.

Proof. This follows from the Rummler-Sullivan characterization of minimality,[3] which asserts that minimality holds when a volume form along the leaves, $\chi_{\mathcal{F}}$, defines a class in $E_2^{0,l}(\mathcal{F})$. □

3. Singular Riemannian foliations

Naturally, the setting of Proposition 1.1 arises from a singular Riemannian foliation on a compact manifold. It induces two stratifications on the manifold, one given by the dimension of the closures of the leaves, and the other by the dimension of the leaves. On the regular stratum for the first one, the foliation has the good properties of the proposition. If we consider the regular stratum for the second stratification, the corollary still holds because, by continuity, the condition on $\chi_{\mathcal{F}}$ extends to the greater stratum. Saralegi, Royo and Wolak prove[2] the above minimality characterization with a different method.

For a singular Riemannian foliation on a compact manifold, the proposition can be understood as a duality theorem for the intersection cohomology of the foliation with extreme perversities.

References

1. Masa, X. Comment. Math. Helv. **67** (1992), 17–27.
2. Saralegi-Aranguren, M., Royo Prieto, J.I. and Wolak, R. Manuscripta Math. **126** (2008), 177-200
3. Sullivan, D., Comment. Math. Helv. **54** (1979), 218–223.

OPEN PROBLEMS ON FOLIATIONS

These problems were collected from the problem session on foliations held at the ICDG 2008.

Problem 0.1 (Takashi Tsuboi). *Let M be a compact and connected manifold of dimension $2m$. Is $\mathrm{Diff}(M)_0$ uniformly perfect?*

The answer is not known for example when M is $\mathbb{T}^2$, the Möbius band, $\mathbb{S}^2 \times \mathbb{S}^2$, $\mathbb{C}P^2$ or $\mathbb{R}P^2$.

If M has a handle body decomposition without handles of index m then the answer is known to be affirmative. It seems that some new ideas are needed for the general case (T. Tsuboi).

Suppose now that M is an open manifold of arbitrary dimension. Let $K \subset M$ be a compact deformation retract of M such that there is a compactly supported diffeomorphism $f : M \to M$ so that $f^k(K) \cap K = \emptyset$ for $k \in \mathbb{Z}$. Then it is known that any element of $\mathrm{Diff}(M)_0$ can be expressed as a composition of two commutators.

(Added in December 2008: The answer is affirmative if $2m \geq 6$. But the cases where $2m = 2$ and 4 are very complicated and they look far from being settled.)

Problem 0.2 (Remi Langevin). *Find, or prove that there are no foliations of $\mathbb{S}^3$, $\mathbb{T}^3$ or $\mathbb{H}^3/\Gamma$ (a compact hyperbolic 3-manifold) whose leaves satisfy some property (P), or that there are only some simple examples.*

Here, (P) can be for example "totally geodesic", "Dupin" or "canal surfaces". For the totally geodesic and Dupin cases, the answer is that there are essentially no such foliations (R. Langevin and P. Walczak).

Let $\mathcal{F}$ be a foliation on $\mathbb{H}^3/\Gamma$, and let $\widetilde{\mathcal{F}}$ denote its lift to $\mathbb{H}^3$. A main point is that $\widetilde{L} \cap S_\infty^2$ is not so bad when L is a Dupin leaf, where $\widetilde{L}$ is a leaf of $\widetilde{\mathcal{F}}$ over L. Thus we may ask how is $\widetilde{L} \cap S_\infty^2$ when L is a canal leaf. We hope that $\widetilde{L} \cap S_\infty^2$ is something like an envelope of circles in S_∞^2.

Here, Reeb components may exist, and a suggestion could be to know whether there are such foliations without Reeb components.

In the same spirit, Zeghib has asked about the existence of foliations $\mathcal{F}$

on $\mathbb{H}^3/\Gamma$ so that the leaves of $\widetilde{\mathcal{F}}$ are analytic and extend through S_∞^2. If the answer is no, then the end of the proof could be like in the Dupin case: we get a partition of S_∞^2, and the density of the orbits of Γ on S_∞^2 is used.

The condition "leaves are canal surfaces" is more flexible. Foliations of $\mathbb{S}^3$ by canal surfaces are now completely understood (Langevin-Walczak).

If the answer for foliations by canal surfaces of $\mathbb{H}^3/\Gamma$ is no, we may ask how to relax the condition to get a few (just a few) examples.

Another property that can be considered for (P) is "elasticity"; foliations with elastic leaves is a natural generalization of foliations by minimal submanifolds.

Problem 0.3 (Pawel Walczak). On $\mathbb{S}^3$ or $\mathbb{H}^3/\Gamma$, does there exist a family of three orthogonal foliations of codimension one? Here, orthogonality means that the unit normal vector fields N_1, N_2 and N_3 of those foliations satisfy $\langle N_i, N_j \rangle = \delta_{ij}$.

In the 1890's, it was proved that there are no local obstructions: on any Riemannian manifold there is a chart around each point so that the matrix (g_{ij}) of metric coefficients is diagonal.

In the 1970's (Hardorp), it was proved that on any compact oriented 3-manifold there exist three foliations of codimension one transverse to each other, so there are no topological obstructions.

Therefore, the only obstructions are global geometric. It is also known that three pairwise orthogonal foliations intersect along common lines of curvature, So, existence of umbilical points could serve as such obstruction.

Problem 0.4 (Pawel Walczak). Describe complete connected Riemannian manifolds of bounded geometry which are not quasi-isometric to (or, Gromov-Hausdorff far from) the leaves or generic leaves of compact manifolds. Here, we refer to the Gromov-Hausdorff distance of arbitrary (compact or not) metric spaces with base points.

Problem 0.5 (Paul Schweitzer). Consider the manifold of dimension 3 (Ghys and Inaba-Nishimori-Takamura-Tsuchiya)

$$L = L_3 \natural L_5 \natural L_7 \natural \cdots \natural L_{p_n} \natural \cdots,$$

where p_n denotes the sequence of prime numbers, and where $\pi_1(L_p) = \mathbb{Z}/p\mathbb{Z}$. Can this L be (homeomorphic or diffeomorphic to) a leaf of a foliation of codimension $q > 1$ on a compact manifold?

Let N be a complete Riemannian manifold of dimension 2 and bounded geometry. Find conditions on N to be quasi-isometric to a leaf of a foliation on a compact manifold (of codimension 1 or q). For instance, what about a Liouville ladder (a cylinder with handles spaced like the nonzero digits in a Liouville decimal number)?

Problem 0.6 (Vladimir Rovenski). Let M be a compact Riemannian manifold of dimension $n + \nu$, and let $\mathcal{F}$ be a totally geodesic foliation of dimension ν on M. Then consider the mixed curvature K_σ, where σ is the plane generated by a vector tangent to the a leaf and a vector normal to that leaf at the same point. We would like to find examples with ν large and $K_\sigma > 0$ for all such σ. For instance we have the Hopf fibrations of spheres. To be more precise, let $\rho(n)$ be the maximum number of linearly independent vector fields on $\mathbb{S}^{n-1}$; it is known that $\rho(n) \leq 2\log_2 n + 1$. Does there exist any example with $\nu > \rho(n)$ and $K_\sigma > 0$ for all mixed σ?

For $K_\sigma = 1$, it was proved that $\nu \leq \rho(n)$ with a local method (Ferus, 1970).

Problem 0.7 (Hiraku Nozawa). Let M be a closed manifold, let $\{\mathcal{F}_t\}_{t \in I = [0,1]}$ be a smooth family of Riemannian foliations on M, and let $H_B^k(\mathcal{F}_t)$ denote their basic cohomology of degree k. Is $\dim H_B^k(\mathcal{F}_t)$ constant on t?

Now let $\{(\mathcal{F}_t, J_t)\}_{t \in I}$ be a smooth family of transversely holomorphic Riemannian foliations on M. Then we can consider the coresponding bigrading of the basic cohomology with complex coefficients, $H_B^{r,s}(\mathcal{F}_t)$. Is $\dim H_B^{r,s}(\mathcal{F}_t)$ constant on t?

Problem 0.8 (Victor Kleptsyn). Let G be a finitely generated group acting minimally on $\mathbb{S}^1$. A point $x \in \mathbb{S}^1$ is called non-expandable when $g'(x) \leq 1$ for all $g \in G$. Let NE denote the set of non-expandable points. Is it true that $NE \neq \emptyset$ implies $\lambda_{\exp} = 0$? To answer this question, we may need the following condition denoted by (*): any non-expandable point x is a fixed point, and thus there exist $g_+, g_- \in G$ such that $g_+(x) = g_-(x) = x$, $\text{Fix}(g_+) \cap (x, x + \epsilon) = \text{Fix}(g_-) \cap (x - \epsilon, x) = \emptyset$ for some $\epsilon > 0$. Does (*) imply that we get a Markov pseudogroup? Is (*) always true?

For instance, the action of $\text{PSL}_2(\mathbb{Z})$ on $\mathbb{S}^1 = \text{PR}(\mathbb{R}^2)$ satisfies (*), as well as the action of the Thomson group.

Problem 0.9 (Gilbert Hector). Consider the manifold of dimension 3

$$L = L_2 \natural L_3 \natural L_5 \natural L_7 \natural \cdots$$

like in Problem 0.5. Is L a leaf of a codimension one foliation in $\mathbb{R}^4$? As variants of this question, we can ask whether L can be realized as a closed submanifold of $\mathbb{R}^4$, or as a fiber of a submersion $\mathbb{R}^4 \to \mathbb{R}$.

If L embeds as a closed submanifold of $\mathbb{R}^4$, then it should be proper. Also, the triviality of the tangent bundle is the first obstruction to be a leaf in $\mathbb{R}^4$, but L is parallelizable because of its dimension is 3. The Chern-Simons classes may be relevant for this problem because they detect lens spaces.

If there is no answer for $\mathbb{R}^4$, what about $\mathbb{R}^5$?

Problem 0.10 (Etienne Ghys (recalled by Bertrand Deroin)).
There is an example of a lamination of a compact space by Riemann surfaces with one parabolic leaf and all other leaves hyperbolic (Ghys and Kenyon). Does there exist any foliation by Riemann surfaces on a closed manifold whose leaves are of such a mixed type? Is it possible to find an example with one parabolic plane and one hyperbolic plane as leaves?

PART B

Riemannian geometry

GRAPHS WITH PRESCRIBED MEAN CURVATURE

Marcos Dajczer

IMPA, Estrada Dona Castorina, 110
22460-320, Rio de Janeiro, Brazil
E-mail: marcos@impa.br

In this note we survey several recent results on graphs with prescribed mean curvature function in Riemannian manifolds endowed with a (conformal) Killing field.

Keywords: Mean curvature, conformal Killing field, Killing graph

1. Introduction

A well known theorem on elliptic PDE due to J. Serrin[19] establishes existence and uniqueness of a solution $u \in C^{2,\alpha}(\bar{\Omega})$ to the Dirichlet problem

$$\begin{cases} \text{div}\left(\dfrac{\nabla z}{\sqrt{1+|\nabla z|^2}}\right) + 2H = 0 \\ z|_{\partial\Omega} = \varphi \end{cases} \quad (1)$$

for any given $\varphi \in C^{2,\alpha}(\partial\Omega)$. Here $H \geq 0$ is a real number and Ω a bounded $C^{2,\alpha}$ domain in the plane $z = 0$ satisfying the geometric condition $k_\Gamma \geq 2H$, being k_Γ the curvature of the boundary curve $\Gamma = \partial\Omega$. In addition, the divergence and gradient are the usual operators in the Euclidean plane. It follows that the graph of a solution z of (1) is a surface in $\mathbb{R}^3$ with constant mean curvature H when oriented with a normal vector pointing downward.

Serrin's classical theorem has been extended to other types of graphs and other ambient spaces. For instance, for results in the hyperbolic space see[11,13,15–17] and.[18] The spherical case was considered in[10] and.[14] Recently, several results for hypersurfaces with prescribed mean curvature function in very general ambient spaces have been given in[1,3–6,8] and.[12] Our goal is to describe the general results in[5] and[6] that extend and generalize almost all previous existence theorems. It seems that the only exceptions are Theorem 3 in[1] and the result in[15] where a larger range for H is allowed.

The first section below is devoted to an existence theorem for Riemannian manifolds endowed with a Killing field with integral orthogonal distribution. In the remaining sections, we generalize this initial result by either removing the integrability condition or by considering (not necessarily closed) conformal Killing fields.

2. Killing graphs

Let M^{n+1} denote a Riemannian manifold endowed with a complete Killing vector field Y such that its orthogonal distribution is integrable. Then, the leaves are totally geodesic hypersurfaces. Fix an integral leaf $\mathbb{P}^n$, and denote by

$$\Phi \colon \mathbb{R} \times \mathbb{P}^n \to M^{n+1}$$

the flow generated by Y. Given a bounded domain $\Omega \subset \mathbb{P}$, we define the *Killing graph* associated to a function z on $\bar{\Omega}$ as the hypersurface

$$\Sigma^n = \{\Phi(z(u), u) : u \in \bar{\Omega}\}.$$

We want to assure the existence and uniqueness of a Killing graph with prescribed mean curvature H and boundary data φ. Here, the functions H and φ are defined on $\bar{\Omega}$ and $\Gamma = \partial\Omega$ respectively. As in the case of Serrin's result and in the sequel, the problem of existence of such a graph is formulated in terms of a Dirichlet problem for a divergence form elliptic PDE.

The *Killing cylinder* K over Γ is ruled by the flow lines of Y through Γ. Thus,

$$K = \{\Phi(t, u) : u \in \Gamma\}.$$

The mean curvature of K pointing inward is denoted by H_K and Ric_M stands for the Ricci tensor of the ambient space.

The following result was obtained in[4] and generalizes that one in.[8] As is the case of all theorems presented below, its proof uses the continuity method for quasilinear elliptic PDE as discussed in.[9] In order to obtain a priori height and gradient estimates, essential to this method, we use the Killing cylinders as barriers.

Theorem 2.1. *Let $\Omega \subset \mathbb{P}^n$ be a bounded domain with $C^{2,\alpha}$ boundary Γ. Assume $H_K \geq 0$ and $\mathrm{Ric}_M \geq -n \inf_\Gamma H_K^2$. Let $H \in C^\alpha(\Omega)$ and $\varphi \in C^{2,\alpha}(\Gamma)$ be given. If*

$$\sup_\Omega |H| \leq \inf_\Gamma H_K,$$

then there exists an unique function $z \in C^{2,\alpha}(\bar{\Omega})$ satisfying $z|_\Gamma = \varphi$ whose Killing graph has mean curvature H.

In a more general case, we will see below that M^{n+1} as above has a Riemannian structure as a warped product Riemannian manifold

$$M^{n+1} = \mathbb{P}^n \times_\varrho \mathbb{R}$$

such that

$$\Sigma^n = \{(z(u), u) : u \in \bar{\Omega}\}.$$

Moreover, we verify that the Killing field is $Y = d/dt$, where t is the parameter for $\mathbb{R}$. Thus, the warped function $\varrho \in C^\infty(\mathbb{P})$ is $\varrho = \|Y\|$.

3. Riemannian submersions

We discuss next the case when we do not necessarily have the integrability of the normal distribution to the Killing field. It turns that a natural setting of the Dirichlet problem for graphs with prescribed mean curvature is to consider these graphs as leaves in a Riemannian submersion transversal to a Killing cylinder.

Given a Riemannian submersion $\pi\colon M^{n+1} \to \mathbb{P}^n$ and a bounded domain $\Omega \subset \mathbb{P}^n$ with boundary Γ, the Killing cylinder K over Γ is the subset $\pi^{-1}(\Gamma)$. The following result was obtained in.[5]

Theorem 3.1. Let $\Omega \subset \mathbb{P}^n$ be a domain with compact closure and $C^{2,\alpha}$ boundary Γ. Assume that $H_{\text{cyl}} > 0$ and $\inf_M \text{Ric} \geq -n \inf_\Gamma H_{\text{cyl}}^2$, where H_{cyl} stands for the mean curvature of the Killing cylinder over Γ. Let $H \in C^\alpha(\bar{\Omega})$ and $\varphi \in C^{2,\alpha}(\Gamma)$ be given. Assume that there exists a $C^{2,\alpha}$ immersion $\iota\colon \bar{\Omega} \to M^{n+1}$ transverse to the vertical fibers. If

$$\sup_\Omega |H| \leq \inf_\Gamma H_K, \qquad (2)$$

then there exists an unique function $z \in C^{2,\alpha}(\bar{\Omega})$ satisfying $z|_\Gamma = \varphi$ whose Killing graph has mean curvature H.

The hypothesis on the existence of the immersion ι is used simultaneously to introduce a set of coordinates well suited to the problem and to define properly the notion of graph. Also $\iota(\bar{\Omega})$ is used as barrier to producing an initial minimal graph by the direct method in Calculus of Variations.

In the case of higher-dimensional Heisenberg spaces there exists a minimal leaf transverse to the flow lines of the vertical vector field. Thus, there is no need of the hypothesis in this particular case. By contrast, if we consider the example of odd-dimensional spheres submersed in the complex projective spaces, it is not guaranteed that always exist such minimal graphs with respect to the Hopf fibers.

We remark that submersions with totally geodesic fibers constitute an important example where we may construct initial Killing graphs. In fact, if we also assume that the solid cylinder over $\bar{\Omega}$ is complete, then geodesic cones with boundary in K and vertex at the mean convex side of K may be taken as initial Killing graphs after smoothing around the vertex. Thus, in this case we may rule out the hypothesis.

The above result has a very interesting application even for Euclidean space as we discuss next.

Let $\mathbb{P}^n$ be a Riemannian manifold endowed with a Killing vector field S_0. Given a positive function $\varrho \in C^\infty(\mathbb{P})$ so that $S_0(\varrho) = 0$, consider the warped product manifold

$$M^{n+1} = \mathbb{P}^n \times_\varrho \mathbb{R}.$$

Then, the lift S of S_0 by the projection

$$(u,t) \in M^{n+1} \mapsto u \in \mathbb{P}^n$$

is a Killing vector field in M^{n+1}. Thus, given constants $a, b \in \mathbb{R}$ with $b \neq 0$, it is easy to see that

$$Y = aS + bT$$

is a Killing field in M^{n+1}, where $T = \partial_t$. The following theorem was obtained in.[7]

Theorem 3.2. *Let Ω be a $C^{2,\alpha}$ bounded domain in M^n. Assume that $H_K \geq 0$ and $\mathrm{Ric}_M|_{T\Omega} \geq -n \inf_\Gamma H_K^2$. Let $H \in C^\alpha(\Omega)$ and $\varphi \in C^{2,\alpha}(\Gamma)$ be given such that*

$$|H| \leq \inf_\Gamma H_K.$$

Then, there exists an unique function $z \in C^{2,\alpha}(\bar{\Omega})$ satisfying $z|_\Gamma = \varphi$ whose (helicoidal) graph has mean curvature function H and boundary data φ.

Assume that $\mathbb{P}^2 = \mathbb{R}^2$ is endowed with a rotationally invariant metric. More precisely, we have polar coordinates r, θ such that the metric is written as

$$ds^2 = dr^2 + \psi^2(r)\, d\theta^2$$

for some positive smooth function ψ. Given the Killing vector field ∂_θ in $\mathbb{P}^2$ and $a, b \in \mathbb{R}$ with $b \neq 0$, it follows that

$$Y = a\partial_\theta + b\partial_t$$

is a Killing vector field in M^3.

We define a submersion $\pi\colon M^3 \to \mathbb{P}^2$ by identifying points in the same orbit through a point of $\mathbb{P}^2$. Then, it turns out that $\pi\colon M^3 \to \mathbb{P}^2$ is a Riemannian submersion if we consider in $\mathbb{P}^2$ the metric

$$ds^2 = dr^2 + \frac{b^2\psi^2}{a^2\psi^2 + b^2}\, d\theta^2$$

which coincides with the Euclidean metric when $a = 0$ and $\psi(r) = r$.

Corollary 3.1. *Let $\Omega \subset \mathbb{R}^2$ be a $C^{2,\alpha}$ bounded domain with boundary Γ such that $H_K \geq 0$. Let $H \in C^\alpha(\Omega)$ and $\varphi \in C^{2,\alpha}(\Gamma)$ be given such that*

$$|H| \leq \inf_\Gamma H_K.$$

Then, there exists an unique function $z \in C^{2,\alpha}(\bar{\Omega})$ satisfying $z|_\Gamma = \varphi$ whose helicoidal graph in $\mathbb{R}^3$ has mean curvature H and boundary data φ.

Notice that the Corollary 4 above reduces to Serrin's theorem if we take $a = 0$. Similar results may be stated for helicoidal graphs in hyperbolic spaces and spheres with respect to linear combinations of Killing vector fields generating translations along a geodesic and rotations.

4. Conformal Killing graphs

Next, we deal with conformal Killing vector fields and the corresponding notion of graph. The result extends Theorem 1 in[4] and gives a substantial improvement of the results proved in.[1] In fact, the method used here allows us to discard or weaken several assumptions that appear in Theorem 2 and Theorem 3 in.[1] In particular, the conformal Killing field we consider does not have to be closed.

Let M^{n+1} denote a Riemannian manifold endowed with a conformal Killing vector field Y whose orthogonal distribution is integrable. Thus, there exists a function $\rho \in C^\infty(M)$ such that we have the conformal Killing equation

$$\langle \bar{\nabla}_V Y, W \rangle + \langle \bar{\nabla}_W Y, V \rangle = 2\rho \langle V, W \rangle, \tag{3}$$

where $V, W \in T\bar{M}$. If the conformal Killing field Y is closed we have

$$\langle \bar{\nabla}_V Y, W \rangle = \rho \langle V, W \rangle.$$

The leaves of the orthogonal distribution are totally umbilical hypersurfaces. In particular, if Y is closed, the leaves are spherical.

We denote the flow generated by Y by

$$\Phi : \mathbb{I} \times \mathbb{P}^n \to M^{n+1},$$

where $\mathbb{I} = (-\infty, a)$ is an interval with $a > 0$ and $\mathbb{P}^n$ is an arbitrarily fixed integral leaf which we label as $t = 0$. It may happen that $a = +\infty$, i.e., that the vector field Y is complete. Notice that this situation occurs when the trajectories of Y are circles and we pass to the universal cover.

Since $\Phi_t = \Phi(t, \cdot)$ is a conformal map for any fixed $t \in \mathbb{I}$, there exists a positive function $\lambda \in C^\infty(\mathbb{I})$ such that $\lambda(0) = 1$ and

$$\Phi_t^* \bar{g} = \lambda^2(t)\bar{g}.$$

It is a standard fact that the functions λ and ρ are related by

$$\rho(t) = (\lambda_t/\lambda)(t). \tag{4}$$

Denoting

$$\|Y(t,u)\|^2 = 1/\bar{\gamma}(t,u)$$

and $\gamma(u) = \bar{\gamma}(0, u)$, we have from (3) and (4) that

$$\bar{\gamma}(t, u) = \gamma(u)/\lambda^2(t).$$

It follows from (3) and the integrability of the orthogonal distribution that

$$\langle \bar{\nabla}_X Y, Z \rangle = \rho(t)\langle X, Z \rangle \tag{5}$$

for $X, Z \perp Y$. Thus, the leaves $\mathbb{P}_t = \Phi_t(\mathbb{P})$ are totally umbilical and the principal curvature $k(t, u)$ of $\mathbb{P}_t$ with respect to the unit normal vector field $-Y/|Y|$ is

$$k(t, u) = \frac{\rho}{\|Y\|} = \frac{\lambda_t \sqrt{\gamma}}{\lambda^2}.$$

The metric in M^{n+1} has the form

$$ds^2 = \lambda^2(t)(\gamma^{-1}(u)dt^2 + d\sigma^2) \tag{6}$$

where (σ_{ij}) is the local expression for the metric $d\sigma^2$ in $\mathbb{P}^n$. Hence, M^{n+1} is conformal with conformal factor $\lambda(t)$ to the Riemannian warped product manifold $\mathbb{P}^n \times_{1/\sqrt{\gamma}} \mathbb{I}$. Moreover, after the change of variable

$$r = r(t) = \int_0^t \lambda(\tau)d\tau$$

we see that (6) takes the form of a Riemannian twisted product
$$ds^2 = \gamma^{-1}(u)dr^2 + \psi^2(r)d\sigma^2$$
where $\psi(r) = \lambda(t(r))$.

Example 4.1. *Let $\alpha \in C^\infty(\mathbb{P})$ be any positive function.*
(i) Then,
$$\bar{M}^{n+1} = \mathbb{R}_+ \times \mathbb{P}^n, \ \ ds^2 = \alpha^2(u)\,dr^2 + r^2 d\sigma^2$$
is isometric to
$$\tilde{M}^{n+1} = \mathbb{R} \times \mathbb{P}^n, \ \ d\tilde{s}^2 = e^{2t}(\alpha^2(u)\,dt^2 + d\sigma^2)$$
by means of the change of variable $e^t = r$.
(ii) Then,
$$M^{n+1} = \mathbb{R} \times \mathbb{P}^n, \ \ ds^2 = \alpha^2(u)\,dr^2 + e^{2r}d\sigma^2$$
is isometric to
$$\tilde{M}^{n+1} = (-\infty, 1) \times \mathbb{P}^n, \ \ d\tilde{s}^2 = \frac{1}{(1-t)^2}(\alpha^2(u)\,dt^2 + d\sigma^2)$$
by means of the change of variable $t = 1 - e^{-r}$.
(iii) If $c > 0$ and $b^{-1} = \tanh(c/2)$, then
$$M^{n+1} = \mathbb{R}_+ \times \mathbb{P}^n, \ \ ds^2 = \alpha^2(u)\,dr^2 + (\sinh t)^2 d\sigma^2$$
is isometric to
$$\tilde{M}^{n+1} = (-\infty, c + \log b) \times \mathbb{P}^n,$$
$$d\tilde{s}^2 = (\sinh(2\,\mathrm{argtanh}\, b^{-1} e^{s-c}))^2(\alpha^2(u)\,dt^2 + d\sigma^2)$$
by means of the change of variable $t = c + \log(b \tanh(t/2))$.

If the conformal Killing field Y is closed, i.e.,
$$\langle \bar{\nabla}_V Y, W \rangle = \rho \langle V, W \rangle,$$
we may assume that $\gamma(u) = 1$. Thus M^{n+1} has a warped product structure and is conformal to the Riemannian product manifold $\mathbb{R} \times M^n$ with conformal factor $\lambda(t)$. Observe the leaves of $\mathcal{D}$ are now spherical, that is, they have constant mean curvature $k(t)$.

The partial differential equation for a prescribed mean curvature function H in $\bar{\Omega}$ to be solved is the quasilinear elliptic equation of divergence form
$$\mathrm{div}\left(\frac{\nabla z}{\sqrt{\gamma + |\nabla z|^2}}\right) - \frac{1}{\sqrt{\gamma + |\nabla z|^2}}\left(\frac{\langle \nabla z, \nabla \gamma\rangle}{2\gamma} + \frac{n\gamma\lambda_t}{\lambda}\right) + n\lambda H = 0.$$

If the Killing field is closed ($\gamma = $ constant, say $\gamma = 1$) the equation reduces to

$$\operatorname{div}\left(\frac{\nabla z}{\sqrt{1+|\nabla z|^2}}\right) + n\left(\lambda H - \frac{\lambda_t}{\lambda\sqrt{1+|\nabla z|^2}}\right) = 0.$$

At this respect, the following result was obtained in.[5]

Let Ω_0 denote the largest open subset of points of Ω that can be joined to Γ by a *unique* minimizing geodesic. At points of Ω_0, we denote

$$\operatorname{Ric}_{\bar{M}}^{rad}(x) = \operatorname{Ric}_{\bar{M}}(\eta, \eta)$$

where $\operatorname{Ric}_{\bar{M}}$ is the ambient Ricci tensor and $\eta \in T_x M$ is a unit vector tangent to the the uniqueminimizing geodesic from $x \in \Omega_0$ to Γ.

Theorem 4.1. *Let $\Omega \subset M$ be a $C^{2,\alpha}$ bounded domain such that $\operatorname{Ric}_{\bar{M}}^{rad} \geq -n \inf_\Gamma H_K^2$. Assume $\lambda_t \geq 0$ and $(\lambda_t/\lambda)_t \geq 0$. Let $H \in C^\alpha(\Omega)$ and $\phi \in C^{2,\alpha}(\Gamma)$ be such that $\inf_\Gamma H_K \geq H \geq 0$ and $\phi \leq 0$. Then, there exists a unique function $z \in C^{2,\alpha}(\bar{\Omega})$ whose conformal Killing graph has mean curvature function H and boundary data ϕ.*

Theorem 4.1 has the following consequence.

Corollary 4.1. *Theorem 4.1 holds if the Ricci curvature assumption is replaced by*

$$n \operatorname{Ric}_{\bar{M}}^{rad} \geq -(n-1)^2 \inf_\Gamma H_\Gamma^2$$

when the conformal Killing field Y is closed.

In comparison with the Theorem 2 in,[1] we observe that Corollary 4.1 above allows us to remove the restriction on the Ricci curvature to be minimal in the direction of the conformal Killing field and rule out conditions on λ. Moreover, it permits to attain equality in $|H| \leq \inf_\Gamma H_K$ thus applying successfully to product ambient spaces. In this sense, Corollary 4.1 generalizes the main result in[1] which applies in turn to graphs of constant mean curvature and initial condition Γ.

References

1. L. Alías and M. Dajczer, *J. Diff. Geometry* **75** (2007), 387–401.
2. L. Alías and M. Dajczer, *Geom. Dedicata* **131** (2008), 173–179.
3. L. Alias, M. Dajczer and H. Rosenberg, *Calc. Var. Partial Differential Equations* **30** (2007), 511–522.
4. M. Dajczer, P. Hinojosa and J. H. de Lira, *Calc. Var. Partial Differential Equations* **33** (2008), 231–248.
5. M. Dajczer and J. H. de Lira, Killing graphs with prescribed mean curvature and Riemannian submersions. *Ann. l'Institut Henri Poincaré - Analyse non linéaire*, to appear.
6. M. Dajczer and J. H. de Lira, Conformal Killing graphs with prescribed mean curvature. Preprint.
7. M. Dajczer and J. H. de Lira, Helicoidal graphs with prescribed mean curvature. *Proc. Am. Math. Soc.*, to appear.
8. M. Dajczer and J. Ripoll, *J. Geom. Anal.* **15** (2005), 193–205.
9. D. Gilbarg and N. Trudinger, *Elliptic partial differential equations of second order*, Springer Verlag Berlin Heidelberg 2001.
10. P. Fusieger and J. Ripoll, *Ann. Global Anal. Geom.* **23** (2003), 373–400.
11. E. Guio and R. Sa Earp, *Commun. Pure Appl. Anal.* **4** (2005), 549–568.
12. D. Hoffman, J. H. de Lira and H. Rosenberg, *Trans. Amer. Math. Soc.* **358** (2006), 491–507.
13. J. H. de Lira, *Geom. Dedicata* **93** (2002), 11–23.
14. R. López, *Manuscripta Math.* **110** (2003), 45–54.
15. R. López and S. Montiel, *Calc. Var* **8**, (1999) 177–190.
16. B. Nelli and R. Sa Earp, *Bull. Sci. Math.* **120** (1996), 537–553.
17. B. Nelli and J. Spruck, On the existence and uniqueness of constant mean curvature hypersurfaces in hyperbolic space, Geometric analysis and the calculus of variations, 253–266, Internat. Press, Cambridge, MA, 1996.
18. P. Nitsche, *Manuscripta Math.* **108** (2002), 349–367.
19. J. Serrin, *Philos. Trans. Roy. Soc. London Ser. A* **264** (1969), 413–496.

GENUINE ISOMETRIC AND CONFORMAL DEFORMATIONS OF SUBMANIFOLDS

Ruy Tojeiro

Universidade Federal de São Carlos, Via Washington Luiz km 235
São Carlos, Brazil
E-mail: tojeiro@dm.ufscar.br

We survey on classical results and some recent developments on the study of isometric and conformal deformations of Euclidean submanifolds, with emphasis on the theory of *genuine* deformations, introduced by M. Dajczer and L. Florit for the isometric case, and recently extended to the conformal realm by L. Florit and the author.

Keywords: Isometric rigidity, conformal rigidity, isometric deformation, conformal deformation, genuine deformation

Mathematics Subject Classification 2000: 53C24, 53C40, 53C42

1. Introduction

Let $f\colon M^n \to \mathbb{R}^{n+p}$ be an isometric immersion with codimension p of an n-dimensional Riemannian manifold into Euclidean space. A basic problem in submanifold theory is to understand when f fails to be unique. Clearly, if $T\colon \mathbb{R}^{n+p} \to \mathbb{R}^{n+p}$ is a rigid motion then $\hat{f} = T \circ f\colon M^n \to \mathbb{R}^{n+p}$ is also an isometric immersion, so uniqueness should be understood up to this trivial equivalence relation. The usual terminology for uniqueness in this sense is *rigidity*. Thus, f is said to be *rigid* if any isometric immersion $\hat{f}\colon M^n \to \mathbb{R}^{n+p}$ is given in this way. Otherwise, f is said to be *isometrically deformable*. Hence, the problem is (at least to start with) to classify the isometrically deformable isometric immersions $f\colon M^n \to \mathbb{R}^{n+p}$, and possibly to describe the set of its isometric deformations.

A similar problem can be posed for *conformal* immersions $f\colon M^n \to \mathbb{R}^{n+p}$ of manifolds with a conformal structure, with rigid motions of $\mathbb{R}^{n+p}$ replaced by conformal (Möbius) transformations.

Our aim in this article is to survey on classical results and some recent developments on this topic, with emphasis on the theory of *genuine* deformations of submanifolds, introduced by Dajczer and Florit[6] for the isometric case, and recently extended to the conformal realm by Florit and the author.[19]

2. The hypersurface case

Let us start with a brief discussion of the only case in which our problem is well understood, namely, the case of hypersurfaces.

2.1. *Local theory for the isometric case*

In the study of isometric deformability of a hypersurface $f\colon M^n \to \mathbb{R}^{n+1}$, a crucial rôle is played by the rank of its shape operator, the so-called *type number* of f. Recall that the shape operator A_N of f with respect to a unit normal vector field N is the endomorphism of the tangent bundle TM given by $A_N X = -\nabla_X N$ for every $X \in TM$, where ∇ stands for the derivative in Euclidean space. Then, the type number $\tau(x)$ of f at x is the rank of A_N at x.

The study of isometric deformability of Euclidean hypersurfaces $f\colon M^n \to \mathbb{R}^{n+1}$ of dimension $n \geq 3$ started with the following classical result due to Beez and Killing.

Theorem 2.1. *Any hypersurface $f\colon M^n \to \mathbb{R}^{n+1}$ with $\tau \geq 3$ everywhere is rigid.*

On the other hand, hypersurfaces with $\tau \leq 1$ are flat, and hence, highly deformable. Thus, the interesting case in the study of local isometric deformations of hypersurfaces occurs when $\tau \equiv 2$.

Locally isometrically deformable hypersurfaces with $\tau \equiv 2$ were classified by Sbrana[25] and Cartan,[2] and are now called *Sbrana-Cartan hypersurfaces*. They split into 4 classes. The simplest one consists of hypersurfaces that are cylinders $L^2 \times \mathbb{R}^{n-2}$ over surfaces $L^2 \subset \mathbb{R}^3$ or products $CL^2 \times \mathbb{R}^{n-3}$, where $CL^2 \subset \mathbb{R}^4$ denotes the cone over a surface $L^2 \subset \mathbb{S}^3$ in the sphere $\mathbb{S}^3 \subset \mathbb{R}^4$. The deformations of such hypersurfaces are given by deformations of the corresponding surface L^2. The second class is that of ruled hypersurfaces, i.e., hypersurfaces that admit a codimension one foliation by (open subsets of) affine subspaces of $\mathbb{R}^{n+1}$. Hypersurfaces in this class are highly deformable, the deformations being parameterized by smooth real functions on an open interval.

The most interesting classes are the two remaining ones. Both can be described as envelopes of certain two-parameter families of affine subspaces, but whereas hypersurfaces in one class admit a one-parameter family of isometric deformations, those in the other have exactly one such deformation.

A modern account of the classification of Sbrana-Cartan hypersurfaces was given in,[12] where also some questions left open in the works by Sbrana and Cartan were answered, as the existence of hypersurfaces of the discrete type and the possibility of gluing together hypersurfaces of distinct classes.

2.2. *Global theory of isometrically deformable hypersurfaces*

A fundamental result in the study of global deformability of hypersurfaces was obtained by Sacksteder.[24]

Theorem 2.2. *A compact hypersurface* $f\colon M^n \to \mathbb{R}^{n+1}$, $n \geq 3$, *is rigid unless the subset of totally geodesic points of f disconnects M^n.*

Recall that totally geodesic points are those where the shape operator vanishes identically. Sacksteder theorem was later extended to *complete* hypersurfaces by Dajczer and Gromoll:[11]

Theorem 2.3. *Let* $f\colon M^n \to \mathbb{R}^{n+1}$, $n \geq 3$, *be an isometric immersion of a complete Riemannian manifold M^n that does not contain an open subset $L^3 \times \mathbb{R}^{n-3}$ with L^3 unbounded. Then f admits (nondiscrete) isometric deformations only along ruled strips. Furthermore, if f is nowhere completely ruled and the subset of totally geodesic points of f does not disconnect M^n, then f is rigid.*

An isometric immersion $f\colon M^n \to \mathbb{R}^{n+p}$ is said to be *completely ruled* if it is ruled with complete rulings. In this case, the leaves in each connected component of M (called a *ruled strip*) form an afine vector bundle over a curve with or without end points. An important open problem in this topic is whether there exists a complete nonruled deformable hypersurface M^3 in $\mathbb{R}^4$ with type number $\tau \equiv 2$ almost everywhere.

2.3. *Digression: Extending intrinsic isometries*

An interesting problem that is closely related to isometric rigidity is whether an intrinsic isometry of an Euclidean submanifold can be extended to a rigid motion of Euclidean space. More precisely, given an isometric immersion $f\colon M^n \to \mathbb{R}^{n+p}$ and an isometry g of M^n, does there exist a rigid motion G of $\mathbb{R}^{n+p}$ such that $f \circ g = G \circ f$?

It is easy to produce examples showing that the answer is "No", in general. On the other hand, the answer is clearly "Yes" whenever f is rigid: it suffices to apply the rigidity of f to the pair of isometric immersions f and $f \circ g$.

It was recently shown in[20,23] that, for compact Euclidean hypersurfaces of dimension $n \geq 3$, the answer is in fact "Yes" for any intrinsic isometry that belongs to the identity component of the isometry group, regardless of the hypersurface being rigid or not:

Theorem 2.4. *Let $f\colon M^n \to \mathbb{R}^{n+1}$, $n \geq 3$, be a compact hypersurface. Then the identity component $Iso^0(M^n)$ of the isometry group of M^n admits an orthogonal representation $\Phi\colon Iso^0(M^n) \to SO(n+1)$ such that $f \circ g = \Phi(g) \circ f$ for all $g \in Iso^0(M^n)$.*

2.4. Local theory of conformal deformations of hypersurfaces

The local theory of conformally deformable hypersurfaces turns out to be quite similar to that of the isometric case, starting with the following analogous due to Cartan[3] of the Beez-Killing theorem.

Theorem 2.5. *A hypersurface $f\colon M^n \to \mathbb{R}^{n+1}$, $n \geq 5$, is conformally rigid if all principal curvatures have multiplicity less than $n-2$ everywhere.*

At the other extreme, hypersurfaces with a principal curvature of multiplicity at least $n-1$ everywhere are locally conformally flat, that is, locally conformal to an open subset of Euclidean space, hence, highly conformally deformable.

Conformally deformable hypersurfaces with a principal curvature of constant multiplicity $n-2$ were classified by Cartan.[3] As in the isometric case, they split into four classes: surface-like, conformally ruled, the ones having a 1–parameter family of deformations and those that admit only one deformation.

3. Higher codimensions: Rigidity results

An important contribution to the study of isometric rigidity of Euclidean submanifolds of higher codimensions was given by Allendoerfer,[1] who proved the following extension of Beez-Killing's criterion for hypersurfaces.

Theorem 3.1. *Any isometric immersion $f\colon M^n \to \mathbb{R}^{n+p}$ with type number $\tau \geq 3$ everywhere is rigid.*

Here, the *type number* of an isometric immersion $f\colon M^n \to \mathbb{R}^{n+p}$ at a given $x \in M^n$ is the integer $0 \le r \le n$ defined as follows. Take any orthonormal basis $\xi_1, \dots, \xi_p$ of $T_x^\perp M$ and consider the maximum number r of linearly independent vectors $X_1, \dots, X_r \in T_x M$ such that the set $\{A_{\xi_j} X_i : 1 \le i \le r,\ 1 \le j \le p\}$ is also linearly independent. This is easily seen to be independent of the choice of the orthonormal normal basis. Obviously, for $p = 1$ the type number is just the rank of the shape operator, so in this case Allendoerfer theorem reduces to the Beez-Killing criterion. Notice also that the assumption $\tau \ge 3$ imposes the restriction $p \le n/3$ on the codimension of f. A conformal version of Allendorfer theorem was obtained by Silva[26] by means of a suitable extension of the notion of type number to the conformal realm.

3.1. *The s-nullity and the conformal s-nullity*

For submanifolds of low codimension, weaker conditions insuring isometric and conformal rigidity of Euclidean submanifolds were given by do Carmo and Dajczer[4] in terms of the so-called s-nullities and the conformal s-nullities of the submanifold.

The *s-nullity* $\nu_s(x)$ of an immersion $f\colon M^n \to \mathbb{R}^{n+p}$ at a point $x \in M^n$ is defined for each $1 \le s \le p$ by

$$\nu_s(x) = \max\{\dim \ker \left(\alpha^f_{V^s}(x)\right) : V^s \subset T_x^\perp M\},$$

where $\alpha^f_{V^s}(x)$ denotes the composition of the second fundamental form α^f of f at x with the orthogonal projection onto V^s. Equivalently, $\nu_s(x)$ is the maximal dimension of a subspace $W \subset T_x M$ for which there exists an s-dimensional subspace $V^s \subset T_x^\perp M$ such that $A_\xi|_W = 0$ for all $\xi \in V^s$. In particular, $\nu_p(x) = \nu(x) = \dim \ker \alpha^f(x)$ is the index of relative nullity of f at x.

Similarly, the *conformal s-nullity* $\nu_s^c(x)$ of an immersion $f\colon M^n \to \mathbb{R}^{n+p}$ at a point $x \in M^n$ is given for any $1 \le s \le p$ by

$$\nu_s^c(x) = \max\{\dim \ker \left(\alpha^f_{V^s}(x) - \langle\,,\,\rangle_f \zeta\right) : V^s \subset T_x^\perp M,\ \zeta \in V_s\},$$

where $\langle\,,\,\rangle_f$ stands for the metric on M^n induced by f.

In other words, $\nu_s^c(x)$ is the maximal dimension of a subspace $W \subset T_x M$ for which there exists an s-dimensional subspace $V^s \subset T_x^\perp M$ and a normal vector field $\zeta \in V^s$ such that $A_\xi|_W = \langle \xi, \zeta \rangle I$ for all $\xi \in V^s$. In particular, $\nu_p^c(x) = \nu^c(x)$ is the conformal nullity of f at x.

Theorem 3.2. (i) *An isometric immersion* $f\colon M^n \to \mathbb{R}^{n+p}$, $p \leq 5$, *is rigid if*

$$\nu_s \leq n - 2s - 1 \text{ for all } 1 \leq s \leq p.$$

(ii) *A conformal immersion* $f\colon M^n \to \mathbb{R}^{n+p}$, $p \leq 4$, *is conformally rigid if*

$$\nu_s^c \leq n - 2s - 1 \text{ for all } 1 \leq s \leq p.$$

For instance, an immersion $f\colon M^n \to \mathbb{R}^{n+2}$, $n \geq 5$, is isometrically rigid whenever $\nu \leq n - 5$ and $\mathrm{rank} A_\xi \geq 3$ for any normal direction ξ. Accordingly, f is conformally rigid whenever $\nu^c \leq n - 5$ and there exists no normal direction ξ such that A_ξ has a principal curvature of multiplicity greater than $n - 3$.

3.2. A main tool: flat bilinear forms

Let $f\colon M^n \to \mathbb{R}^{n+p}$ and $\hat{f}\colon M^n \to \mathbb{R}^{n+q}$ be isometric immersions. Endow the vector bundle $T_f^\perp M \oplus T_{\hat{f}}^\perp M$ with the indefinite metric of type (p,q) given by $\langle\!\langle\ ,\ \rangle\!\rangle_{T_f^\perp M \oplus T_{\hat{f}}^\perp M} = \langle\ ,\ \rangle_{T_f^\perp M} - \langle\ ,\ \rangle_{T_{\hat{f}}^\perp M}$. Set

$$\beta = \alpha \oplus \hat{\alpha}\colon TM \times TM \to T_f^\perp M \oplus T_{\hat{f}}^\perp M,$$

where α and $\hat{\alpha}$ denote the second fundamental forms of f and $\hat{f}$, respectively. The Gauss equations for f and $\hat{f}$ imply that β is *flat* with respect to $\langle\!\langle\ ,\ \rangle\!\rangle$:

$$\langle\!\langle \beta(X,Y), \beta(Z,T) \rangle\!\rangle - \langle\!\langle \beta(X,T), \beta(Z,Y) \rangle\!\rangle = 0$$

for all $X, Y, Z, T \in TM$. Flat bilinear forms were introduced by J. D. Moore[21] as an outgrowth of Cartan's exteriorly orthogonal forms. The fundamental result on flat bilinear forms is the following lemma, whose most general form appears in,[10] extending several previous cases.

Lemma 3.1. *Let* $\beta\colon V^n \times V^n \to W^{p,q}$ *be a flat symmetric bilinear form. If* $\min\{p,q\} \leq 5$ *and the subspace* $\mathcal{S}(\beta)$ *spanned by the image of* β *is non-degenerate, then* $\dim \ker(\beta) \geq n - \dim \mathcal{S}(\beta)$.

It has been conjectured by Dajczer for some time that the lemma should hold without the restrictions on $\min\{p,q\}$. However, in[10] a counterexample was given showing that the lemma is actually false if $\min\{p,q\} \geq 6$.

As an illustration of how the lemma is used to derive rigidity results, let us consider the simplest case of Beez-Killing theorem. Namely, given another isometric immersion $\hat{f}\colon M^n \to \mathbb{R}^{n+1}$, the assumption $\mathrm{rank}\,\alpha^f \geq 3$

implies that $\beta = \alpha^f \oplus \alpha^{\hat{f}}$ satisfies $\dim \ker(\beta) \leq n - 3$. Hence $\mathcal{S}(\beta)$ is degenerate by Lemma 3.1. Choosing unit normal vectors N and $\hat{N}$ to f and $\hat{f}$, respectively, so that $N + \hat{N}$ spans $\mathcal{S}(\beta) \cap \mathcal{S}(\beta)^{\perp}$, we obtain from $\langle\!\langle \alpha^f \oplus \alpha^{\hat{f}}, N + \hat{N} \rangle\!\rangle = 0$ that $A_N = \hat{A}_{\hat{N}}$, and the Beez-Killing criterion follows from the fundamental theorem of hypersurfaces.

3.3. Conformal geometry in the light cone

A basic tool for studying conformal deformations of Euclidean submanifolds, in particular for proving conformal rigidity theorems, is the model of Euclidean space $\mathbb{R}^n$ as a paraboloid in the light-cone

$$\mathbb{V}^{n+q+1} = \{x \in \mathbb{L}^{n+q+2} : \langle x, x \rangle = 0\}$$

of the $(n+q+2)$-dimensional Lorentz space $\mathbb{L}^{n+q+2}$. Namely, for a fixed $e_0 \in \mathbb{V}^{n+q+1}$, let $\mathcal{H}$ be the degenerate affine hyperplane

$$\mathcal{H} = \{x \in \mathbb{L}^{n+q+2} : \langle x, e_0 \rangle = 1\}.$$

Then, one can define an explicit isometric embedding $\Psi \colon \mathbb{R}^{n+q} \to \mathbb{L}^{n+q+2}$ such that $\Psi(\mathbb{R}^{n+q}) = \mathbb{E}^{n+q} := \mathbb{V}^{n+q+1} \cap \mathcal{H}$. The usefulness of the model is summarized by the following facts:

- If $f \colon M^n \to \mathbb{R}^{n+q}$ is a conformal immersion with conformal factor $\varphi \in C^{\infty}(M)$ of a Riemannian manifold, then

$$\mathcal{I}(f) := \varphi^{-1} \Psi \circ f \colon M^n \to \mathbb{V}^{n+q+1} \subset \mathbb{L}^{n+q+2}$$

is an *isometric* immersion, called the *isometric light cone representative* of f.

- If $g \colon M^n \to \mathbb{V}^{n+q+1}$ is an isometric immersion such that $g(M^n) \subset \mathbb{V}^{n+q+1} \setminus \mathbb{R}_{e_0}$, where $\mathbb{R}_{e_0} = \{te_0 : t > 0\}$, then $\mathcal{C}(g) \colon M^n \to \mathbb{R}^{n+q}$ given by

$$\Psi \circ \mathcal{C}(g) = \langle g, e_0 \rangle^{-1} g$$

is a conformal immersion with conformal factor $\langle g, e_0 \rangle^{-1}$.

- Two conformal immersions $f, g \colon M^n \to \mathbb{R}^{n+q}$ are conformaly congruent if and only if their isometric light cone representatives $\mathcal{I}(f), \mathcal{I}(g) \colon M^n \to \mathbb{V}^{n+q+1} \subset \mathbb{L}^{n+q+2}$ are isometrically congruent in $\mathbb{L}^{n+q+2}$.

4. The general deformation problem

All results we have discussed so far, concerning submanifolds of codimension greater than one, were sufficient criteria in terms of algebraic conditions on the second fundamental form that insure rigidity or conformal rigidity of a submanifold with low codimension. One might foresee higher-dimensional versions of the Sbrana and Sbrana-Cartan theories of isometrically and conformally deformable hypersurfaces, at least in the next simplest case $p = 2$. However, such a theory is far from being fully developed (see[7] for some partial results in the isometric case).

A basic observation when one goes into that problem is the following. Start with a (nontrivial) isometric (conformal) deformation $\hat{F}\colon N^{n+1} \to \mathbb{R}^{n+2}$ of a Sbrana-Cartan (Cartan) hypersurface $F\colon N^{n+1} \to \mathbb{R}^{n+2}$. Take any isometric (conformal) immersion $j\colon M^n \to N^{n+1}$. Then $\hat{f} = \hat{F} \circ i$ is an isometric (conformal) deformation of $f = F \circ i$.

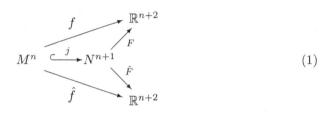
(1)

In other words, any hypersurface of a deformable hypersurface is also deformable. Thus, deformable submanifolds that arise in this way should be regarded as "nongenuine", and one should only pay attention to submanifolds that admit "genuine" deformations. Before one can turn this observation into a precise definition, it is convenient to put our deformation problem in a more general form, by allowing deformations to take place in possibly different codimensions:

The general deformation problem: For fixed $p > 0$ and $q \geq 0$, classify isometric (conformal) immersions $f\colon M^n \to \mathbb{R}^{n+p}$ that admit isometric (conformal) deformations $g\colon M^n \to \mathbb{R}^{n+q}$ (and possibly describe, in some sense, such deformations).

Notice that the problem makes sense even for $q = 0$: in the isometric (conformal) case, it amounts to classifying flat (conformally flat) submanifolds in $\mathbb{R}^{n+p}$.

A much more inspiring example, however, is the following: consider the standard umbilical inclusion $i\colon U \subset \mathbb{S}^n \to \mathbb{R}^{n+1}$ and let $h\colon W \supset i(U) \to \mathbb{R}^{n+p}$ be any isometric immersion. Then one can produce "trivial" isometric immersions $f\colon U \subset \mathbb{S}^n \to \mathbb{R}^{n+p}$ as compositions $f = h \circ i$:

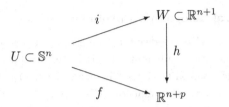

A natural and interesting problem is whether every isometric immersion $f: U \subset \mathbb{S}^n \to \mathbb{R}^{n+p}$ is given in this way. The answer turns out to be "Yes" whenever $p \leq n-2$, at least on connected components of an open and dense subset of U. This was shown in,[14] extending a previous result for $p=2$ due to Erbacher.[18]

On the other hand, it was shown in[21] and[15] that the answer is "No" for $p = n-1$. More precisely, it was proved in[15] that the set of "genuine" isometric deformations in $\mathbb{R}^{2n-1}$ of the umbilic inclusion $i: U \subset \mathbb{S}^n \to \mathbb{R}^{n+1}$ is in correspondence with the set of solutions of a system of nonlinear PDE's called the Generalized elliptic sinh-Gordon equations. Thus, the umbilic inclusion $i: U \subset \mathbb{S}^n \to \mathbb{R}^{n+1}$ is "genuinely rigid" in $\mathbb{R}^{n+p}$ for $p \leq n-2$, but not for $p = n-1$.

5. Genuine deformations

The ideas in the preceding section can be made precise by the following definitions, introduced by Dajczer and Florit[6] in the isometric case.

Definition 5.1. *A pair $\{f, \hat{f}\}$ of isometric (conformal) immersions $f: M^n \to \mathbb{R}^{n+p}$ and $\hat{f}: M^n \to \mathbb{R}^{n+q}$ is said to extend isometrically (conformally) when there exist an isometric (conformal) embedding $j: M^n \to N^{n+r}$, with $r \geq 1$, and isometric (conformal) immersions $F: N^{n+r} \to \mathbb{R}^{n+p}$ and $\hat{F}: N^{n+r} \to \mathbb{R}^{n+q}$ such that $f = F \circ j$ and $\hat{f} = \hat{F} \circ j$.*

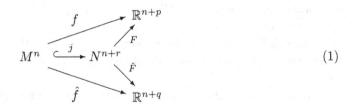 (1)

Definition 5.2. (Genuine deformation) *We say that $\hat{f}: M^n \to \mathbb{R}^{n+q}$ is a genuine isometric (conformal) deformation of $f: M^n \to \mathbb{R}^{n+p}$ if there*

is no open subset $U \subset M^n$ along which the restrictions $f|_U$ and $\hat{f}|_U$ extend isometrically (conformally).

Definition 5.3. (Genuine rigidity) *An isometric (conformal) immersion $f\colon M^n \to \mathbb{R}^{n+p}$ is genuinely isometrically (conformally) rigid in $\mathbb{R}^{n+q}$ for a fixed integer $q > 0$ if, for any given isometric (conformal) immersion $\tilde{f}\colon M^n \to \mathbb{R}^{n+q}$, there is an open dense subset $U \subset M^n$ such that the pair $\{f|_U, \tilde{f}|_U\}$ extends isometrically (conformally).*

The deformation problem revisited: For fixed $p > 0$ and $q \geq 0$, classify isometric (conformal) immersions $f\colon M^n \to \mathbb{R}^{n+p}$ that admit *genuine* isometric (conformal) deformations $g\colon M^n \to \mathbb{R}^{n+q}$.

The relevant geometric property related to existence of genuine isometric deformations turns out to be the following.

Definition 5.4. (Ruled immersions) *An immersion $f\colon M^n \to \mathbb{R}^{n+p}$ is D^d- ruled if M^n carries an integrable d-dimensional distribution $D^d \subset TM$ whose leaves are mapped diffeomorphically by f onto open subsets of affine subspaces of $\mathbb{R}^{n+p}$.*

Given a D^d- *ruled* isometric immersion $f\colon M^n \to \mathbb{R}^{n+p}$, set

$$L_D(x) = L_D(f)(x) = \text{span}\,\{\alpha^f(Z,X) : Z \in D^d(x) \text{ and } X \in T_x M\}.$$

Dajczer and Florit[6] have shown that admitting genuine isometric deformations imposes strong restrictions on a submanifold.

Theorem 5.1. *Let $\hat{f}\colon M^n \to \mathbb{R}^{n+q}$ be a genuine isometric deformation of $f\colon M^n \to \mathbb{R}^{n+p}$ with $p + q < n$ and $\min\{p,q\} \leq 5$. Then, along each connected component of an open dense subset of M^n, the immersions f and $\hat{f}$ are mutually D^d-ruled with*

$$d \geq n - p - q + 3\,\ell_D,$$

and $D^d = \ker \alpha^f_{L^\perp} \cap \ker \alpha^{\hat{f}}_{\hat{L}^\perp}$, where $L := L_D(f)$, $\hat{L} := L_D(\hat{f})$ and $\ell_D := \mathrm{rank}\, L = \mathrm{rank}\, \hat{L}$. Moreover, there exists a parallel vector bundle isometry $\mathcal{T}\colon L \to \hat{L}$ such that $\alpha^{\hat{f}}_{\hat{L}} = \mathcal{T} \circ \alpha^f_L$.

Theorem 5.1 yields the following sufficient conditions for genuine rigidity.

Corollary 5.1. *Let $f\colon M^n \to \mathbb{R}^{n+p}$ be an isometric immersion and $q \in \mathbb{N}$ with $p + q < n$ and $\min\{p,q\} \leq 5$. Then the following holds:*

(i) If f is nowhere $(n–p–q)$-ruled, f is genuinely rigid in $\mathbb{R}^{n+q}$.
(ii) If $Ric_M > 0$, then f is genuinely rigid in $\mathbb{R}^{n+q}$.

5.1. Ruled extensions

The proof of Theorem 5.1 follows by a careful study of when a pair of isometric immersions $f: M^n \to \mathbb{R}^{n+p}$ and $\hat{f}: M^n \to \mathbb{R}^{n+q}$ admits isometric ruled extensions.

Given a pair $\{f, \hat{f}\}$ of isometric immersions as in the statement, one can show (after hard work!) that there exist vector subbundles $L^\ell \subset T_f^\perp M$ and $\hat{L}^\ell \subset T_{\hat{f}}^\perp M$ and a vector bundle isometry $\mathcal{T}: L^\ell \to \hat{L}^\ell$ such that

- $D^d = \ker \alpha^f_{L^\perp} \cap \ker \alpha^{\hat{f}}_{\hat{L}^\perp}$ defines a smooth integrable subbundle of TM.
- $\mathcal{T}$ is parallel and preserves second fundamental forms.
- The subbundles L and $\hat{L}$ are parallel along D in the normal connections.
- $d \geq n - p - q + 3\ell$.

Then, one proves that if f and $\hat{f}$ are not D-ruled, they admit non-trivial Δ-ruled isometric extensions $F: N \to \mathbb{R}^{n+p}$ and $\hat{F}: N \to \mathbb{R}^{n+q}$, with $\Delta \cap TM = D$, contradicting the fact that $\hat{f}$ is a genuine isometric deformation of f. The estimate on d follows since $L_D \subset L$ by the definition of D.

5.2. Genuine conformal deformations of submanifolds

The conformal version of the property of a submanifold being ruled is the following.

Definition 5.5. *An immersion $f: M^n \to \mathbb{R}^{n+p}$ is D^d-conformally ruled if M^n carries an integrable d-dimensional distribution $D^d \subset TM$ whose leaves are mapped diffeomorphically by f onto open subsets of affine subspaces or round spheres of $\mathbb{R}^{n+p}$.*

Given a D^d-conformally ruled submanifold $f: M^n \to \mathbb{R}^{n+p}$, one now defines $\beta^f = \beta^f(x): T_xM \times T_xM \to T_x^\perp M$ by

$$\beta^f(Z, X) := \alpha^f(Z, X) - \langle Z, X \rangle \eta(x),$$

where $\eta(x)$ is the normal component of the mean curvature vector of the (image by f of) the leaf of D through x, and set

$$L_D(x) = L_D(f)(x) = \operatorname{span} \{\beta^f(Z, X) : Z \in D^d(x) \text{ and } X \in T_xM\}.$$

Then, one has the following conformal version of Theorem 5.1.[19]

Theorem 5.2. *Let $\bar{f}\colon M^n \to \mathbb{R}^{n+q}$ be a genuine conformal deformation of $f\colon M^n \to \mathbb{R}^{n+p}$, with $p+q \leq n-3$ and $\min\{p,q\} \leq 5$. Then, along each connected component of an open dense subset of M^n, the immersions f and $\bar{f}$ are mutually conformally D^d-ruled, with*

$$d \geq n - p - q + 3\ell_D^c,$$

and $D^d = \ker \beta_{L^\perp}^f \cap \ker \beta_{\bar{L}^\perp}^{\bar{f}}$, where $L := L_D^c(f)$, $\bar{L} := L_D^c(\bar{f})$ and $\ell_D^c := \operatorname{rank} L = \operatorname{rank} \bar{L}$. Moreover, there exists a parallel vector bundle isometry $\mathcal{T}\colon L \to \bar{L}$ such that $\beta_{\bar{L}}^{\bar{f}} = \varphi \mathcal{T} \circ \beta_L^f$, where φ is the conformal factor relating the metrics induced by f and $\bar{f}$.

Theorem 5.2 yields an immediate criterion for genuine conformal rigidity.

Corollary 5.2. *Let $f\colon M^n \to \mathbb{R}^{n+p}$ be a conformal immersion and let q be a positive integer with $p + q \leq n - 3$ and $\min\{p,q\} \leq 5$. If f is not $(n-p-q)$-conformally ruled on any open subset of M^n, then f is genuinely conformally rigid in $\mathbb{R}^{n+q}$.*

To get a better understanding of the content of Theorem 5.2, let us see how it implies several previous results on conformal deformations of submanifolds, starting with Cartan's conformal rigidity theorem.

Namely, for $p = 1 = q$, the estimate on d in Theorem 5.2 implies that $\ell_D^c = 0$ and that $d \geq n-2$. Thus, if $\{f, \bar{f}\}$ is a pair of (nowhere conformally congruent) conformal hypersurfaces, then f and $\bar{f}$ must carry principal curvatures of multiplicity at least $n-2$ (with common eigenspaces).

For $p \leq n-3$ and $q = 0$ (hence $\ell_D^c = 0$), the estimate gives $d \geq n-p$. This yields Moore's result[22] that a conformally flat n-dimensional Euclidean submanifold of codimension $p \leq n-3$ must have $\nu_c^f \geq n-p$.

The estimate on d for $p = q = 2$ gives $\ell_D^c \leq 1$. This yields $d \geq n-4$ if $\ell_D^c = 0$ and $d \geq n-1$ if $\ell_D^c = 1$. In both cases, the conformal nullity $\nu^c = \nu_2^c$ of both immersions satisfies $\nu^c \geq n-4$. Therefore, under the assumptions that $n \geq 7$ and $\nu_f^c \leq n-5$ everywhere, we conclude that f is genuinely conformally rigid.

Here, this means that if $\bar{f}$ is a conformal deformation of f, then there exists an open and dense subset of M^n on each connected component of which either $\bar{f}$ is conformally congruent to f, or f can be extended to either a conformally flat or a Cartan hypersurface and $\bar{f}$ is induced by a conformal deformation of such hypersurface. This was first proved in.[17]

One of the main applications of Theorem 5.2, however, is an extension of do Carmo and Dajczer conformal rigidity theorem.

Corollary 5.3. *Let $f\colon M^n \to \mathbb{R}^{n+p}$ be an immersion and let q be a positive integer with $p \leq q \leq n-p-3$. Suppose that $p \leq 5$ and that f satisfies*

$$\nu_s^c \leq n+p-q-2s-1 \text{ for all } 1 \leq s \leq p.$$

For $q \geq p+5$ assume further that $\nu_1^c \leq n-2(q-p)+1$. Then, any immersion $\bar{f}\colon M^n \to \mathbb{R}^{n+q}$ conformal to f is locally a composition, i.e., there exists an open dense subset $V \subset M^n$ such that the restriction $\bar{f}$ to any connected component U of V satisfies $\bar{f}|_U = h \circ f|_U$, where $h\colon W \subset \mathbb{R}^{n+p} \to \mathbb{R}^{n+q}$ is a conformal immersion of an open subset $W \supset f(U)$.

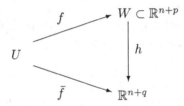

5.3. Constructing conformal pairs from isometric ones

The discussion in this section sheds some light on the reason behind the similarity between the theories of isometric and conformal deformations of submanifolds.

Let N^{n+1} be a *Riemannian* manifold that admits an isometric immersion $F'\colon N^{n+1} \to \mathbb{R}^{n+p}$ and an isometric embedding $\hat{F}\colon N^{n+1} \to \mathbb{L}^{n+q+2}$ transversal to the light cone $\mathbb{V}^{n+q+1}$. Then, set

$$M^n := \hat{F}^{-1}(\hat{F}(N^{n+1}) \cap \mathbb{V}^{n+q+2}),$$

$f = F' \circ i$ and $\bar{f} = \mathcal{C}(\hat{F} \circ i)$, where $i\colon M^n \to N^{n+1}$ is the inclusion map.

The following result from[19] states that any *genuine* conformal pair $(f, \bar{f})$ can be constructed in this way, as soon as the codimensions of both immersions are low enough and $\bar{f}$ is not conformally congruent to an isometric deformation of f on any open subset.

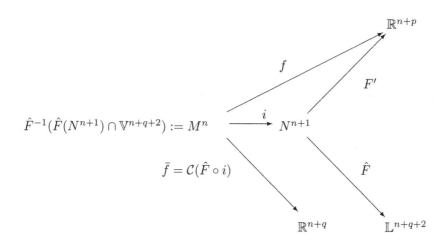

Theorem 5.3. *Let $f\colon M^n \to \mathbb{R}^{n+p}$, $p \geq 1$, and $\bar{f}\colon M^n \to \mathbb{R}^{n+q}$ form a genuine conformal pair, with $p + q \leq n - 3$ and $\min\{p,q\} \leq 5$. Suppose further that $\bar{f}$ is nowhere conformally congruent to an immersion that is isometric to f. Then (locally on an open dense subset of M^n) there exist a Riemannian manifold N^{n+1} that admits an isometric immersion $F'\colon N^{n+1} \to \mathbb{R}^{n+p}$ and an isometric embedding $\hat{F}\colon N^{n+1} \to \mathbb{L}^{n+q+2}$ transversal to the light cone $\mathbb{V}^{n+q+1}$, and a conformal diffeomorphism $i\colon M^n \to \hat{F}^{-1}(\hat{F}(N^{n+1}) \cap \mathbb{V}^{n+q+1})$ such that $\{F', \hat{F}\}$ is a genuine isometric pair, $f = F' \circ i$, and $\bar{f} = \mathcal{C}(\hat{F} \circ i)$.*

The case $p = 1$ is particularly interesting. Roughly speaking, it says that any hypersurface $f\colon M^n \to \mathbb{R}^{n+1}$ that admits a genuine conformal (but not isometric) deformation in $\mathbb{R}^{n+q}$ can be locally produced as the intersection of an $(n+1)$-dimensional flat submanifold of $\mathbb{L}^{n+q+2}$ with the light cone.

Corollary 5.4. *Let $f\colon M^n \to \mathbb{R}^{n+1}$ and $\bar{f}\colon M^n \to \mathbb{R}^{n+q}$ form a conformal pair, with $q \leq n - 4$. Assume that $\bar{f}$ is neither conformally congruent to an isometric deformation of f nor a composition on any open subset. Then, (locally on an open dense subset of M^n) there exist an isometric embedding $\hat{F}\colon U \subset \mathbb{R}^{n+1} \to \mathbb{L}^{n+q+2}$ transversal to the light cone $\mathbb{V}^{n+q+1}$ and a conformal diffeomorphism $\tau\colon M^n \to \bar{M}^n := \hat{F}^{-1}(\hat{F}(U) \cap \mathbb{V}^{n+q+1}) \subset U$ such that $f = i \circ \tau$ and $\bar{f} = \mathcal{C}(\hat{F} \circ \tau)$, where $i\colon \bar{M}^n \to U$ is the inclusion map.*

$$\hat{F}^{-1}(\hat{F}(U)\cap \mathbb{V}^{n+q+1}):=M^n \xrightarrow{i} U\subset \mathbb{R}^{n+1} \xrightarrow{\hat{F}} \mathbb{L}^{n+q+2}$$

In the special case $q = 1$, Corollary 5.4 was proved in,[13] and can be regarded as a nonparametric description of Cartan's conformally deformable hypersurfaces.

For $q = 0$, Theorem 5.3 reduces to a geometric construction of conformally flat submanifolds $f\colon M^n \to \mathbb{R}^{n+p}$ of codimension $p \leq n-3$ that are free of flat points, first obtained in.[9]

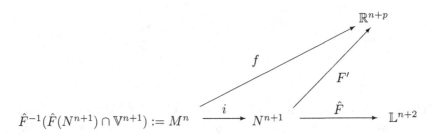

Acknowledgment

This survey is based on a lecture given at the VIII International Colloquium on Differential Geometry (E. Vidal Abascal Centenniun Congress) at Santiago de Compostela on July 2007. We would like to thank the people of Universidade de Santiago de Compostela, specially Eduardo García-Río and Luis Cordero for the warm hospitality during our stay at Santiago.

References

1. C. B. Allendoerfer, *Amer. J. Math.* **61** (1939), 633–644.
2. E. Cartan, *Bull. Soc. Math. France* **44** (1916), 65–99.
3. E. Cartan, *Bull. Soc. Math. France* **45** (1917), 57–121.
4. M. do Carmo and M. Dajczer, *Amer. J. Math.* **109** (1987), 963–985.
5. M. Dajczer et al., "Submanifolds and Isometric Immersions", Math. Lecture Ser. 13, Publish or Perish Inc. Houston, 1990.
6. M. Dajczer and L. Florit, *Comm. Anal. Geometry* **12** (2004), 1105–1129.
7. M. Dajczer and L. Florit, *Geom. Dedicata* **106** (2004), 195–210.
8. M. Dajczer and L. Florit, *Manuscripta Math.* **105** (2001), 507–517. Erratum, *Manuscripta Math.* **110** (2003), 135.
9. M. Dajczer and L. Florit, *Comm. Anal. Geometry* **4** (1996), 261–284.
10. M. Dajczer and L. Florit, *Proc. Amer. Math. Soc.* **132** (2004), 3703–3704.
11. M. Dajczer and D. Gromoll, *J. Diff. Geometry* **31** (1990), 401–416.

12. M. Dajczer, L. Florit and R. Tojeiro, *Ann. Mat. Pura Appl.* **174** (1998), 361–390.
13. M. Dajczer and R. Tojeiro, *Michigan Math. J.* **47** (2000), 529–557.
14. M. Dajczer and R. Tojeiro, *J. Diff. Geometry* **36** (1992), 1–18.
15. M. Dajczer and R. Tojeiro, *J. Reine Angew. Math.* **467** (1995), 109–147.
16. M. Dajczer and R. Tojeiro, *Indiana Univ. Math. J.* **46** (1997), 491–504.
17. M. Dajczer and R. Tojeiro, *J. Math. Soc. Japan* **52** (2000), 41–50.
18. J. Erbacher, *Nagoya Math. J.* **45** (1972), 139–165.
19. L. Florit and R. Tojeiro, Genuine deformations of submanifolds II: the conformal case, Preprint.
20. F. Mercuri, F. Podestà, J. A. Seixas and R. Tojeiro, *Comment. Math. Helv.* **81** (2) (2006), 471-479.
21. J. Moore, *Duke Math. J.* **44** (1977), 449–484.
22. J. Moore, *Math. Ann.* **225** (1977), 89-97.
23. I. Moutinho and R. Tojeiro, *Ann. Glob. Anal. Geom.* **33** (4) (2008), 323-336.
24. R. Sacksteder, *J. of Math. Mech.* **11** (1962), 929-939.
25. V. Sbrana, *Rend. Circ. Mat. Palermo* **27** (1909), 1–45.
26. S. Silva, *Pacific J. Math.* **199** (2001), 227–247.
27. R. Tojeiro, *J. Reine Angew. Math.* **598** (2006), 1-24.

TOTALLY GEODESIC SUBMANIFOLDS IN RIEMANNIAN SYMMETRIC SPACES

Sebastian Klein

Universität Mannheim, Lehrstuhl für Mathematik III, Seminargebäude A5, 68131 Mannheim, Germany
E-mail: mail@sebastian-klein.de

In the first part of this expository article, the most important constructions and classification results concerning totally geodesic submanifolds in Riemannian symmetric spaces are summarized. In the second part, we describe the results of our classification of the totally geodesic submanifolds in the Riemannian symmetric spaces of rank 2.

1. Totally geodesic submanifolds

A submanifold M' of a Riemannian manifold M is called *totally geodesic*, if every geodesic of M' is also a geodesic of M. In this article, we will discuss totally geodesic submanifolds in Riemannian symmetric spaces; in such spaces, a connected, complete submanifold is totally geodesic if and only if it is a symmetric subspace.

There are several important construction principles for totally geodesic submanifolds in Riemannian symmetric spaces M. First, we note that the connected components of the fixed point set of any isometry f of M are totally geodesic submanifolds (this is in fact true in any Riemannian manifold, see Ref. 11, Theorem II.5.1, p. 59). This construction principle is especially important in the case where f is involutive (i.e. $f \circ f = \mathrm{id}_M$); the totally geodesic submanifolds resulting in this case are called *reflective* submanifolds; they have been studied extensively, for example by LEUNG (see below).

Further constructions of totally geodesic submanifolds in Riemannian symmetric spaces of compact type M were introduced by CHEN and NAGANO:[2,4] For $p \in M$, the connected components $\neq \{p\}$ of the fixed point set of the geodesic reflection of M at p are called *polars* or M_+-*submanifolds* of M; note that they are in particular reflective submanifolds of M. A *pole* of M is a polar which is a singleton. It has been shown by Chen/Nagano[4] that for every polar M_+ of M and every $q \in M_+$ there exists another reflective submanifold M_- of M with $q \in M_-$ and $T_q M_- = (T_q M_+)^\perp$; M_- is called a *meridian* or M_--*submanifold* of M. For $p_1, p_2 \in M$, a point $q \in M$ is called a *center point* between p_1 and p_2 if there exists a geodesic joining p_1 with p_2 so that q is the middle point on that geodesic. If p_2 is a pole of p_1, then the set $C(p_1, p_2)$ of center points between p_1 and p_2 is called the *centrosome* of p_1 and p_2; its connected components are totally geodesic submanifolds of M (see Ref. 2, Proposition 5.1).

Moreover, every symmetric space of compact type can be embedded in its transvection group as a totally geodesic submanifold: Let $M = G/K$ be such a space, then there exists an involutive automorphism σ of G so that $\text{Fix}(\sigma)^0 \subset K \subset \text{Fix}(\sigma)$. Because of this property, the *Cartan map*

$$f : G/K \to G, \; gK \mapsto \sigma(g) \cdot g^{-1}$$

is a well-defined covering map onto its image, which turns out to be a totally geodesic submanifold of G. If M is a "bottom space", i.e. there exists no non-trivial symmetric covering map with total space M, then we have $K = \text{Fix}(\sigma)$, and therefore f is an embedding. In this setting f is called the *Cartan embedding* of M.

It is a significant and interesting problem to determine all totally geodesic submanifolds in a given symmetric space. Because totally geodesic submanifolds are rigid (i.e. if M'_1, M'_2 are connected, complete totally geodesic submanifolds of M with $p \in M'_1 \cap M'_2$ and $T_p M'_1 = T_p M'_2$, then we already have $M_1 = M_2$), they can be classified by determining those linear subspaces $U \subset T_p M$ which occur as tangent spaces of totally geodesic submanifolds of M.

The elementary answer to the latter problem is the following: There exists a totally geodesic submanifold of M with a given tangent space $U \subset T_p M$ if and only if U is *curvature invariant* (i.e. we have $R(u,v)w \in U$ for all $u, v, w \in U$, denoting by R the Riemannian curvature tensor of M).

Therefore the classification of totally geodesic submanifolds of M reduces to the purely algebraic problem of the classification of curvature invariant subspaces of $T_p M$. However, because of the algebraic complexity of the curvature tensor, classifying the curvature invariant subspaces is by no means an easy task, and therefore also the classification of totally geodesic submanifolds remains a significant problem.

This problem has been solved for the Riemannian symmetric spaces of rank 1 by WOLF in Ref. 19, §3. CHEN/NAGANO claimed a classification for the complex quadrics (which are symmetric spaces of rank 2) in Ref. 3, and then for all symmetric spaces of rank 2 in Ref. 4, using their construction of polars and meridians described above. However, it turns out that their classifications are incorrect: For several spaces of rank 2, totally geodesic submanifolds have been missed, and also some other details are faulty. In my papers Refs. 7–10 I discuss these shortcomings and give a full classification of the totally geodesic submanifolds in all irreducible symmetric spaces of rank 2; Section 2 of the present exposition contains a summary of these results. For symmetric spaces of rank ≥ 3, the full classification problem is still open.

However, there are several results concerning the classification of special classes of totally geodesic submanifolds. Probably the most significant result of this kind is the classification of reflective submanifolds in all Riemannian symmetric spaces due to LEUNG; his results are found in final form in Ref. 14, but also see Refs. 12,13. Another important problem of this kind is the classification of the totally geodesic submanifolds M' of $M = G/K$ with maximal rank (i.e. $\mathrm{rk}(M') = \mathrm{rk}(M)$); this problem has been solved for the symmetric spaces with $\mathrm{rk}(M) = \mathrm{rk}(G)$ by IKAWA/TASAKI,[6] and then for all irreducible symmetric spaces by ZHU/LIANG.[20]

Further important classification results concern Hermitian symmetric spaces M: In them, the complex totally geodesic submanifolds have been classified by IHARA.[5] Moreover, the real forms of M (i.e. the totally real, totally geodesic submanifolds M' of M with $\dim_{\mathbb{R}}(M') = \dim_{\mathbb{C}}(M)$) are all reflective; due to this fact LEUNG was able to derive a classification of the real forms of all Hermitian symmetric spaces from his classification of reflective submanifolds.[15]

Finally we mention a result by WOLF concerning totally geodesic submanifolds M' of (real, complex or quaternionic) Grassmann manifolds $G_r(\mathbb{K}^n)$ with the property that any two distinct elements of M' have zero intersection, regarded as r-dimensional subspaces of $\mathbb{K}^n$. Wolf showed[18,19] that any such totally geodesic submanifold is isometric either to a sphere or to a projective space over $\mathbb{R}$, $\mathbb{C}$ or $\mathbb{H}$; he was also able to describe embeddings for these submanifolds explicitly and to calculate their maximal dimension depending on r and n.

2. Maximal totally geodesic submanifolds in the Riemannian symmetric spaces of rank 2

In the following, I list the isometry types corresponding to all the congruence classes of totally geodesic submanifolds which are maximal (i.e. not strictly contained in another connected totally geodesic submanifold) in all Riemannian symmetric spaces of rank 2. In many cases I also briefly describe totally geodesic embeddings corresponding to these submanifolds. This is a summary of my work in Refs. 7–10, where it is proved that the lists given here are complete, and where the totally geodesic embeddings are described in more detail.

The invariant Riemannian metric of an irreducible Riemannian symmetric space is unique only up to a positive constant. In the sequel, we use the following notations to describe the metric which is induced on the totally geodesic submanifolds: For $\ell \in \mathbb{N}$ and $r > 0$ we denote by $\mathbb{S}^\ell_r$ the ℓ-dimensional sphere of radius r, and for $\varkappa > 0$ we denote by $\mathbb{R}P^\ell_\varkappa$, $\mathbb{C}P^\ell_\varkappa$, $\mathbb{H}P^\ell_\varkappa$ and $\mathbb{O}P^2_\varkappa$ the respective projective spaces, their metric being scaled in such a way that the *minimal* sectional curvature is $\varkappa$. ($\mathbb{R}P^\ell_\varkappa$ is then of constant sectional curvature $\varkappa$, $\mathbb{C}P^\ell_\varkappa$ is of constant holomorphic sectional curvature $4\varkappa$, and we have the inclusions $\mathbb{R}P^\ell_\varkappa \subset \mathbb{C}P^\ell_\varkappa \subset \mathbb{H}P^\ell_\varkappa$ of totally geodesic submanifolds). For symmetric spaces of rank 2, we describe the appropriate metric by stating the length a of the shortest restricted root of the space as a subscript $_{\text{srr}=a}$. For the three infinite families of Grassmann manifolds $G_2^+(\mathbb{R}^n)$, $G_2(\mathbb{C}^n)$ and $G_2(\mathbb{H}^n)$, we also use the notation $_{\text{srr}=1*}$ to denote the metric scaled in such a way that the shortest root occurring *for large n* has length 1, disregarding the fact that this root might vanish for certain small values of n.

The spaces in which the totally geodesic submanifolds are classified below are always taken with $_{\text{srr}=1*}$ (for the Grassmann manifolds) or $_{\text{srr}=1}$ (for all others).

2.1. $G_2^+(\mathbb{R}^{n+2})$

(a) $G_2^+(\mathbb{R}^{n+1})_{\text{srr}=1*}$
The linear isometric embedding $\mathbb{R}^{n+1} \to \mathbb{R}^{n+2}$, $(x_1, \ldots, x_{n+1}) \mapsto (x_1, \ldots, x_{n+1}, 0)$ induces a totally geodesic, isometric embedding $G_2^+(\mathbb{R}^{n+1}) \to G_2^+(\mathbb{R}^{n+2})$.

(b) $\mathbb{S}^n_{r=1}$
Fix a unit vector $v_0 \in \mathbb{R}^{n+2}$, and let $\mathbb{S} := \{ v \in \mathbb{R}^{n+2} \,|\, \langle v, v_0 \rangle = 0, \|v\| = 1 \} \cong \mathbb{S}^n_{r=1}$. Then the map $\mathbb{S} \to G_2^+(\mathbb{R}^{n+2})$, $v \mapsto \mathbb{R}v \oplus \mathbb{R}v_0$ is a totally geodesic, isometric embedding.

(c) $(\mathbb{S}^\ell_{r=1} \times \mathbb{S}^{\ell'}_{r=1})/\mathbb{Z}_2$, where $\ell + \ell' = n$
The map $\mathbb{S}^\ell_{r=1} \times \mathbb{S}^{\ell'}_{r=1} \to G_2^+(\mathbb{R}^{n+2})$ given by

$$((x_0, \ldots, x_\ell), (y_0, \ldots, y_{\ell'})) \mapsto \mathbb{R}\,(x_0, \ldots, x_\ell, 0, \ldots, 0) \oplus \mathbb{R}\,(0, \ldots, 0, y_0, \ldots, y_{\ell'})$$

is a totally geodesic, isometric immersion, and a two-fold covering map onto its image in $G_2^+(\mathbb{R}^{n+2})$.

(d) For $n \geq 4$ even: $\mathbb{C}P^{n/2}_{\varkappa=1/2}$

Let us fix a complex structure J on $\mathbb{R}^{n+2}$. Then the complex-1-dimensional linear subspaces of $(\mathbb{R}^{n+2}, J)$ are in particular real-2-dimensional oriented linear subspaces of $\mathbb{R}^{n+2}$. Therefore the complex projective space $\mathbb{P} \cong \mathbb{C}P^{n/2}$ over $(\mathbb{R}^{n+2}, J)$ is contained in $G_2^+(\mathbb{R}^{n+2})$; it turns out to be a totally geodesic submanifold.

(e) For $n = 2$: $\mathbb{C}P^1_{\varkappa=1/2} \times \mathbb{R}P^1_{\varkappa=1/2}$

The image of the Segré embedding $\mathbb{C}P^1 \times \mathbb{C}P^1 \to \mathbb{C}P^3$ (see for example Ref. 17, p. 55f.) is a 2-dimensional complex quadric in $\mathbb{C}P^3$; such a quadric is isometric to $G_2^+(\mathbb{R}^4)$. Thereby we see that $G_2^+(\mathbb{R}^4)$ is isometric to $\mathbb{C}P^1_{\varkappa=1/2} \times \mathbb{C}P^1_{\varkappa=1/2}$. Let C be the trace of a (closed) geodesic in $\mathbb{C}P^1_{\varkappa=1/2}$; then C is isometric to $\mathbb{R}P^1_{\varkappa=1/2}$, and $\mathbb{C}P^1_{\varkappa=1/2} \times C$ is a totally geodesic submanifold of $\mathbb{C}P^1_{\varkappa=1/2} \times \mathbb{C}P^1_{\varkappa=1/2} \cong G_2^+(\mathbb{R}^4)$.

(f) For $n = 3$: $\mathbb{S}^2_{r=\sqrt{5}}$

To describe this totally geodesic submanifold, as well as similar totally geodesic submanifolds occurring in $G_2(\mathbb{C}^6)$ and $G_2(\mathbb{H}^7)$ (see Sections 2.2(g) and 2.3(f) below), we note that there is exactly one irreducible, 14-dimensional, quaternionic representation of $Sp(3)$ (see Ref. 1, Chapter VI, Section (5.3), p. 269ff.). It can be constructed as follows: The vector representation of $Sp(3)$ on $\mathbb{C}^6$ induces a representation of $Sp(3)$ on $\bigwedge^3 \mathbb{C}^6$. This 20-dimensional representation decomposes into two irreducible components: One, 6-dimensional, is equivalent to the vector representation of $Sp(3)$; the other, acting on a 14-dimensional linear space V, is the irreducible representation we are interested in.

It turns out that the restriction of the representation of $Sp(3)$ on V to an $SO(3)$ embedded in $Sp(3)$ in the canonical way, is a real representation, and that in any real form $V_{\mathbb{R}}$ of V, two linear independent vectors are left invariant. By splitting off the subspace of V' spanned by these vectors, we get a real-5-dimensional representation $V'_{\mathbb{R}}$ of $SO(3)$, which turns out to be irreducible (and equivalent to the Cartan representation $SO(3) \times End_+(\mathbb{R}^3)_0 \to End_+(\mathbb{R}^3)_0$, $(B,X) \mapsto BXB^{-1}$). It turns out that the corresponding action of $SO(3)$ on $G_2^+(V'_{\mathbb{R}}) \cong G_2^+(\mathbb{R}^5)$ has exactly one totally geodesic orbit; this orbit is isometric to $\mathbb{S}^2_{r=\sqrt{5}}$.

2.2. $G_2(\mathbb{C}^{n+2})$

(a) $G_2(\mathbb{C}^{n+1})_{\mathrm{srr}=1*}$
The linear isometric embedding $\mathbb{C}^{n+1} \to \mathbb{C}^{n+2}$, $(z_1,\ldots,z_{n+1}) \mapsto (z_1,\ldots,z_{n+1},0)$ induces a totally geodesic, isometric embedding $G_2(\mathbb{C}^{n+1}) \to G_2(\mathbb{C}^{n+2})$.

(b) $G_2(\mathbb{R}^{n+2})_{\mathrm{srr}=1*}$
The map $G_2(\mathbb{R}^{n+2}) \to G_2(\mathbb{C}^{n+2})$, $\Lambda \mapsto \Lambda \oplus i\Lambda$ is a totally geodesic, isometric embedding.

(c) $\mathbb{C}P^n_{\varkappa=1}$
Fix a unit vector $v_0 \in \mathbb{C}^{n+2}$. Then the map $\mathbb{C}P((\mathbb{C}v_0)^\perp) \to G_2(\mathbb{C}^{n+2})$, $\mathbb{C}v \mapsto \mathbb{C}v \oplus \mathbb{C}v_0$ is a totally geodesic, isometric embedding.

(d) $\mathbb{C}P^\ell_{\varkappa=1} \times \mathbb{C}P^{\ell'}_{\varkappa=1}$ with $\ell + \ell' = n$
Let $\mathbb{C}^{n+2} = W \oplus W'$ be a splitting of $\mathbb{C}^{n+2}$ into complex-linear subspaces of dimension $\ell+1$ resp. $\ell'+1$; here $\oplus$ denotes an orthogonal direct sum. Then $\mathbb{C}P(W) \times \mathbb{C}P(W') \to G_2(\mathbb{C}^{n+2})$, $(\mathbb{C}v, \mathbb{C}v') \mapsto \mathbb{C}v \oplus \mathbb{C}v'$ is a totally geodesic, isometric embedding.

(e) For n even: $\mathbb{H}P^{n/2}_{\varkappa=1/2}$
Let us fix a quaternionic structure τ on $\mathbb{C}^{n+2}$ (i.e. $\tau : \mathbb{C}^{n+2} \to \mathbb{C}^{n+2}$ is anti-linear with $\tau^2 = -\mathrm{id}$). Then the quaternionic-1-dimensional linear subspaces of $(\mathbb{C}^{n+2}, \tau)$ are in particular complex-2-dimensional linear subspaces of $\mathbb{C}^{n+2}$. Therefore the quaternionic projective space $\mathbb{P} \cong \mathbb{H}P^{n/2}$ over $(\mathbb{C}^{n+2}, \tau)$ is contained in $G_2(\mathbb{C}^{n+2})$; it turns out that $\mathbb{P}$ is a totally geodesic submanifold of $G_2(\mathbb{C}^{n+2})$.

(f) For $n=2$: $G_2^+(\mathbb{R}^5)_{\mathrm{srr}=\sqrt{2}}$ and $(\mathbb{S}^3_{r=1/\sqrt{2}} \times \mathbb{S}^1_{r=1/\sqrt{2}})/\mathbb{Z}_2$
Note that $G_2(\mathbb{C}^4)_{\mathrm{srr}=1*}$ is isometric to $G_2^+(\mathbb{R}^6)_{\mathrm{srr}=\sqrt{2}}$. This isometry can be exhibited via the Plücker map $G_2(\mathbb{C}^m) \to \mathbb{C}P(\bigwedge^2 \mathbb{C}^m)$, $\mathbb{C}u \oplus \mathbb{C}v \mapsto \mathbb{C}(u \wedge v)$, which is an isometric embedding for any m; for $m=4$ its image in $\mathbb{C}P(\bigwedge^2 \mathbb{C}^4) \cong \mathbb{C}P^5$ turns out to be a 4-dimensional complex quadric; such a quadric is isomorphic to $G_2^+(\mathbb{R}^6)$. Thus $G_2(\mathbb{C}^4)$ is isometric to $G_2^+(\mathbb{R}^6)$, hence its maximal totally geodesic submanifolds are those given in Section 2.1 for $n=4$, namely: $G_2^+(\mathbb{R}^5)_{\mathrm{srr}=\sqrt{2}}$, $(\mathbb{S}^3_{r=1/\sqrt{2}} \times \mathbb{S}^1_{r=1/\sqrt{2}})/\mathbb{Z}_2$, $(\mathbb{S}^2_{r=1/\sqrt{2}} \times \mathbb{S}^2_{r=1/\sqrt{2}})/\mathbb{Z}_2$, $\mathbb{S}^4_{r=1/\sqrt{2}}$, $\mathbb{C}P^2_{\varkappa=1}$. The first two of these submanifolds are those which are listed under this point; the remaining submanifolds have already been listed above (note that $(\mathbb{S}^2 \times \mathbb{S}^2)/\mathbb{Z}_2$ and $\mathbb{S}^4$ are isometric to $G_2(\mathbb{R}^4)$ and $\mathbb{H}P^1$, respectively).

(g) For $n=4$: $\mathbb{C}P^2_{\varkappa=1/5}$
Let us consider the 14-dimensional quaternionic, irreducible represen-

tation V of Sp(3) described in Section 2.1(f). The restriction of that representation to a SU(3) canonically embedded in Sp(3) leaves a totally complex 6-dimensional linear subspace $V_{\mathbb{C}}$ of V invariant; the resulting 6-dimensional representation $V_{\mathbb{C}}$ of SU(3) is irreducible. It turns out that the induced action of SU(3) on $G_2(V_{\mathbb{C}}) \cong G_2(\mathbb{C}^6)$ has exactly one totally geodesic orbit; this orbit is isometric to $\mathbb{C}P^2_{\varkappa=1/5}$.

2.3. $G_2(\mathbb{H}^{n+2})$

(a) $G_2(\mathbb{H}^{n+1})_{\mathrm{srr}=1*}$
 The linear isometric embedding $\mathbb{H}^{n+1} \to \mathbb{H}^{n+2}$, $(q_1,\ldots,q_{n+1}) \mapsto (q_1,\ldots,q_{n+1},0)$ induces a totally geodesic, isometric embedding $G_2(\mathbb{H}^{n+1}) \to G_2(\mathbb{H}^{n+2})$.

(b) $G_2(\mathbb{C}^{n+2})_{\mathrm{srr}=1*}$
 We fix two orthogonal imaginary unit quaternions i and j, and let $\mathbb{C} = \mathbb{R} \oplus \mathbb{R}i$. Then the map $G_2(\mathbb{C}^{n+2}) \to G_2(\mathbb{H}^{n+2})$, $\Lambda \mapsto \Lambda \oplus \Lambda j$ is a totally geodesic, isometric embedding.

(c) $\mathbb{H}P^n_{\varkappa=1}$
 Fix a unit vector $v_0 \in \mathbb{H}^{n+2}$. Then the map $\mathbb{H}P((v_0\mathbb{H})^\perp) \to G_2(\mathbb{H}^{n+2})$, $v\mathbb{H} \mapsto v\mathbb{H} \oplus v_0\mathbb{H}$ is a totally geodesic, isometric embedding.

(d) $\mathbb{H}P^\ell_{\varkappa=1} \times \mathbb{H}P^{\ell'}_{\varkappa=1}$ with $\ell + \ell' = n$
 Let $\mathbb{H}^{n+2} = W \oplus W'$ be a splitting of $\mathbb{H}^{n+2}$ into quaternionic-linear subspaces of dimension $\ell+1$ resp. $\ell'+1$. Then $\mathbb{H}P(W) \times \mathbb{H}P(W') \to G_2(\mathbb{H}^{n+2})$, $(v\mathbb{H}, v'\mathbb{H}) \mapsto v\mathbb{H} \oplus v'\mathbb{H}$ is a totally geodesic, isometric embedding.

(e) For $n=2$: $\mathrm{Sp}(2)_{\mathrm{srr}=\sqrt{2}}$ and $(\mathbb{S}^5_{r=1/\sqrt{2}} \times \mathbb{S}^1_{r=1/\sqrt{2}})/\mathbb{Z}_2$
 Let $U \in G_2(\mathbb{H}^4)$ be given, then $U^\perp$ is the only pole corresponding to U in $G_2(\mathbb{H}^4)$. The centrosome between this pair of poles is a totally geodesic submanifold of $G_2(\mathbb{H}^4)$ which is isometric to $\mathrm{Sp}(2)$. This $\mathrm{Sp}(2)$ is also a reflective submanifold of $G_2(\mathbb{H}^4)$, the complementary reflective submanifold is isometric to $(\mathbb{S}^5_{r=1/\sqrt{2}} \times \mathbb{S}^1_{r=1/\sqrt{2}})/\mathbb{Z}_2$.

(f) For $n=5$: $\mathbb{H}P^2_{\varkappa=1/5}$
 We again consider the irreducible, quaternionic 14-dimensional representation V of Sp(3) introduced in Section 2.1(f); we now view V as a quaternionic-7-dimensional linear space. The representation of Sp(3) on V induces an action of Sp(3) on the quaternionic 2-Grassmannian $G_2(V) \cong G_2(\mathbb{H}^7)$; again it turns out that this action has exactly one totally geodesic orbit, which is isometric to $\mathbb{H}P^2_{\varkappa=1/5}$.

(g) For $n=4$: $\mathbb{S}^3_{r=2\sqrt{5}}$
According to the present list, two of the maximal totally geodesic submanifolds of the 2-Grassmannian $G_2(\mathbb{H}^7)$ are isometric to $G_2(\mathbb{H}^6)$ and $\mathbb{H}P^2_{\varkappa=1/5}$, respectively. The intersection of these two totally geodesic submanifolds is a totally geodesic submanifold of $G_2(\mathbb{H}^6)$, which turns out to be isometric to $\mathbb{S}^3_{r=2\sqrt{5}}$.

2.4. SU(3)/SO(3)
(a) $(\mathbb{S}^2_{r=1} \times \mathbb{S}^1_{r=\sqrt{3}})/\mathbb{Z}_2$ (b) $\mathbb{R}P^2_{\varkappa=1/4}$

2.5. SU(6)/Sp(3)
(a) $\mathbb{H}P^2_{\varkappa=1/4}$ (b) $\mathbb{C}P^3_{\varkappa=1/4}$
(c) $SU(3)_{srr=1}$
The map $SU(3) \to SU(6)/Sp(3)$, $B \mapsto \begin{pmatrix} B & 0 \\ 0 & B^{-1} \end{pmatrix} \cdot Sp(3)$ is a totally geodesic embedding of this type.
(d) $(\mathbb{S}^5_{r=1} \times \mathbb{S}^1_{r=\sqrt{3}})/\mathbb{Z}_2$

2.6. SO(10)/U(5)
In the descriptions of the embeddings for this symmetric space, we consider both $U(5)$ and $SO(10)$ as acting on $\mathbb{C}^5 \cong \mathbb{R}^{10}$; in the latter case, this action is only $\mathbb{R}$-linear.
(a) $\mathbb{C}P^4_{\varkappa=1}$ (b) $G_2(\mathbb{C}^5)_{srr=1}$
(c) $\mathbb{C}P^3_{\varkappa=1} \times \mathbb{C}P^1_{\varkappa=1}$
Let $G := SO(6) \times SO(4)$ be canonically embedded in $SO(10)$ in such a way that its intersection with $U(5)$ is maximal. Then $G/(G \cap U(5))$ is a totally geodesic submanifold of $SO(10)/U(5)$ which is isometric to $(SO(6)/U(3)) \times (SO(4)/U(2)) \cong \mathbb{C}P^3 \times \mathbb{C}P^1$.
(d) $G_2^+(\mathbb{R}^8)_{srr=\sqrt{2}}$
Let $G := SO(8)$ be canonically embedded in $SO(10)$ in such a way that its intersection with $U(5)$ is maximal. Then $G/(G \cap U(5))$ is a totally geodesic submanifold of $SO(10)/U(5)$ which is isometric to $SO(8)/U(4) \cong G_2^+(\mathbb{R}^8)$.
(e) $SO(5)_{srr=1}$
The map $SO(5) \to SO(10)/U(5)$, $B \mapsto \begin{pmatrix} B & 0 \\ 0 & B^{-1} \end{pmatrix} \cdot U(5)$ is a totally geodesic embedding of this type.

2.7. $E_6/(U(1) \cdot Spin(10))$
(a) $\mathbb{O}P^2_{\varkappa=1/2}$ (b) $\mathbb{C}P^5_{\varkappa=1} \times \mathbb{C}P^1_{\varkappa=1}$ (c) $G_2^+(\mathbb{R}^{10})_{srr=\sqrt{2}}$
(d) $G_2(\mathbb{C}^6)_{srr=1}$ (e) $(G_2(\mathbb{H}^4)/\mathbb{Z}_2)_{srr=1}$ (f) $SO(10)/U(5)_{srr=1}$

2.8. E_6/F_4
(a) $\mathbb{O}P^2_{\varkappa=1/4}$
(b) $\mathbb{H}P^3_{\varkappa=1/4}$
(c) $((SU(6)/Sp(3))/\mathbb{Z}_3)_{srr=1}$
(d) $(\mathbb{S}^9_{r=1} \times \mathbb{S}^1_{r=\sqrt{3}})/\mathbb{Z}_4$

2.9. $G_2/SO(4)$
(a) $SU(3)/SO(3)_{srr=\sqrt{3}}$
(b) $(\mathbb{S}^2_{r=1} \times \mathbb{S}^2_{r=1/\sqrt{3}})/\mathbb{Z}_2$
(c) $\mathbb{C}P^2_{\varkappa=3/4}$
(d) $\mathbb{S}^2_{r=\frac{2}{3}\sqrt{21}}$

2.10. $SU(3)$
(a) $SU(3)/SO(3)_{srr=1}$
  The Cartan embedding $f : SU(3)/SO(3) \to SU(3)$ is a totally geodesic embedding of this type.
(b) $(\mathbb{S}^3_{r=1} \times \mathbb{S}^1_{r=\sqrt{3}})/\mathbb{Z}_2$
(c) $\mathbb{C}P^2_{\varkappa=1/4}$
  The Cartan embedding $f : SU(3)/S(U(2) \times U(1)) \to SU(3)$ is a totally geodesic embedding of this type.
(d) $\mathbb{R}P^3_{\varkappa=1/4}$

2.11. $Sp(2)$
(a) $G_2^+(\mathbb{R}^5)_{srr=1}$
  The Cartan embedding $f : Spin(5)/(Spin(2) \times Spin(3)) \to Spin(5) \cong Sp(2)$ is a totally geodesic embedding of this type.
(b) $Sp(1) \times Sp(1)$
  The canonically embedded $Sp(1) \times Sp(1) \subset Sp(2)$ is a totally geodesic submanifold of this type.
(c) $\mathbb{H}P^1_{\varkappa=1/2}$
  The Cartan embedding $f : Sp(2)/(Sp(1) \times Sp(1)) \to Sp(2)$ is a totally geodesic embedding of this type.
(d) $\mathbb{S}^3_{r=\sqrt{5}}$

2.12. G_2
(a) $G_2/SO(4)_{srr=1}$
  The Cartan embedding $f : G_2/SO(4) \to G_2$ is a totally geodesic embedding of this type.
(b) $(\mathbb{S}^3_{r=1} \times \mathbb{S}^3_{r=1/\sqrt{3}})/\mathbb{Z}_2$
(c) $SU(3)_{srr=\sqrt{3}}$
  Regard G_2 as the automorphism group of the division algebra of the octonions $\mathbb{O}$ and fix an imaginary unit octonion i. Then the subgroup

$\{g \in G_2 \,|\, g(i) = i\}$ is isomorphic to $SU(3)$ and a totally geodesic submanifold of this type.

(d) $\mathbb{S}^3_{r=\frac{2}{3}\sqrt{21}}$

Acknowledgments

This work was supported by a fellowship within the Postdoc-Programme of the German Academic Exchange Service (DAAD).

References

1. T. Bröcker, T. tom Dieck, *Representations of compact Lie groups*, Springer, New York, 1985.
2. B.-Y. Chen, *A new approach to compact symmetric spaces and applications*, Katholieke Universiteit Leuven, 1987.
3. B.-Y. Chen, T. Nagano, *Duke Math. J.* **44** (1977), 745–755.
4. B.-Y. Chen, T. Nagano, *Duke Math. J.* **45** (1978), 405–425.
5. S. Ihara, *J. Math. Soc. Japan* **19** (1967), 261–302.
6. O. Ikawa, H. Tasaki, *Japan. J. Math.* **26** (2000), 1–29.
7. S. Klein, *Differential Geom. Appl.* **26** (2008), 79–96.
8. S. Klein, to appear in *Trans. Amer. Math. Soc.*, arXiv:0709.2644.
9. S. Klein, to appear in *Geom. Dedicata*, arXiv:0801.4127.
10. S. Klein, arXiv:0809.1319.
11. S. Kobayashi, *Transformation groups in Differential geometry*, Springer-Verlag Berlin 1972.
12. D. S. P. Leung, *Indiana Math. J.* **24** (1974), 327–339.
13. D. S. P. Leung, *Indiana Math. J.* **24** (1975), 1199.
14. D. S. P. Leung, *J. Diff. Geom.* **14** (1979), 167–177.
15. D. S. P. Leung, *J. Diff. Geom.* **14** (1979), 179–185.
16. O. Loos, *Symmetric spaces II: Compact spaces and classification*, W. A. Benjamin Inc., New York, 1969.
17. I. R. Shafarevich, *Basic algebraic geometry I*, Springer, Berlin, 1994.
18. J. A. Wolf, *Illinois J. Math.* **7** (1963), 425–446.
19. J. A. Wolf, *Illinois J. Math.* **7** (1963), 447–462.
20. F. Zhu, K. Liang, *Science in China Ser. A* **47** (2004), 264–271.

THE ORBITS OF COHOMOGENEITY ONE ACTIONS ON COMPLEX HYPERBOLIC SPACES

José Carlos Díaz-Ramos

Department of Mathematics, University College Cork
Cork, Ireland
E-mail: jc.diazramos@ucc.ie

We study cohomogeneity one actions on complex hyperbolic spaces. We provide characterizations of their orbits in terms of their second fundamental form.

Keywords: Complex hyperbolic space, homogeneous submanifolds, real hypersurfaces, constant principal curvatures

A *cohomogeneity one action* is an isometric action on a Riemannian manifold whose principal orbits are hypersurfaces. A submanifold is *homogeneous* if it is an orbit of an isometric action. Thus, the study of cohomogeneity one actions is essentially equivalent to the study of homogeneous hypersurfaces.

The investigation of the orbits of cohomogeneity one actions goes back to Segre[12] who proved that any isoparametric hypersurface in the Euclidean space is homogeneous. Then, any such hypersurface is a tube around an affine subspace of the Euclidean space. Cartan[8] generalized this result to the real hyperbolic space. A homogeneous hypersurface of the real hyperbolic space is a horosphere or a tube around a totally geodesic real hyperbolic space of lower dimension. A result by Hsiang and Lawson[10] implies that a hypersurface in the sphere S^n is homogeneous if and only if it is a principal orbit of the isotropy representation of an $(n+1)$-dimensional semisimple symmetric space of rank two. Hence, the classification of homogeneous hypersurfaces in S^n follows from Cartan's classification of symmetric spaces.

Takagi[13] showed that a real hypersurface of $\mathbb{C}P^n$ is homogeneous if and only if it is a principal orbit of an action induced via the Hopf map $S^{2n+1} \to \mathbb{C}P^n$ by a cohomogeneity one action on a sphere. A remarkable consequence is that each homogeneous hypersurface in $\mathbb{C}P^n$ is a Hopf hypersurface. Berndt[1] classified the homogeneous Hopf hypersurfaces in $\mathbb{C}H^n$, but surprisingly, a homogeneous hypersurface in the complex hyperbolic space is not necessarily Hopf. Lohnherr gave the first counterexample.

Recently, Berndt and Tamaru[6] obtained the classification of homogeneous hypersurfaces in $\mathbb{C}H^n$ (see Section 1 for explicit definitions).

Theorem 0.1. *A connected real hypersurface in $\mathbb{C}H^n$, $n \geq 2$, is homogeneous if and only if it is holomorphically congruent to one of the following:*

(A) a tube around a totally geodesic $\mathbb{C}H^k$ for some $k \in \{0, \ldots, n-1\}$;
(B) a tube around a totally geodesic $\mathbb{R}H^n$;
(H) a horosphere in $\mathbb{C}H^n$;
(S) the real hypersurface W^{2n-1} or one of its equidistant hypersurfaces;
(W) a tube around the minimal ruled submanifold W_φ^{2n-k} for some $\varphi \in (0, \pi/2]$ and $k \in \{2, \ldots, n-1\}$, where k is even if $\varphi \neq \pi/2$.

The first aim of this paper is to study the geometry of these hypersurfaces and the corresponding cohomogeneity one actions on $\mathbb{C}H^n$. We accomplish this task in Section 1.

It is a major problem in submanifold geometry to characterize homogeneous hypersurfaces of a manifold. The motivation for our investigations originates from the question: Is every real hypersurface with constant principal curvatures in $\mathbb{C}H^n$ an open part of a homogeneous hypersurface? Segre[12] solved this problem for the Euclidean space and Cartan[8] gave an affirmative answer in the real hyperbolic space. Remarkably, this is not true in spheres (see *e.g.* Ref. 9). The problem in $\mathbb{C}P^n$ has been addressed by Takagi[14] who proved that a hypersurface of the complex hyperbolic space with 2 or 3 constant principal curvatures is homogeneous. There are homogeneous hypersurfaces in $\mathbb{C}P^n$ with 5 principal curvatures. However, it is not known whether the number of constant principal curvatures of a hypersurface in $\mathbb{C}P^n$ is 2, 3 or 5. No examples of inhomogeneous hypersurfaces with constant principal curvatures are known in the complex projective space. In Section 2 of this paper we review the known facts about hypersurfaces with constant principal curvatures in the complex hyperbolic space.

1. The geometry of the orbits

We first summarize some basic facts about the complex hyperbolic space. For details we refer to Ref. 7.

Let $\mathbb{C}H^n$ be the n-dimensional complex hyperbolic space equipped with the Bergman metric $\langle \cdot, \cdot \rangle$ of constant holomorphic sectional curvature -1 and let J be its complex structure. We denote by $\mathbb{C}H^n(\infty)$ the ideal boundary of $\mathbb{C}H^n$. Each $x \in \mathbb{C}H^n(\infty)$ is an equivalence class of asymptotic geodesics in $\mathbb{C}H^n$. For each $o \in \mathbb{C}H^n$ and $x \in \mathbb{C}H^n(\infty)$ there exists a unique unit speed geodesic γ_{ox} such that $\gamma_{ox}(0) = o$ and $\lim_{t\to\infty} \gamma_{ox}(t) = x$. The connected component of the isometry group of $\mathbb{C}H^n$ is the special unitary group $G = SU_{1,n}$. We fix $o \in \mathbb{C}H^n$ and let K be the isotropy group of G at o. Then $K \cong S(U_1 \times U_n)$ and $\mathbb{C}H^n = G/K$. Let $\mathfrak{g} = \mathfrak{k} + \mathfrak{p}$ be the Cartan decomposition of the Lie algebra $\mathfrak{g}$ of G with respect to o. We now fix $x \in \mathbb{C}H^n(\infty)$. Let $\mathfrak{a}$ be the 1-dimensional linear subspace of $\mathfrak{p}$ spanned by $\dot{\gamma}_{ox}(0)$. Let $\mathfrak{g} = \mathfrak{g}_{-2\alpha} + \mathfrak{g}_{-\alpha} + \mathfrak{g}_0 + \mathfrak{g}_\alpha + \mathfrak{g}_{2\alpha}$ be the root space decomposition of $\mathfrak{g}$ induced by $\mathfrak{a}$. Then, $\mathfrak{n} = \mathfrak{g}_\alpha + \mathfrak{g}_{2\alpha}$ is a 2-step nilpotent subalgebra isomorphic to the $(2n-1)$-dimensional Heisenberg algebra; its center is the 1-dimensional subalgebra $\mathfrak{g}_{2\alpha}$. Moreover, $\mathfrak{g} = \mathfrak{k} + \mathfrak{a} + \mathfrak{n}$ is an Iwasawa decomposition of $\mathfrak{g}$. We denote by A and N the connected subgroups of G with Lie algebras $\mathfrak{a}$ and $\mathfrak{n}$, respectively. The solvable subgroup AN acts simply transitively on $\mathbb{C}H^n$. Thus we can identify $\mathfrak{a}+\mathfrak{n}$ with $T_o\mathbb{C}H^n$. The metric on $\mathbb{C}H^n$ induces an inner product $\langle \cdot, \cdot \rangle$ on $\mathfrak{a}+\mathfrak{n}$, and we may identify $\mathbb{C}H^n$ with the solvable Lie group AN equipped with the left-invariant metric induced from $\langle \cdot, \cdot \rangle$. The complex structure J induces a complex structure on $\mathfrak{a} + \mathfrak{n}$ that we also denote by J. We define $B = \dot{\gamma}_{ox}(0) \in \mathfrak{a}$ and $Z = JB \in \mathfrak{g}_{2\alpha}$. Note that $\mathfrak{g}_\alpha$ is J-invariant. The Lie algebra structure on $\mathfrak{a} + \mathfrak{n}$ is given by the trivial skew-symmetric bilinear extension of the relations $[B, Z] = Z$, $2[B, U] = U$ and $[U, V] = \langle JU, V \rangle Z$, with $U, V \in \mathfrak{g}_\alpha$.

It is clear that a homogeneous hypersurface has constant principal curvatures. Let M be a hypersurface of $\mathbb{C}H^n$ with constant principal curvatures. We denote by T_λ the distribution associated with eigenspaces of a principal curvature λ. If $\xi \in \Gamma(\nu M)$, then $J\xi$ is called the Hopf vector field. If $J\xi$ is a principal curvature vector, then M is said to be a *Hopf hypersurface*.

(A) *The action of* $S(U_{1,k} \times U_{n-k})$

The group $S(U_{1,k} \times U_{n-k}) \subset SU_{1,n}$ ($k \in \{0, \ldots, n-1\}$) acts on $\mathbb{C}H^n$ with cohomogeneity one. This action has exactly one singular orbit: a totally geodesic $\mathbb{C}H^n[k] \subset \mathbb{C}H^n$. If M is one principal orbit of this action then M is a tube of certain radius $r > 0$ around the singular orbit. Standard Jacobi vector field theory shows that M has three principal curvatures

$$\lambda_1 = \frac{1}{2}\tanh\frac{r}{2}, \qquad \lambda_2 = \frac{1}{2}\coth\frac{r}{2}, \qquad \lambda_3 = \coth r,$$

with multiplicities $m_1 = 2(n-k-1)$, $m_2 = 2k$ and $m_3 = 1$. Furthermore, T_{λ_1} and T_{λ_2} are complex distributions and the Hopf vector field is a principal curvature vector of λ_3. Thus, M is a Hopf hypersurface. If $k = 0$, the singular orbit is a point and the principal orbits are geodesic spheres with two eigenvalues $\lambda_1 = \frac{1}{2}\tanh\frac{r}{2}$ and $\lambda_3 = \coth r$ and multiplicities $2(n-1)$ and 1. Similarly, if $k = n-1$, M has two constant principal curvatures $\lambda_2 = \frac{1}{2}\coth\frac{r}{2}$ and $\lambda_3 = \coth r$ with multiplicities $2(n-1)$ and 1.

(B) *The action of $SO^0_{1,n}$*

The group $SO^0_{1,n} \subset SU_{1,n}$ acts on $\mathbb{C}H^n$ with cohomogeneity one. This action has one singular orbit which is a totally geodesic $\mathbb{R}H^n \subset \mathbb{C}H^n$. Let M be a principal orbit of this action. Then, M is a tube of certain radius $r > 0$ around the singular orbit. It has three principal curvatures

$$\lambda_1 = \frac{1}{2}\tanh\frac{r}{2}, \qquad \lambda_2 = \frac{1}{2}\coth\frac{r}{2}, \qquad \lambda_3 = \tanh r,$$

with multiplicities $m_1 = n-1$, $m_2 = n-1$ and $m_3 = 1$. The distributions T_{λ_1} and T_{λ_2} are real and the Hopf vector field is a principal curvature vector of λ_3. If $r = \log(2+\sqrt{3})$, there are two principal curvatures $\lambda_1 = \sqrt{3}/6$ and $\lambda_2 = \lambda_3 = \sqrt{3}/2$ with multiplicities $n-1$ and n.

(H) *The horosphere foliation*

Let $\mathfrak{k} + \mathfrak{a} + \mathfrak{n}$ be the Iwasawa decomposition of the isometry group of $\mathbb{C}H^n$ with respect to $o \in \mathbb{C}H^n$ and $x \in \mathbb{C}H^n(\infty)$. The group N acts on $\mathbb{C}H^n$ with cohomogeneity one and all its orbits are principal. The resulting foliation is called the horosphere foliation. A horosphere is constructed as follows. Let γ be a unit speed geodesic with $\gamma(0) = o$. The Busemann function with respect to γ, $\mathcal{B}_\gamma : \mathbb{C}H^n \to \mathbb{R}$, is defined by $\mathcal{B}_\gamma(p) = \lim_{t\to\infty}(d(p,\gamma(t)) - t)$, where d is the Riemannian distance function. The level sets of this function are horospheres centered at the point of infinity x.

A horosphere has exactly two distinct principal curvatures

$$\lambda_1 = 1/2 \qquad \text{and} \qquad \lambda_2 = 1,$$

with corresponding multiplicities $m_1 = 2(n-1)$ and $m_2 = 1$. The distribution T_{λ_1} is complex and each horosphere is a Hopf hypersurface.

(S) *The solvable foliation*

Let $\mathfrak{k} + \mathfrak{a} + \mathfrak{n}$ be the Iwasawa decomposition of the isometry group of $\mathbb{C}H^n$ with respect to some $o \in \mathbb{C}H^n$ and $x \in \mathbb{C}H^n(\infty)$. Let $\mathfrak{w}$ be a linear hyperplane in $\mathfrak{g}_\alpha$. Then $\mathfrak{s} = \mathfrak{a} + \mathfrak{w} + \mathfrak{g}_{2\alpha}$ is a Lie subalgebra of $\mathfrak{a} + \mathfrak{n}$ of codimension one. If S is the connected Lie subgroup whose Lie algebra is $\mathfrak{s}$, then S acts on $\mathbb{C}H^n$ with cohomogeneity one. This action has no singular orbits and therefore induces a foliation on $\mathbb{C}H^n$. We call it the *solvable foliation* (Ref. 2). Different choices of $\mathfrak{w}$ lead to congruent actions.

The orbit $S \cdot o$ is a minimal homogeneous ruled real hypersurface which we denote by W^{2n-1}. It can be seen as a particular case of the submanifolds W_φ^{2n-k} described below. Any other orbit of the action of S is an equidistant hypersurface to this one. Let $M(r)$ denote an orbit of S at a distance $r \geq 0$ from $S \cdot o$. The shape operator of $M(r)$ has exactly three eigenvalues

$$\lambda_{1/2} = \frac{3}{4}\tanh\frac{r}{2} \pm \frac{1}{2}\sqrt{1 - \frac{3}{4}\tanh^2\frac{r}{2}}, \quad \lambda_3 = \frac{1}{2}\tanh\frac{r}{2},$$

with corresponding multiplicities $m_1 = 1$, $m_2 = 1$ and $m_3 = 2n - 3$. The Hopf vector field $J\xi$ has nontrivial projection onto T_{λ_1} and T_{λ_2}.

The subspace $\mathfrak{a} + \mathfrak{w}^\perp + J\mathfrak{w}^\perp + \mathfrak{g}_{2\alpha}$ is a subalgebra of $\mathfrak{a} + \mathfrak{n}$, and the orbit through o of the corresponding connected subgroup of AN is a totally geodesic $\mathbb{C}H^2$. The action of the connected subgroup of S with Lie algebra $\mathfrak{a} + J\mathfrak{w}^\perp + \mathfrak{g}_{2\alpha}$ induces the solvable foliation on this totally geodesic $\mathbb{C}H^2$. The "slice" $\mathbb{C}H^2$ contains the relevant information of the solvable foliation.

Theorem 1.1. *The leaves of the solvable foliation on $\mathbb{C}H^2$ are diffeomorphic to $\mathbb{R}^3$ and are foliated orthogonally by a 1-dimensional totally geodesic foliation and a 2-dimensional foliation whose leaves are Euclidean planes.*

(W) The ruled submanifolds W_φ^{2n-k} and their tubes

We follow Ref. 5. Let $\mathfrak{k} + \mathfrak{a} + \mathfrak{n}$ be the Iwasawa decomposition of $SU_{1,n}$ with respect to $o \in \mathbb{C}H^n$ and $x \in \mathbb{C}H^n(\infty)$. Let $\mathfrak{w}$ be a linear subspace of $\mathfrak{g}_\alpha$ such that $\mathfrak{w}^\perp = \mathfrak{g}_\alpha \ominus \mathfrak{w}$ has *constant Kähler angle* $\varphi \in [0, \pi/2]$, i.e. for each $0 \neq v \in \mathfrak{g}_\alpha$ the angle between $\mathfrak{w}^\perp$ and the real span of v is constantly equal to φ. Then $\mathfrak{s} = \mathfrak{a} + \mathfrak{w} + \mathfrak{g}_{2\alpha}$ is a subalgebra of $\mathfrak{a} + \mathfrak{n}$. Denote by S the connected subgroup of AN with Lie algebra $\mathfrak{s}$ and by $N_K^0(S)$ the identity component of its normalizer in K. Then $N_K^0(S)S$ acts with cohomogeneity one. The orbit $W_\varphi^{2n-k} = N_K^0(S)S \cdot o = S \cdot o$ is a $(2n-k)$-dimensional submanifold of $\mathbb{C}H^n$, where $k = \dim \mathfrak{w}^\perp$. As $\mathbb{C}H^n$ is a two-point homogeneous space, the construction of the submanifolds W_φ^{2n-k} does not depend on the choice of $o \in \mathbb{C}H^n$ and $x \in \mathbb{C}H^n(\infty)$.

If $\varphi = 0$, then W_0^{2n-k} is a totally geodesic $\mathbb{C}H^{n-k'}$, where $k = 2k'$. If $\varphi = \pi/2$, then $\mathfrak{w}^\perp$ is a k-dimensional real subspace of $\mathfrak{g}_\alpha$. If $k = 1$, then $W_{\pi/2}^{2n-1}$ is the hypersurface W^{2n-1} described above. If $k > 1$, $W_{\pi/2}^{2n-k}$ is a $(2n-k)$-dimensional homogeneous submanifold of $\mathbb{C}H^n$ with totally real normal bundle of rank k. We denote $W^{2n-k} = W_{\pi/2}^{2n-k}$.

We investigate the geometry of the submanifolds W_φ^{2n-k}, $0 < \varphi \leq \pi/2$. Note that k is even if $0 < \varphi < \pi/2$. Let $\mathbb{C}\mathfrak{w}^\perp$ be the complex span of $\mathfrak{w}^\perp$ and $\mathfrak{d} = \mathbb{C}\mathfrak{w}^\perp \ominus \mathfrak{w}^\perp$. As $\varphi > 0$, we have $k = \dim_\mathbb{C} \mathbb{C}\mathfrak{w}^\perp = \dim \mathfrak{w}^\perp = \dim \mathfrak{d}$. For each $\xi \in \mathfrak{w}^\perp$ we decompose $J\xi \in \mathbb{C}\mathfrak{w}^\perp = \mathfrak{d} + \mathfrak{w}^\perp$ into $J\xi = P\xi + F\xi$ with $P\xi \in \mathfrak{d}$ and $F\xi \in \mathfrak{w}^\perp$. Then $\mathfrak{d}$ has constant Kähler angle φ as well. We denote by $\mathfrak{c}$ the maximal complex subspace of $\mathfrak{s}$. We have the orthogonal decomposition $\mathfrak{a} + \mathfrak{n} = \mathfrak{c} + \mathfrak{d} + \mathfrak{w}^\perp$. We denote by $\mathfrak{A}$, $\mathfrak{C}$, $\mathfrak{D}$ and $\mathfrak{W}^\perp$ the left-invariant distributions on $\mathbb{C}H^n$ along W_φ^{2n-k} induced by $\mathfrak{a}$, $\mathfrak{c}$, $\mathfrak{d}$ and $\mathfrak{w}^\perp$, respectively. By construction, $\mathfrak{C} + \mathfrak{D} = TW_\varphi^{2n-k}$ and $\mathfrak{W}^\perp = \nu W_\varphi^{2n-k}$.

Proposition 1.1. *The submanifold W_φ^{2n-k}, $0 < \varphi \leq \pi/2$, satisfies:*

(i) The maximal holomorphic subbundle $\mathfrak{C}$ of TW_φ^{2n-k} is autoparallel and the leaves of the induced foliation on W_φ^{2n-k} are totally geodesic $\mathbb{C}H^{n-k} \subset \mathbb{C}H^n$. Hence W_φ^{2n-k} is a ruled submanifold of $\mathbb{C}H^n$.

(ii) The following statements are equivalent:

  (a) the distribution $\mathfrak{D}$ on W_φ^{2n-k} is integrable;
  (b) the distribution $\mathfrak{A} + \mathfrak{D}$ on W_φ^{2n-k} is integrable;
  (c) the normal bundle $\mathfrak{W}^\perp$ is flat;
  (d) $\varphi = \pi/2$.

In this case, the leaves of the foliation induced by $\mathfrak{A} + \mathfrak{D}$ are totally geodesic $\mathbb{R}H^{k+1} \subset \mathbb{C}H^n$ and the leaves of the foliation induced by $\mathfrak{D}$ are horospheres with center x in these totally geodesic $\mathbb{R}H^{k+1}$.

(iii) For each $0 \neq \xi \in \mathfrak{w}^\perp$ the distribution $\mathfrak{A} + \mathbb{R}P\xi$ is autoparallel and the leaves of the induced foliation are totally geodesic $\mathbb{R}H^2 \subset \mathbb{C}H^n$.

(iv) For each $0 \neq \xi \in \mathfrak{w}^\perp$ the distribution $\mathbb{R}P\xi$ is integrable and the leaves of the induced foliation are horocycles with center x in the totally geodesic $\mathbb{R}H^2 \subset \mathbb{C}H^n$ given by the distribution $\mathfrak{A} + \mathbb{R}P\xi$.

(v) Let $\xi, \eta \in \mathfrak{w}^\perp$, $U, V \in \mathfrak{c} \ominus (\mathfrak{a} + \mathfrak{g}_{2\alpha})$ and $a, b, x, y \in \mathbb{R}$. Then, the second fundamental form II of W_φ^{2n-k} is given by

$$2II(aB + U + P\xi + xZ, bB + V + P\eta + yZ) = (\sin^2 \varphi)(y\xi + x\eta).$$

As a consequence, W_φ^{2n-k} is a minimal ruled submanifold of $\mathbb{C}H^n$. The second fundamental form characterizes the submanifolds W_φ^{2n-k}:

Theorem 1.2 (Rigidity of the submanifold W_φ^{2n-k}). *Let M be a $(2n-k)$-dimensional connected submanifold in $\mathbb{C}H^n$, $n \geq 2$, with normal bundle $\nu M \subset T\mathbb{C}H^n$ of constant Kähler angle $\varphi \in (0, \pi/2]$. Assume that there exists a unit vector field Z tangent to the maximal complex distribution on M such that the second fundamental form II of M is given by*

the trivial symmetric bilinear extension of
$$2II(Z, P\xi) = (\sin^2 \varphi) \xi, \quad \text{for all } \xi \in \nu M,$$
where $P\xi$ is the tangential component of $J\xi$. Then M is holomorphically congruent to an open part of the ruled minimal submanifold W_φ^{2n-k}.

In order to investigate the geometry of orbits of the cohomogeneity one actions of type (W) we deal with two possibilities: $\varphi = \pi/2$ or $\varphi \in (0, \pi/2)$.

Constant Kähler angle $\varphi = \pi/2$

In this case $\mathfrak{w}^\perp$ is real and νW^{2n-k} is totally real. The eigenvalues of the shape operator of W^{2n-k} with respect to a unit vector $\xi \in \nu W^{2n-k}$ are $1/2, -1/2$ and 0, with multiplicities $1, 1$ and $2n-2-k$, and corresponding principal curvature spaces $\mathbb{R}(Z+J\xi), \mathbb{R}(Z-J\xi)$ and $TW^{2n-k} \ominus (\mathbb{R}Z + \mathbb{R}J\xi)$.

Let $M(r)$ be the tube of radius $r > 0$ around W^{2n-k}. Take a unit speed geodesic γ with $\dot\gamma(0) = \xi \in \nu W_\varphi^{2n-k}$. For $X \in T\mathbb{C}H^n$ let $\mathcal{P}_X$ be the parallel transport of X along γ. The shape operator of $M(r)$ with respect to $-\dot\gamma(r)$ and the orthogonal decomposition $T_{\gamma_\xi(r)}M(r) = \mathcal{P}_{\mathbb{R}Z+\mathbb{R}J\xi}(r) + \mathcal{P}_{TW^{2n-k}\ominus(\mathbb{R}Z+\mathbb{R}J\xi)}(r) + \mathcal{P}_{\nu W^{2n-k}\ominus \mathbb{R}\xi}(r)$ is given by

$$\mathcal{S}(r) = \frac{1}{2}\begin{pmatrix} \tanh^3 \frac{r}{2} & -\operatorname{sech}^3 \frac{r}{2} & & \\ -\operatorname{sech}^3 \frac{r}{2} & 2\left(1+\frac{1}{2}\operatorname{sech}^2 \frac{r}{2}\right)\tanh \frac{r}{2} & & \\ \hline & & (\tanh \frac{r}{2})I_{2n-2-k} & \\ \hline & & & (\coth \frac{r}{2})I_{k-1} \end{pmatrix}$$

Then $M(r)$ has 4 principal curvatures
$$\lambda_{1/2} = \frac{3}{4}\tanh \frac{r}{2} \pm \frac{1}{2}\sqrt{1 - \frac{3}{4}\tanh^2 \frac{r}{2}}, \quad \lambda_3 = \frac{1}{2}\tanh \frac{r}{2}, \quad \lambda_4 = \frac{1}{2}\coth \frac{r}{2}.$$
The Hopf vector field on M has nontrivial orthogonal projection onto T_{λ_1} and T_{λ_2}. If $r = \ln(2 + \sqrt{3})$ there are 3 principal curvatures $\lambda_1 = 0$, $\lambda_2 = \lambda_4 = \sqrt{3}/2$ and $\lambda_3 = \sqrt{3}/6$ with multiplicities $1, k$ and $2n-k-2$.

The interesting part of the shape operator of both W^{2n-k} and $M(r)$ concerns the vectors Z and $J\xi$. More precisely, consider the subalgebra $\tilde{\mathfrak{g}} = \mathfrak{a} + \mathbb{R}\xi + \mathbb{R}J\xi + \mathfrak{g}_{2\alpha}$ and let $\tilde{G}$ be the connected subgroup of AN with Lie algebra $\tilde{\mathfrak{g}}$. The orbit $\tilde{G} \cdot o$ is a totally geodesic $\mathbb{C}H^2$ in $\mathbb{C}H^n$. This $\mathbb{C}H^2$ defines a "slice" of the action of $N_K^0(S)S$ through o. Next, $\tilde{\mathfrak{s}} = \mathfrak{s} \cap \tilde{\mathfrak{g}}$ is a subalgebra of $\tilde{\mathfrak{g}}$ of codimension one. Let $\tilde{S}$ be the connected subgroup of $\tilde{G}$ with Lie algebra $\tilde{\mathfrak{s}}$. Then $\tilde{S}$ acts on $\mathbb{C}H^2 \cong \tilde{G} \cdot o$ with cohomogeneity one and gives the solvable foliation of $\mathbb{C}H^2$ described above. The action of G on $\mathbb{C}H^n$ on the slice $\mathbb{C}H^2$ is equivalent to the action of $\tilde{H}$ on $\mathbb{C}H^2$.

Constant Kähler angle $\varphi \in (0, \pi/2)$

The eigenvalues of the shape operator with respect to a unit $\xi \in \nu W_\varphi^{2n-k}$ are $(\sin\varphi)/2$, $-(\sin\varphi)/2$ and 0, with multiplicities 1, 1 and $2n-k-2$. If $\bar{P}\xi = P\xi/\sin(\varphi)$ and $\bar{F}\xi = F\xi/\cos(\varphi)$, then, the corresponding eigenspaces are $\mathbb{R}(Z+\bar{P}\xi)$, $\mathbb{R}(-Z+\bar{P}\xi)$ and $TW_\varphi^{2n-k} \ominus (\mathbb{R}Z + \mathbb{R}\bar{P}\xi)$.

The shape operator $\mathcal{S}(r)$ of the orbit $M(r)$ at a distance r from W_φ^{2n-k} in the direction $-\dot{\gamma}(r)$, where γ is a geodesic with $\dot{\gamma}(0) = \xi \in \nu W_\varphi^{2n-k}$, is

$$\mathcal{S}(r) = \begin{pmatrix} s(r) & & \\ & \frac{1}{2}\tanh\frac{r}{2} I_{2n-k-2} & \\ & & \frac{1}{2}\coth\frac{r}{2} I_{k-2} \end{pmatrix},$$

with respect to $T_{\gamma(r)}M(r) = \mathcal{P}_{\mathbb{R}Z+\mathbb{R}\bar{P}\xi+\mathbb{R}\bar{F}\xi}(r) + \mathcal{P}_{TW_\varphi^{2n-k}\ominus(\mathbb{R}Z+\mathbb{R}\bar{P}\xi)}(r) + \mathcal{P}_{\nu W_\varphi^{2n-k}\ominus(\mathbb{R}\xi+\mathbb{R}\bar{F}\xi)}(r)$. Here $s(r)$ is a 3×3 matrix whose characteristic polynomial has 3 distinct real roots for any r. We define $\beta_{r,\varphi} = 27(\sin^2\varphi)\tanh^2\frac{r}{2}\text{sech}^4\frac{r}{2} - 2$. Let $u_{r,\varphi}^i$, $i \in \{1,2,3\}$, denote each cubic root of the complex number $(\beta_{r,\varphi} + \sqrt{\beta_{r,\varphi}^2 - 4})/2$. Then, the eigenvalues of $s(r)$ are

$$\lambda_i(r) = -\frac{1}{6}\left(\left(u_{r,\varphi}^i + \frac{1}{u_{r,\varphi}^i}\right)\coth\frac{r}{2} - \text{csch}\frac{r}{2}\text{sech}\frac{r}{2} - 4\tanh\frac{r}{2}\right),$$

$i \in \{1,2,3\}$. Neither $\frac{1}{2}\tanh\frac{r}{2}$ nor $\frac{1}{2}\coth\frac{r}{2}$ are eigenvalues of $s(r)$, so $M(r)$ has 5 distinct constant principal curvatures when $k > 2$ and 4 when $k = 2$.

Let $\mathfrak{v}_0 \subset \mathfrak{g}_\alpha$ be a two-dimensional subspace with constant Kähler angle φ. Then, $\tilde{\mathfrak{g}} = \mathfrak{a} + \mathbb{C}\mathfrak{v}_0 + \mathfrak{g}_{2\alpha}$ is a Lie subalgebra of $\mathfrak{a} + \mathfrak{n}$. Let $\tilde{G}$ be the connected Lie subgroup of AN whose Lie algebra is $\tilde{\mathfrak{g}}$. Then, $\tilde{G}\cdot o$ is a totally geodesic $\mathbb{C}H^3$ in $\mathbb{C}H^n$ containing o. The Lie subalgebra $\tilde{\mathfrak{s}} = \mathfrak{a} + \mathfrak{v}_0 + \mathfrak{g}_{2\alpha}$ of $\tilde{\mathfrak{g}}$ has codimension two. Denote by $\tilde{S}$ the connected Lie subgroup of $\tilde{G}$ whose Lie algebra is $\tilde{\mathfrak{s}}$. We know that $N_K^0(\tilde{S})\tilde{S}$ acts on $\tilde{G}\cdot o \cong \mathbb{C}H^3$ with cohomogeneity one and its orbit through o is exactly $\tilde{S}\cdot o \cong W_\varphi^4$. This cohomogeneity one action is of the type described in this subsection.

Let $M(r)$ denote the tube around $W_\varphi^4 \subset \mathbb{C}H^3$ at distance $r > 0$. For each $p \in M(r)$ there exists a unique unit vector $\xi(p) \in \nu W_\varphi^4$ such that $p = \exp(r\,\xi(p))$. Let $\gamma_{\xi(p)}(t) = \exp(t\,\xi(p))$ be the unique geodesic perpendicular to W_φ^4 that joins W_φ^4 and p. For any $X \in T_{\gamma_{\xi(p)}(0)}\mathbb{C}H^3$ we denote by $\mathcal{P}_X^p(r)$ the parallel displacement of X to p along $\gamma_{\xi(p)}$. We have

Theorem 1.3. *The following two statements hold.*

(i) Let $\mathcal{D}$ be the distribution on $M(r)$ defined by $\mathcal{P}_B^p(r)$, $p \in M(r)$, and denote by $\mathcal{D}^\perp$ its orthogonal complement in $TM(r)$. Then $\mathcal{D}$ and $\mathcal{D}^\perp$

are integrable and $\mathcal{D}$ is autoparallel. If $p \in M(r)$ and $\mathbb{R}H^2$ is the totally geodesic real hyperbolic plane determined by $\xi(p)$ and B_o, where $o \in W_\varphi^4$ is the footpoint of $\xi(p)$, then the leaf of $\mathcal{D}$ through p is parametrized by the parallel curve through p in $\mathbb{R}H^2$ of the geodesic in $\mathbb{R}H^2$ through o and in direction B_o.

(ii) Let $\mathcal{E}$ be the distribution on $M(r)$ defined by $\mathbb{R}\mathcal{P}_B^p(r) + \mathbb{R}\mathcal{P}_{PF\xi(p)}^p(r)$. Then $\mathcal{E}$ is autoparallel and each integral manifold has constant curvature $-(1/4)\mathrm{sech}(r/2)$. If $p \in M(r)$ and $\mathbb{R}H^3$ is the totally geodesic real hyperbolic space determined by $\xi(p)$, $PF\xi(p)$ and B_o, where $o \in W_\varphi^4$ is the footpoint of $\xi(p)$, then the leaf of $\mathcal{E}$ through p is the parallel surface through p in $\mathbb{R}H^3$ of the totally geodesic $\mathbb{R}H^2$ in $\mathbb{R}H^3$ through o determined by $PF\xi(p)$ and B_o.

2. Hypersurfaces with constant principal curvatures

In this section we summarize the known facts about real hypersurfaces of $\mathbb{C}H^n$ with constant principal curvatures.

Theorem 2.1. *Let M be a connected real hypersurface with constant principal curvatures in $\mathbb{C}H^n$ and g its number of principal curvatures. Then:*

(i) There are no umbilical hypersurfaces in $\mathbb{C}H^n$, i.e. $g = 1$ is not possible.

(ii) If $g = 2$ then M is an open part of one of:

  (a) A geodesic sphere.
  (b) A tube around a totally geodesic $\mathbb{C}H^{n-1}$.
  (c) A tube of radius $r = \log(2 + \sqrt{3})$ around a totally geodesic $\mathbb{R}H^n$.
  (d) A horosphere in $\mathbb{C}H^n$.

(iii) If $g = 3$ then M is an open part of one of:

  (a) A tube around a totally geodesic $\mathbb{C}H^k$ for some $k \in \{1, \ldots, n-2\}$.
  (b) A tube of radius $r \ne \log(2 + \sqrt{3})$ around a totally geodesic $\mathbb{R}H^n$.
  (c) The hypersurface W^{2n-1} or one of its equidistant hypersurfaces.
  (d) A tube of radius $r = \log(2 + \sqrt{3})$ around the minimal ruled submanifold W^{2n-k} for some $k \in \{2, \ldots, n-1\}$.

Part (i) was proved by Tashiro and Tachibana.[15] Part (ii) follows from the work by Montiel[11] for $n \geq 3$ and from Ref. 4 for $n = 2$. Part (iii) follows from Ref. 3 for $n \geq 3$ and from Ref. 4 for $n = 2$. In particular we have:

Corollary 2.1. *Any complete real hypersurface with constant principal curvatures in the complex hyperbolic plane is homogeneous.*

For $g \geq 4$ not much is known. However, Theorem 1.2 characterizes the singular orbits of cohomogeneity one actions of type (W) in terms of their second fundamental form. The classification of Hopf hypersurfaces with constant principal curvatures is due to Berndt.[1]

Theorem 2.2. *A connected Hopf real hypersurface of $\mathbb{C}H^n$ with constant principal curvatures is holomorphically congruent to an open part of a Hopf homogeneous hypersurface of type (A), (B) or (H) in Theorem 0.1.*

To conclude this section we give a list of open problems

(a) Determine the possible number of principal curvatures for a real hypersurface with constant principal curvatures in $\mathbb{C}H^n$. If there were no non-homogeneous examples the answer would be 2, 3, 4 or 5.
(b) Find examples of non-homogenous hypersurfaces with constant principal curvatures in $\mathbb{C}H^n$ or prove that such examples do not exist.
(c) Classify hypersurfaces with constant principal curvatures in $\mathbb{C}H^n$. An answer for 4 or 5 distinct principal curvatures would be of interest.

Acknowledgements

The author has been supported by a Marie Curie Intra-European Fellowship (MEIF-CT-2006-038754) and by PGIDIT06PXIB207054PR (Spain).

References

1. J. Berndt, *J. Reine Angew. Math.* **395**, 132–141 (1989).
2. J. Berndt, *Math. Z.* **229**, 589–600 (1998).
3. J. Berndt and J. C. Díaz-Ramos, *J. London Math. Soc.* **74**, 778–798 (2006).
4. J. Berndt and J. C. Díaz-Ramos, *Proc. Amer. Math. Soc.* **135**, 3349–3357 (2007).
5. J. Berndt and J. C. Díaz-Ramos, *Geom. Dedicata*, DOI 10.1007/s10711-008-9303-8.
6. J. Berndt and H. Tamaru, *Trans. Amer. Math. Soc.* **359**, 3425–3438 (2007).
7. J. Berndt, F. Tricerri and L. Vanhecke, *Lecture Notes in Math.* **1598** (Springer–Verlag, Berlin, 1995).
8. É. Cartan, *Ann. Mat. Pura Appl. IV. Ser.* **17**, 177–191 (1938).
9. D. Ferus, H. Karcher and H. F. Münzner, *Math. Z.* **177**, 479–502 (1981).
10. W.-Y. Hsiang and H. B. Lawson Jr., *J. Differential Geom.* **5**, 1–38 (1971).
11. S. Montiel, *J. Math. Soc. Japan* **37**, 515–535 (1985).
12. P. Segre, *Atti. Accad. Naz. Lincei Rend. Cl. SCi. Fis. Mat. Natur.* (6). **27**, 203–207 (1938).
13. R. Takagi, *Osaka J. Math.* **10**, 495–506 (1973).
14. R. Takagi, *J. Math. Soc. Japan* **27**, 43–53 (1975).
15. Y. Tashiro and S. I. Tachibana, *Kōdai Math. Sem. Rep.*, **15**, 176–183 (1963).

RIGIDITY RESULTS FOR GEODESIC SPHERES IN SPACE FORMS

Julien Roth

Laboratoire d'Analyse et de Mathématiques Appliquées, Université Paris-Est
5, boulevard Descartes, Cité Descartes
77454 Marne-la-Vallée cedex 2, France
E-mail: Julien.Roth@univ-mlv.fr

We prove that a hypersurface of a space form with almost constant mean curvature and almost constant scalar curvature is close to a geodesic sphere. In the case of Euclidean space, we deduce new characterizations of geodesic spheres.

Keywords: Hypersurfaces, space forms, rigidity, pinching

1. Introduction

The well-known Alexandrov theorem ([1]) claims that any compact without boundary hypersurface embedded into the Euclidean space $\mathbb{R}^{n+1}$ with constant mean curvature (CMC) is a geodesic sphere. Later, A. Ros ([14]) proved that the result also holds for hypersurfaces of the hyperbolic space $\mathbb{H}^{n+1}$ and the open half sphere $\mathbb{S}^{n+1}_+$.

In these results, the assumption that the hypersurface is embedded is crucial. Indeed, the results are false for immersed hypersurfaces. For instance, the so-called Wente's torus (see[18]) is an example of (non-embedded) immersed surface in $\mathbb{R}^3$ with constant mean curvature which is not a geodesic sphere. Other examples of higher genus are known ([10]).

For surfaces in $\mathbb{R}^3$, Hopf ([8]) proved that CMC immersed spheres are geodesic spheres. Here again, the result is not true in general since, we know examples of CMC spheres in higher dimension which are not geodesic spheres (see[9]).

The goal of this article is to find an alternative assumption to the embedding such that under this assumption, CMC hypersurfaces are geodesic spheres. It is a well-known fact that if the mean curvature H and the scalar curvature Scal are both constant, then the hypersurface is a geodesic sphere. We will show a stability result associated with this assertion. Namely, we have the following result which was proved for $\delta = 0$ in ([16]). The cases $\delta > 0$ and $\delta < 0$ are new.

Theorem 1.1. *Let (M^n, g) be a compact without boundary, connected and oriented Riemannian manifold, isometrically immersed into the simply connected space form $\mathbb{M}_\delta^{n+1}$ of constant sectional curvature δ. Let R be the extrinsic radius of M. If $\delta > 0$, we assume that $R < \frac{\pi}{4\sqrt{\delta}}$. Let $h > 0$ and $\theta \in]0,1[$. Then, there exists $\varepsilon(n, h, R, \theta) > 0$ such that if*

- $|H - h| < \varepsilon$ *and*
- $|\mathrm{Scal} - s| < \varepsilon$ *for a constant s,*

then $\left| h^2 - \frac{s}{n(n-1)} + \delta \right| \leqslant A\varepsilon$, for a positive constant A depending on n, h, δ and R, and M is diffeomorphic and θ-almost isometric to a geodesic sphere in $\mathbb{M}_\delta^{n+1}$ of radius $t_\delta^{-1}\left(\frac{1}{h}\right)$, where t_δ is the function defined in Section 2.

Remark 1.1. By θ-almost isometric, we mean that there exists a diffeomorphism F from M into a geodesic sphere of appropriate radius such that

$$\left| |dF_x(u)|^2 - 1 \right| \leq \theta,$$

for any $x \in M$ and any unit vector $u \in T_x M$.

Remark 1.2. The extrinsic radius of M is the radius of the smallest closed ball in $\mathbb{M}_\delta^{n+1}$ containing M.

Then, from this stability result, we will deduce a new characterization of geodesic spheres in $\mathbb{R}^{n+1}$ with a weaker assumption on the scalar curvature (see Section 4).

2. Preliminaries

Let (M^n, g) be a n-dimensional compact, connected, oriented Riemannian manifold without boundary, isometrically immersed into the $(n + 1)$-dimensional Euclidean space $(\mathbb{R}^{n+1}, can)$, where can is the canonical metric of $\mathbb{R}^{n+1}$. The (real-valued) second fundamental form B of the immersion is the bilinear symmetric form on $\Gamma(TM)$ defined for two vector fields X, Y by

$$B(X, Y) = -g\left(\overline{\nabla}_X \nu, Y\right),$$

where $\overline{\nabla}$ is the Riemannian connection on $\mathbb{R}^{n+1}$ and ν a normal unit vector field on M. When M is embedded, we choose ν as the inner normal field.

From B, we can define the mean curvature,
$$H = \frac{1}{n}\text{tr}(B).$$
Now, we recall the Gauss formula. For $X, Y, Z, W \in \Gamma(TM)$,
$$R(X,Y,Z,W) = \overline{R}(X,Y,Z,W) + \langle AX, Z \rangle \langle AY, W \rangle - \langle AY, Z \rangle \langle AX, W \rangle \quad (1)$$
where R and $\overline{R}$ are respectively the curvature tensor of M and $\mathbb{M}^{n+1}_\delta$, and A is the Weingarten operator defined by $AX = -\overline{\nabla}_X \nu$.

By taking the trace and for $W = Y$, we get
$$\text{Ric}(Y) = \overline{\text{Ric}}(Y) - \overline{R}(\nu, Y, \nu, Y) + nH \langle AY, Y \rangle - \langle A^2 Y, Y \rangle. \quad (2)$$
Since, the ambient space is of constant sectional curvature δ, by taking the trace a seconde time, we have
$$\text{Scal} = n(n-1)\delta + n^2 H^2 - |A|^2, \quad (3)$$
or equivalently
$$\text{Scal} = n(n-1)\left(H^2 + \delta\right) - |\tau|^2, \quad (4)$$
where $\tau = B - H\text{Id}$ is the umbilicity tensor.

Now, we define the higher order mean curvatures, for $k \in \{1, \cdots, n\}$, by
$$H_k = \frac{1}{\binom{n}{k}} \sigma_k(\kappa_1, \cdots, \kappa_n),$$
where σ_k is the k-th elementary symmetric polynomial and $\kappa_1, \cdots, \kappa_n$ are the principal curvatures of the immersion.

From the definition, it is obvious that H_1 is the mean curvature H. We also remark from the Gauss formula (1) that
$$H_2 = \frac{1}{n(n-1)} \text{Scal} - \delta. \quad (5)$$
On the other hand, we have the well-known Hsiung-Minkowski formula
$$\int_M \left(H_{k+1} \langle Z, \nu \rangle + c_\delta(r) H_k \right) = 0, \quad (6)$$
where $r(x) = d(p_0, x)$ is the distance function to a base point p_0, Z is the position vector defined by $Z = s_\delta(r)\overline{\nabla} r$, and the functions c_δ and s_δ are defined by
$$c_\delta(t) = \begin{cases} \cos(\sqrt{\delta}t) & \text{if } \delta > 0 \\ 1 & \text{if } \delta = 0 \\ \cosh(\sqrt{-\delta}t) & \text{if } \delta < 0 \end{cases} \quad \text{and} \quad s_\delta(t) = \begin{cases} \frac{1}{\sqrt{\delta}}\sin(\sqrt{\delta}t) & \text{if } \delta > 0 \\ t & \text{if } \delta = 0 \\ \frac{1}{\sqrt{-\delta}}\sinh(\sqrt{-\delta}t) & \text{if } \delta < 0. \end{cases}$$

Finally, we define the function $t_\delta = \frac{s_\delta}{c_\delta}$.

We finish this section of preliminaries by the following recollection about the first eigenvalue of the Laplacian. In ([7]), Heintze proved the following upper bound for $\lambda_1(\Delta)$

$$\lambda_1(\Delta) \leqslant n(||H||_\infty^2 + \delta), \tag{7}$$

with equality for geodesic spheres. Grosjean ([6]) proved a stability result associated with Heintze's estimate. Precisely, he proved the following

Theorem 2.1 (Grosjean, 2007). *Let M be compact without boundary, connected and oriented hypersurface of $\mathbb{M}_\delta^{n+1}$. If $\delta > 0$, we assume that M is contained in an open ball of $\mathbb{M}_\delta^{n+1}$ of radius $\frac{\pi}{4\sqrt{\delta}}$. Let $\theta \in]0,1[$, then there exitst a constant $C_\theta(n, ||B||_\infty, V(M), \delta) > 0$ such that if*

$$n(||H||_\infty^2 + \delta) - C_\theta < \lambda_1(\Delta),$$

then M is diffeomorphic and θ-almost isometric to a geodesic sphere of radius $\sqrt{\frac{n}{\lambda_1}}$.

Now, we have all the ingredients to prove Theorem 1.1.

3. Proof of Theorem 1.1

We begin the proof of Theorem 1.1 by the following lemma.

Lemma 3.1. *Let h and s be two positive constants. If the mean curvature and the scalar curvature satisfy*

- $|H - h| < \varepsilon$ *and*
- $|\text{Scal} - s| < \varepsilon$,

for some positive ε, then

$$\left| h^2 - \frac{s}{n(n-1)} + \delta \right| \leqslant A\varepsilon,$$

where A is a positive constante depending on h, R, n and δ.

Proof. The proof of this lemma comes directly from the Hisung-Minkowski formula (6). Indeed, the Hisung-Minkowski formula for $k = 1$ is the following

$$\int_M \left(H_2 \langle Z, \nu \rangle + c_\delta(r) H \right) = 0. \tag{8}$$

Since we assume that $|\text{Scal} - s| < \varepsilon$, we get easily from (5) that

$$\left| H_2 - \left(\frac{s}{n(n-1)} - \delta \right) \right| < \frac{1}{n(n-1)} \varepsilon. \tag{9}$$

For more convenience, we will denote $h_2 = \frac{s}{n(n-1)} - \delta$. Then, from (8)

$$0 = \int_M \left(H_2 \langle Z, \nu \rangle + c_\delta(r) H \right)$$

$$= \int_M \left(h_2 \langle Z, \nu \rangle + c_\delta(r) H \right) + \int_M (H_2 - h_2) \langle Z, \nu \rangle$$

$$= \frac{h_2}{h} \int_M h \langle Z, \nu \rangle + \int_M c_\delta(r) H + \int_M (H_2 - h_2) \langle Z, \nu \rangle$$

$$= \frac{h_2}{h} \int_M H \langle Z, \nu \rangle + \frac{h_2}{h} \int_M (h - H) \langle Z, \nu \rangle + \int_M c_\delta(r) h + \int_M c_\delta(r)(H - h)$$

$$+ \int_M (H_2 - h_2) \langle Z, \nu \rangle .$$

Now, we use the Hsiung-Minkowski formula for $k = 0$, that is

$$\int_M \left(H \langle Z, \nu \rangle + c_\delta(r) \right) = 0, \tag{10}$$

to get

$$0 = -\frac{h_2}{h} \int_M c_\delta(r) + \frac{h_2}{h} \int_M (h - H) \langle Z, \nu \rangle + \int_M c_\delta(r) h + \int_M c_\delta(r)(H - h)$$

$$+ \int_M (H_2 - h_2) \langle Z, \nu \rangle$$

$$= \left(h - \frac{h_2}{h} \right) \int_M c_\delta(r) + \frac{h_2}{h} \int_M (h - H) \langle Z, \nu \rangle + \int_M c_\delta(r)(H - h)$$

$$+ \int_M (H_2 - h_2) \langle Z, \nu \rangle .$$

Then, since s_δ is an increasing function, we deduce

$$\left| h - \frac{h_2}{h} \right| \int_M c_\delta(r) \leqslant \frac{h_2}{h} \varepsilon \int_M s_\delta(R) + \varepsilon \int_M c_\delta(r) + \frac{\varepsilon}{n(n-1)} \int_M s_\delta(R) .$$

Using the fact that $|H_2| \leqslant H^2$, we deduce from the assumptions on h and h_2 that

$$|h_2| \leqslant h^2 + A_1(n, h, \delta) \varepsilon,$$

and then we have
$$\left|h - \frac{h_2}{h}\right| \int_M c_\delta(r) \leqslant \varepsilon \int_M c_\delta(r) + A_2(n,h,\delta) \int_M s_\delta(R).$$

We conclude using the fact that $V(M) \leqslant \int_M c_\delta(r) \leqslant A_3(n,\delta,R)V(M)$, which yields
$$|h^2 - h_2| \leqslant A_4(n,h,R,\delta)\varepsilon,$$
which gives the wanted assertion. $\square$

Let $0 < \varepsilon < 1$. We will choose a particular ε later. We assume $|H - h| < \varepsilon$ and $|\text{Scal} - s| < \varepsilon$, then from (3) and Lemma 3.1, we have
$$|\tau|^2 = n(n-1)H^2 - \text{Scal} \tag{11}$$
$$\leqslant A'(n,h,R,\delta)\varepsilon.$$

This means that M is almost umbilical. Moreover, since $|H - h| \leqslant \varepsilon$, Inequality (11) is equivalent to
$$|B - h\text{Id}| \leqslant A''(n,h,R,\delta)\varepsilon. \tag{12}$$

This last inequality combining with the Gauss formula (2) yields
$$\left|\text{Ric}(Y) - (n-1)(h^2 + \delta)|Y|^2\right| \leqslant A'''(n,h,R,\delta)\varepsilon. \tag{13}$$

Now, we use the Lichnerowicz formula (^{11}) to obtain the following lower bound of the first eigenvalue of the Laplacian on M
$$\lambda_1(\Delta) \geqslant n(h^2 + \delta) - \alpha_1(\varepsilon), \tag{14}$$
or equivalently
$$\lambda_1(\Delta) \geqslant n(\|H\|_\infty^2 + \delta) - \alpha_2(\varepsilon), \tag{15}$$
where the positive functions α_1 and α_2 depend on n, h, R and δ and tend to 0 when ε tends to 0. Now, we fix $\theta \in]0,1[$. Since α_2 tends to 0 when ε tends to 0, there exists $\varepsilon_1(n,h,R,\delta) > 0$, such that $\alpha_2(\varepsilon_1) \leqslant \frac{\theta}{2}$. Then, we use Theorem 2.1 to conclude that M is diffeomorphic and $\frac{\theta}{2}$-almost isometric to a geodesic sphere of radius $\sqrt{\frac{n}{\lambda_1}}$. Moreover, because of the pinching of $\lambda_1(\Delta)$, the radii $\frac{1}{h}$ and $\sqrt{\frac{n}{\lambda_1}}$ are close. So, there exists $\varepsilon_2(n,h,R,\delta) > 0$ such that geodesic spheres of radii $\frac{1}{h}$ and $\sqrt{\frac{n}{\lambda_1}}$ are $\frac{\theta}{2}$-almost isometric. Then, we take $\varepsilon = \inf\{\varepsilon_1, \varepsilon_2\}$ and then M is θ-quasi-isometric to a geodesic sphere of radius $\frac{1}{h}$. $\square$

4. Rigidity results in the Euclidean space

For the Euclidean case, that is $\delta = 0$, we can obtain from Theorem 1.1 new characterizations of geodesic spheres. Namely, we have the following

Corollary 4.1. *Let (M^n, g) be a compact, connected and oriented Riemannian manifold without boundary isometrically immersed into $\mathbb{R}^{n+1}$. Let h be a positive constant. Then, there exists $\varepsilon(n, h) > 0$ such that if*

(a) $H = h$ and
(b) $|\text{Scal} - s| \leqslant \varepsilon$,

for some constant s, then M is a geodesic sphere of radius $\frac{1}{h}$.

Remark 4.1. Note that in this result, ε does not depend on the extrinsic radius R. Indeed, since the mean curvature is constant, from (10), we have $\int_M \langle Z, \nu \rangle = \frac{1}{h} V(M)$. So we do not have to control the extrinsic radius.

Proof. This corollary is a direct consequence of Theorem 1.1. Indeed, we know that there exists a diffeomorphism F from M to $\mathbb{S}^n \left(\frac{1}{h}\right)$. But, in the Euclidean space, this diffeomorphism is explicit. Namely,

$$F(x) = \frac{1}{h} \frac{\phi(x)}{|\phi(x)|},$$

where ϕ is the immersion on M into $\mathbb{R}^{n+1}$. See ([3]) to get this expression.

The fact that F is a diffeomorphism implies that the immersion ϕ is necessarily a one-to-one map, that is an embedding. Since M is embedded with constant mean curvature, it is a geodesic sphere by the Alexandrov theorem. □

Now, we state a second characterization of geodesic spheres.

Corollary 4.2. *Let (M^n, g) be a compact, connected and oriented Riemannian manifold without boundary isometrically immersed into $\mathbb{R}^{n+1}$. Let s be a positive constant. Then, there exists $\varepsilon(n, s) > 0$ such that if*

(a) $\text{Scal} = s$
(b) $|H - h| \leqslant \varepsilon$,

for some constant h, then M is a geodesic sphere of radius $\sqrt{\frac{n(n-1)}{s}}$.

Proof. The proof is the same, using an Alexandrov type theorem for the scalar curvature due to Ros ([14]). □

This second corollary gives a partial answer to a conjecture by Yau which states that the only immersed hypersurfaces of the Euclidean space with constant scalar curvature are geodesic spheres.

Acknowledgements

The author would like to thank the organizers of the VIII International Colloquium in Differential Geometry in Santiago de Compostela.

References

1. A. D. Alexandrov, *Vesnik Leningrad Univ.* **11** (1956), 5–17.
2. E. Aubry, Diameter pinching in almost positive Ricci curvature, to appear.
3. B. Colbois and J.F. Grosjean, *Comment. Math. Helv.* **82** (2007), 175–195.
4. A. Fialkow, *Ann. Math.* **39** (1938), 762–783.
5. J.F. Grosjean, *Pacific J. Math.* **206** (2002), 93–111.
6. J.F. Grosjean, arXiv:math.DG/0709.0831.
7. E. Heintze, *Math. Ann.* **280** (1988), 389–402.
8. H. Hopf, *Lecture Notes in Math.* **1000**, Springer-Verlag, Berlin, 1983.
9. W. J. Hsiang, Z. U. Teng and W. C. Yu, *Ann. Math. (2)* **117** (1983), 609–625.
10. M. Jleli and F. Pacard, *Pacific J. Math.* **221** (2005), 81–108.
11. A. Lichnerowicz, *Géométrie des groupes de transformation*, Dunod, 1958.
12. S. Montiel and A. Ros, *Differential geometry*, 279–296, Pitman Monogr. Surveys Pure Appl. Math., **52**, Longman Sci. Tech., Harlow, 1991.
13. R. C. Reilly, *Comment. Math. Helv.* **52** (1977), 525–533.
14. A. Ros, *Revista Mat. Iberoamer.* **3** (1987), 477–483.
15. J. Roth, *Ann. Glob. Anal. Geom.*, to appear.
16. J. Roth, arXiv:Math.DG/0710.5041.
17. T. Y. Thomas, *Amer. J. Math.* **58** (1936), 702–704.
18. H. Wente, *Pacific J. Math.* **121** (1986), 193–243.

MEAN CURVATURE FLOW AND BERNSTEIN-CALABI RESULTS FOR SPACELIKE GRAPHS

Guanghan Li

School of Mathematics and Computer Science, Hubei University
Wuhan, 430062, P. R. China
E-mail: liguanghan@163.com

Isabel M. C. Salavessa

Centro de Física das Interacções Fundamentais, Instituto Superior Técnico
Technical University of Lisbon, Edifício Ciência, Piso 3, Av. Rovisco Pais, 1049-001
Lisboa, Portugal
E-mail: isabel.salavessa@ist.utl.pt

This is a survey of our work on spacelike graphic submanifolds in pseudo-Riemannian products, namely on Heinz-Chern and Bernstein-Calabi results and on the mean curvature flow, with applications to the homotopy of maps between Riemannian manifolds

Keywords: Spacelike submanifolds, Bernstein theorem, mean curvature flow

1. Introduction

It has been an important issue in geometry and in topology to determine when a map $f : \Sigma_1 \to \Sigma_2$ between manifolds can be homotopically deformed to a constant one. If each Σ_i has a Riemannian structure g_i, the curvature of these spaces may give an answer. This is particularly more complex if both Σ_i are compact. For Σ_i noncompact, by a famous result due to Gromov,[8] Σ_i admits a Riemannian metric of negative sectional curvature and also one of positive sectional curvature. In each of these cases, if Σ_i is complete and simply connected, then Σ_i is diffeomorphic to a contractible space, by the Cartan-Hadmard theorem and by a result of Cheeger and Gromoll, respectively (see in [4]). Hence, if this is the case for one of the Σ_i, then f is obviously homotopically trivial.

A deformation problem of an initial map can be handled using some geometric evolution equation, obtaining homotopic deformations of a certain type and with geometrical and analytical meaning, namely, giving at infinite time a solution of a certain partial differential equation. We recall the great discovery of Eells and Sampson,[7] a first example of this kind, on using the heat flow to deform a map to a harmonic one:

Theorem 1.1 (Eells and Sampson (1964)). *If Σ_1 and Σ_2 are closed and Σ_2 has nonpositive sectional curvature then f is homotopic to a harmonic map f_∞. Furthermore, if the Ricci tensor of Σ_1 is nonnegative then f_∞ is totally geodesic and if it is positive somewhere, then f_∞ is constant.*

The last part of this theorem can be seen as a Bernstein-type theorem, and it was obtained from a Weitzenböck formula for the Laplacian of $\|df_\infty\|^2$. We recall that Bernstein-type theorems are theorems that give conditions that ensures that a solution of certain P.D.Es. with geometrical meaning, must be a "trivial" solution, as for example a totally geodesic or a constant map.

In this note, a survey of our main results in [10–12], we will show how to use the mean curvature flow and a Bernstein-Calabi type result for spacelike graphs to obtain a deformation of a map between Riemannian manifolds to a totally geodesic or a constant one.

The Bernstein-Calabi result is obtained by computing the Laplacian of a positive geometric quantity, the hyperbolic cosine of the hyperbolic angle of a spacelike graph, and analyzing the sign of this Laplacian, based on an idea of Chern[5] on computing a similar quantity in the Riemannian case.

Furthermore, we also will show that under somehow more general curvature conditions as in the above theorem, we can obtain a direct proof of the homotopy to a constant map, with no need to use a Bernstein-type result. This approach was started by Wang[14] for the graph Γ_f of the map f, considered as a submanifold of the Riemannian product $\Sigma_1 \times \Sigma_2$ of closed spaces with constant sectional curvature, and take its mean curvature flow and show that under certain conditions the flow preserves the graphic structure of the submanifold and converges to the graph of a constant map. The main difference with our approach is that we consider the pseudo-Riemannian structure on $\Sigma_1 \times \Sigma_2$ instead the Riemannian one. Our assumption on Γ_f to be a spacelike submanifold is essentially identical to the assumptions on the eigenvalues of f^*g_2 imposed in [13,14] in the corresponding Riemannian setting. Our advantage is that the pseudo-Riemannian setting turns out to be a more natural one, since it allows less restrictive assumptions on the curvature tensors (and that include the case of any negative sectional curvature for Σ_2) and on the map f itself after a suitable rescaling of the metric of Σ_2, and long time existence and convergence of the flow are easier obtained. In [14] it is necessary to use a White's regularity theorem, based on a monocity formula due to Huisken, to detect possible singularities of the mean curvature flow, while in the pseudo-Riemannian case, because of good signature in the evolution equations, we have better regularity. This will become clear in equations (1), (3) and (4) below.

Let (Σ_1, g_1) and (Σ_2, g_2) be Riemannian manifolds of dimension $m \geq 2$ and $n \geq 1$ respectively, and of sectional curvatures K_i and Ricci tensors $Ricci_i$. On $\overline{M} = \Sigma_1 \times \Sigma_2$ we consider the pseudo-Riemannian metric $\bar{g} = g_1 - g_2$. We assume Σ_1 oriented. Given a map f, we assume the graph, $\Gamma_f : \Sigma_1 \to \overline{M}$, $\Gamma_f(p) = (p, f(p))$, is a spacelike submanifold that is, $g := \Gamma_f^* \bar{g} = g_1 - f^* g_2$ is a Riemannian metric on Σ_1. Thus, the eigenvalues of $f^* g_2$, at $p \in M$, $\lambda_1^2 \geq \ldots \geq \lambda_m^2 \geq 0$, are bounded from above by $1 - \delta(p)$, where $0 < \delta(p) \leq 1$ is a constant depending on p. The *hyperbolic angle* θ of Γ_f is given by one of the equivalent definitions:

$$\cosh \theta = \frac{1}{\sqrt{\Pi_i (1 - \lambda_i^2)}} = Vol_{\Sigma_1}(\pi_1(e_1), \ldots, \pi_1(e_m)) = \frac{Vol_{(\Sigma_1, g_1)}}{Vol_{(\Sigma_1, g)}}$$

where $\pi_1 : \overline{M} \to \Sigma_1$ is the projection and e_i is a direct o.n. basis of $T\Gamma_f$, and $Vol_{(\Sigma_1, g)}$ is the volume element of (Σ_1, g). Then $\cosh \theta \equiv 1$ iff f is constant, that is Γ_f is a slice.

2. Bernstein-Calabi and Heinz-Chern type results

The classic Bernstein theorem says that an entire minimal graph in $\mathbb{R}^3$ is a plane. This result was generalized to codimension one graphs in $\mathbb{R}^{m+1}$ for $m \leq 7$, and for higher dimensions and codimensions under additional conditions by many other authors. Calabi[2] considered the same problem for the maximal (the mean curvature $H = 0$) spacelike hypersurfaces M in the Lorentz-Minkowski space $\mathbb{R}_1^{m+1}$ with the metric $ds^2 = \sum_{i=1}^m (dx_i)^2 - (dx_{m+1})^2$. If M is given by the graph of a function f on $\mathbb{R}^m$ with $|Df| < 1$, the equation $H = 0$ has the form

$$\sum_{i=1}^m \frac{\partial}{\partial x_i} \left(\frac{\partial f / \partial x_i}{\sqrt{1 - |Df|^2}} \right) = 0.$$

Calabi showed that for $m \leq 4$, the graph of any entire solution to the above equation is a hyperplane. The same conclusion was established by Cheng and Yau[3] for any m. A further generalization of this problem to $\mathbb{R}_n^{m+n}$ has been obtained by some authors (see for instance in [9]).

Another natural generalization is to consider maximal spacelike graphic submanifolds in a non flat ambient space and in higher codimension. We consider a spacelike graph Γ_f, for a map $f : \Sigma_1 \to \Sigma_2$.

We can take a_i an orthonormal basis of $T_p\Sigma_1$ and e_α of $T_{f(p)}\Sigma_2$, $1 \leq i \leq m$, $m+1 \leq \alpha \leq m+n$, such that $df(a_i) = -\lambda_i a_{m+i}$ ($\lambda_i = 0$ if $i > n$). Then $e_i = (1-\lambda_i^2)^{-1/2}(a_i + \lambda_i a_{m+i})$ and $e_{m+i} = (1-\lambda_i^2)^{-1/2}(a_{m+i} + \lambda_i a_i)$, $e_\alpha = a_\alpha$ if $\alpha > 2m$, define o.n.bs. of $T_{(p,f(p))}\Gamma_f$, and of the normal bundle at $(p, f(p))$ respectively. Assuming Γ_f has parallel mean curvature, in this basis we have

$$\Delta \cosh\theta = \cosh\theta \Big\{ \|B\|^2 + 2\sum_k \sum_{i<j} \lambda_i \lambda_j h_{ik}^{m+i} h_{jk}^{m+j}$$
$$- 2\sum_k \sum_{i<j} \lambda_i \lambda_j h_{ik}^{m+j} h_{jk}^{m+i} \qquad (1)$$
$$+ \sum_i \Big(\frac{\lambda_i^2}{(1-\lambda_i^2)} Ricci_1(a_i, a_i) + \sum_{j \neq i} \frac{\lambda_i^2 \lambda_j^2}{(1-\lambda_i^2)(1-\lambda_j^2)}[K_1(P_{ij}) - K_2(P'_{ij})]\Big) \Big\}$$

where $P_{ij} = span\{a_i, a_j\}$ and $P'_{ij} = span\{a_{m+i}, a_{m+j}\}$. Here h_{ij}^α are the components of the second fundamental form B of Γ_f in the basis e_i, e_α.

Theorem 2.1 ([10,12]). *Let $M = \Gamma_f$ be a spacelike graph submanifold of $\overline{M}$ with parallel mean curvature vector. We assume for each $p \in \Sigma_1$, $Ricci_1(p) \geq 0$ and for any two-dimensional planes $P \subset T_p\Sigma_1$, $P' \subset T_{f(p)}\Sigma_2$, $K_1(P) \geq K_2(P')$. We have:*

(i) If $n = 1$ and $\cosh\theta \leq o(r)$ when $r \to +\infty$, where r is the distance function to a point $p \in (\Sigma_1, g_1)$, and Σ_1 is complete, then M is maximal.

(ii) If M is compact, then it is totally geodesic. Moreover, if $Ricci_1(p) > 0$ at some point, then M is a slice, that is f is constant;

(iii) If M is complete, noncompact, and K_1, K_2 and $\cosh\theta$ are bounded, then M is maximal.

(iv) If M is a complete maximal spacelike surface, then M is totally geodesic. Moreover, (a) if $K_1(p) > 0$ at some point $p \in M$, then M is a slice; (b) If $\Sigma_1 = \mathbb{R}^2$ and $\Sigma_2 = \mathbb{R}^n$, then M is a plane; (c) if Σ_1 is flat and $K_2 < 0$ at some point $f(p)$, then either M is a slice or the image of f is a geodesic of Σ_2.

We obtain (i) by applying a Heinz-Chern inequality derived in [12], for the absolute norm of H,

$$m\|H\| \leq \sup_{B_r(p)} \cosh\theta \; \mathfrak{h}(B_r(p)),$$

where $\mathfrak{h}(B_r(p)) = \inf_D V_{m-1}(\partial D)/V_m(D)$, is the Cheeger constant of the open geodesic ball of center p and radius r, where D runs all over the bounded domains of the ball with smooth boundary ∂D. Since $Ricci_1 \geq 0$, $\mathfrak{h}(B_r(p)) \leq C/r$, when $r \to +\infty$, where $C > 0$ is a constant. For Σ_1 the m-hyperbolic space (with non-zero Cheeger constant), we give examples in [12] of foliations of $\mathbb{H}^m \times \mathbb{R}$ by complete spacelike graphic hypersurfaces with bounded hyperbolic angle and with constant mean curvature any real c, the same for all leaves, or parameterized by the leaf.

The proof of (ii) and (iii) consists on showing that, under the curvature conditions, one has $\Delta \ln \cosh \theta \geq \delta \|B\|^2$, where $\delta > 0$ is a constant that does not depend on p and in (iii) showing that the Ricci tensor of M is bounded from below, and applying the Omori-Cheng-Yau maximum principle for noncompact manifolds. (i) and (iii) are obtained by different approaches. If M is a maximal Riemannian surface, (iv) gives a generalization of the Bernstein type theorem of Albujer-Alías[1] for maximal graphic spacelike surfaces in a Lorentzian three manifold $\Sigma_1 \times \mathbb{R}$ to higer codimension. As in [1,5] the proof is based on a parabolicity argument for surfaces with nonnegative Gauss curvature. In fact, in this case, we have that $\Delta\left(\frac{1}{\cosh \theta}\right) \leq 0$ and the Gauss curvature of M satisfies

$$K_M = \tfrac{1}{(1-\lambda_1^2)(1-\lambda_2^2)}[K_1 - \lambda_1^2\lambda_2^2 K_2(a_3,a_4)] + \sum_\alpha [(h_{11}^\alpha)^2 + (h_{12}^\alpha)^2] \geq 0.$$

The conclusion that $B = 0$ comes from analyzing the vanishing of the term involving the components of B in the expression of $\Delta(1/\cosh \theta)$. Our proof for $(iv)(b)$, gives a simpler proof of the same result of Jost and Xin[9] for de case of surfaces, but using their result that any entire maximal graph in $\mathbb{R}_n^{m+n}$ is complete.

We also derive in [11] a Simons' type identity for the absolute norm of the second fundamental form $\|B\|^2$ of a spacelike submanifold M of any pseudo-Riemannian manifold $\overline{M}$,

$$\Delta \|B\|^2 = 2\|\nabla B\|^2 + \sum_{ij\alpha} 2h_{ij}^\alpha H_{,ij}^\alpha - \sum_{ij\alpha} 2h_{ij}^\alpha [\sum_k (\bar{\nabla}_j \bar{R})_{kik}^\alpha + \sum_k (\bar{\nabla}_k \bar{R})_{ijk}^\alpha]$$
$$+ \sum_{ij\alpha\beta} 2\{\sum_k (4\bar{R}_{\beta ki}^\alpha h_{kj}^\beta h_{ij}^\alpha - \bar{R}_{k\beta k}^\alpha h_{ij}^\alpha h_{ij}^\beta) + \bar{R}_{i\beta j}^\alpha H^\beta h_{ij}^\alpha\}$$
$$+ \sum_{ijkl\alpha} 4(\bar{R}_{ijk}^l h_{ij}^\alpha h_{kl}^\alpha + \bar{R}_{kik}^l h_{lj}^\alpha h_{ij}^\alpha) - \sum_{ijk\alpha\beta} 2h_{ij}^\alpha h_{jk}^\beta h_{ki}^\beta H^\beta$$
$$+ 2\sum_{ij\alpha\beta}\left(\sum_k (h_{ik}^\alpha h_{jk}^\beta - h_{ik}^\beta h_{jk}^\alpha)\right)^2 + 2\sum_{\alpha\beta}(\sum_{ij} h_{ij}^\alpha h_{ij}^\beta)^2.$$

3. The mean curvature flow

The mean curvature flow of an immersion $F_0 : M \to \overline{M}$ is a family of immersions $F_t : M \to \overline{M}$ defined in a maximal interval $t \in [0,T)$ evolving according to

$$\begin{cases} \frac{d}{dt}F(x,t) = H(x,t) = \Delta_{g_t}F_t(x) \\ F(\cdot,0) = F_0 \end{cases} \quad (2)$$

where H_t is the mean curvature of $M_t = F_t(M) = (M, g_t = F_t^*\bar{g})$. The mean curvature flow of hypersurfaces in a Riemannian manifold has been extensively studied. Recently, mean curvature flow of submanifolds with higher co-dimensions has been paid more attention. In [14], the graph mean curvature flow is studied in Riemannian product manifolds, and it is proved long-time existence and convergence of the flow under suitable conditions. When $\overline{M}$ is a pseudo-Riemannian manifold, it is considered the mean curvature flow of spacelike submanifolds. This flow for spacelike hypersurfaces has also been largely studied, but very little is known on mean curvature flow in higher codimensions except in a flat space $\mathbb{R}_n^{n+m}$ [15]. In [11] we consider (2) with $\overline{M}$ any pseudo-Riemannian manifold and F_0 any spacelike submanifold, and we derive the evolution of the following quantities at a given point (x,t) with respect to an o.n. frame e_α of the normal bundle of M_t and a coordinate chart x_i of M, normal at x relatively to the metric g_t,

$$\frac{d}{dt}g_{ij} = 2H^\alpha h_{ij}^\alpha$$

$$\frac{d}{dt}Vol_{M_t} = \|H\|^2 Vol_{M_t}$$

$$\frac{d}{dt}\|B\|^2 = \Delta\|B\|^2 - 2\|\nabla B\|^2 + \sum_{ij\alpha}2h_{ij}^\alpha\left(\sum_k(\bar{\nabla}_j\bar{R})_{kik}^\alpha + (\bar{\nabla}_k\bar{R})_{ijk}^\alpha\right)$$
$$-\sum_{ijk\alpha\beta}2(4\bar{R}_{\beta ki}^\alpha h_{kj}^\beta h_{ij}^\alpha - \bar{R}_{k\beta k}^\alpha h_{ij}^\alpha h_{ij}^\beta)$$
$$+\sum_{ijkl\alpha}4(\bar{R}_{ijk}^l h_{ij}^\alpha h_{kl}^\alpha + \bar{R}_{kik}^l h_{lj}^\alpha h_{ij}^\alpha)$$
$$-2\sum_{ij\alpha\beta}\left(\sum_k(h_{ik}^\alpha h_{jk}^\beta - h_{ik}^\beta h_{jk}^\alpha)\right)^2 - 2\sum_{\alpha,\beta}(\sum_{ij}h_{ij}^\alpha h_{ij}^\beta)^2.$$

In this section we assume (Σ_1, g_1) closed and (Σ_2, g_2) complete, and the curvature tensor R_2 of Σ_2 and all its covariant derivatives are bounded. We consider the mean curvature flow on the pseudo-Riemannian manifold $\overline{M} = \Sigma_1 \times \Sigma_2$, when the initial immersed submanifold $F_0 = \Gamma_f : M = \Sigma_1 \to \overline{M}$ is a spacelike graph submanifold. Furthermore, we assume, as in Theorem 2,

$$Ricci_1(p) \geq 0, \quad \text{and} \quad K_1(p) \geq K_2(q) \quad \forall p \in \Sigma_1, q \in \Sigma_2.$$

This means that either $K_1(p) \geq K_2^+(q) = \max\{K_2(q), 0\}$, or $Ricci_1(p) \geq 0$ and $K_1(p)(P) < 0$ for some two-plane P and $K_1(p) \geq K_2(q)$ with $K_2(q) < 0, \forall p, q$.

We recall the mais steps of [11].

For $t > 0$ sufficiently small, F_t is near F_0 and so it is a spacelike graph with $\lambda_i^2(t) \leq 1 - \delta(t)$. We derive the evolution of the hyperbolic angle

$$\frac{d}{dt}\ln(\cosh\theta) = \Delta\ln(\cosh\theta)$$

$$-\underbrace{\left\{\|B\|^2 - \sum_{k,i}\lambda_i^2(h_{ik}^{m+i})^2 - 2\sum_{k,i<j}\lambda_i\lambda_j h_{ik}^{m+j}h_{jk}^{m+i}\right\}}_{\geq \delta(t)\|B\|^2} \quad (3)$$

$$-\sum_i \lambda_i^2\Big(\frac{1}{(1-\lambda_i^2)}\underbrace{Ricci_1(e_i,e_i)}_{\geq 0} + \sum_{i\neq j}\frac{\lambda_j^2}{(1-\lambda_i^2)(1-\lambda_j^2)}\underbrace{[K_1(P_{ij})-K_2(P'_{ij})]}_{\geq 0}\Big).$$

Therefore, $\frac{d}{dt}\ln(\cosh\theta) \leq \Delta\ln(\cosh\theta) - \delta(t)\|B\|^2 \leq \Delta\ln(\cosh\theta)$, and by the maximum principle for parabolic equations, $\max_{\Sigma_1}\cosh\theta_t$ is a nondecreasing function on t, and in particular F_t remains a spacelike graph $F_t = \Gamma_{f_t}$ for a smooth map $f_t : \Sigma_1 \to \overline{M}$. On what follows, c_i denotes positive constants. We may take a uniform bound $\delta = \delta(0)$, such that $\lambda_i^2(t) \leq 1 - \delta$ for all t as long as the flow exists. Consequently $g_t = g_1 - f_t^*g_2$ are uniformly equivalent metrics on Σ_1 and Vol_{M_t} are uniformly bounded, and from the above evolution equations $Vol_{M_t} = e^{\int_0^t \|H_s\|^2 ds}Vol_{M_0}$, what implies $\int_0^T \sup_{\Sigma_1}\|H_t\|^2 dt < c_0$. From the evolution equations one gets

$$\frac{d}{dt}\|B\|^2 \leq \Delta\|B\|^2 + c_1\|B\| + c_2\|B\|^2 - \frac{2}{n}\|B\|^4 \leq \Delta\|B\|^2 - \frac{1}{n}\|B\|^4 + c_3. \quad (4)$$

This is the point where regularity theory is better in the pseudo-Riemannian setting than the Riemannian one (note the negative coefficient of the highest power of $\|B\|$, that holds in the pseudo-Riemannian case and not in the Riemannian case). From the above inequality we may use a result of Ecker and Huisken[6] to conclude that $\|B\|^2$ is uniformly bounded. From this inequality we may apply an interpolation formula due to Hamilton and applying parabolic maximum principles we conclude $\|\nabla^k B\|^2$ is uniformly bounded forall k.

For each t, it is defined on $\overline{M}$ a Riemannian metric $\hat{g}_t = \bar{g}_{|T_pM_t} - \bar{g}_{|T_pM_t^\perp}$ that makes e_i, e_α an orthonormal basis. These metrics are uniformly equivalent to the natural Riemannian metric $\bar{g}^+ = g_1 + g_2$ of $\overline{M} = \Sigma_1 \times \Sigma_2$, for we have some positive constants $c(\delta)$ and $c'(\delta)$, depending only on δ, such that $c(\delta)\bar{g}_+ \leq \hat{g} \leq c'(\delta)\bar{g}_+$ holds. We observe that the Levi-Civita connections of $(\overline{M},\bar{g}_+)$ and of $(\overline{M},\bar{g})$ are the same and $\|\bar{\nabla}B\|_{\hat{g}}^2 \leq c_{22}\|B\|^4 + \|\nabla B\|^2$. By induction on k we see that $\bar{\nabla}^k B$ are $\hat{g}$ and so $\bar{g}_+$-uniformly bounded for all $k \geq 0$, that is all derivatives of B in $\overline{M}$ are also bounded for the Riemannian structure. Then we can apply Schauder theory, by embedding isometrically (Σ_i, g_i) into an Euclidean space $\mathbb{R}^{N_i}$. The spaces $C^{k+\sigma}(\Sigma_1, \overline{M})$, $k \in \mathbb{N}$, $0 \le \sigma < 1$ are Banach manifolds and can be seen as closed subsets of the Banach space $C^{k+\sigma}(\Sigma_1, \mathbb{R}^{N_1+N_2})$ with the Hölder norms. Equation (2) in local coordinates is of the form

$$\sum_{ij} a_{ij} \frac{\partial^2 F^a}{\partial x_i \partial x_j} - \sum_k b_k \frac{\partial F^a}{\partial x_k} = \bar{G}(x,t)^a + \frac{dF^a}{dt}$$

where $a_{ij} = g^{ij}$, $b_k = g^{ij}\Gamma_{ij}^k$, $\bar{G}(x,t)^a = (\bar{\Gamma}_{bc}^a \circ F_t)\frac{\partial F^b}{\partial x_i}\frac{\partial F^c}{\partial x_j}$. From the uniform bounds of $\bar{\nabla}^k B$ and of $\bar{\nabla}^k H$ we have that the coefficients a_{ij}, b_j are $C^{k-1+\sigma}(\Sigma_1)$- uniformly bounded, and if F_t lies on a compact set of $\overline{M}$ then

$$\|F(\cdot,t)\|_{C^{1+\sigma}(\Sigma_1,\overline{M})} \le c_{-1}, \quad \|F(\cdot,t)\|_{C^{2+k+\sigma}(\Sigma_1,\overline{M})} \le c_k, \quad k \ge 0$$

for some positive constants c_i that do not depend on t. Standard use of Ascoli-Arzela's theorem to F_t leads to the conclusion that $T = +\infty$ (by assuming $T < +\infty$ one has $F_t = F_0 + \int_0^t H$ lies in a compact set and gets an extension of the maximal solution F_t to $t = T$, what is a contradiction). We also note that the assumption of R_2 and its derivatives to be bounded is necessary to guarantee the existence of a maximal solution of the flow, as well the trick of DeTurck can also be applied in the pseudo-Riemannian case like in the Riemannian case, to reparametrize F_t in a suitable way to convert the above system in one of strictly parabolic equations (see [16] p. 17). This is necessary since the coefficients b_k also depend on the second derivatives of F_t, and so it can give a degenerated system.

Theorem 3.1 ([11]). *The mean curvature flow of the spacelike graph of f remains a spacelike graph of a map $f_t : \Sigma_1 \to \Sigma_2$ and exists for all time $t \ge 0$.*

Since $\int_0^{+\infty} \sup_{\Sigma_1} \|H_t\|^2 dt \le c_{12}$, then $\exists t_N \to +\infty$ such that $H_{t_N} \to 0$. Assuming f_t lies in a compact set of Σ_2 we obtain a subsequence F_{t_n} that C^∞-converges at infinity to a map $F_\infty \in C^\infty(\Sigma_1, \overline{M})$, necessarily a spacelike graph of a map $f_\infty \in C^\infty(\Sigma_1, \Sigma_2)$, and maximal, for $H_\infty = 0$. From Bernstein theorem 2, we conclude

Theorem 3.2 ([11]). *If Σ_2 is also compact there is a sequence $t_n \to +\infty$ such that the sequence $\Gamma_{f_{t_n}}$ of the flow converges at infinity to a spacelike graph Γ_{f_∞} of a totally geodesic map f_∞, and if $Ricci_1(p) > 0$ at some point $p \in \Sigma_1$, the sequence converges to a slice.*

Finally we consider the case $Ricci_1 > 0$ everywhere. In this case we will see that we can droop the compactness assumption of Σ_2. From (3)

$$\frac{d}{dt}\ln(\cosh\theta) \leq \Delta\ln(\cosh\theta) - c_{15}\sum_i \lambda_i^2,$$

what implies $\frac{d}{dt}\ln(\cosh\theta) \leq \Delta\ln(\cosh\theta) - c_{15}(1 - \frac{1}{\cosh^2\theta})$, and consequently,

$$\begin{cases} 1 \leq \max_{\Sigma_1} \cosh\theta \leq 1 + c_{16}e^{-2c_{15}t} \\ \lambda_i^2(p,t) \leq \frac{c_{16}e^{-2c_{15}t}}{(1+c_{16}e^{-2c_{15}t})} \leq c_{16}e^{-2c_{15}t} =: (1-\delta(t)) \end{cases} \quad (5)$$

that is, we have for each t a constant $\delta(t)$ explicitly defined, and that approaches one in an exponentially decreasing way, and

$$\frac{d}{dt}\cosh\theta \leq \Delta\cosh\theta - \delta(t)\cosh\theta\|B\|^2.$$

Setting $p(t) = \frac{1}{\sqrt{nc_{16}}}e^{c_{15}t}$ and $\psi = e^{\frac{1}{2}c_{15}t}\cosh^{p(t)}\theta\|B\|^2$, we have

$$\frac{d}{dt}\psi \leq \Delta\psi - 2\cosh^{-p}\theta\nabla\cosh^p\theta\nabla\psi - c_{17}\left\{e^{\frac{1}{2}c_{15}t}\psi^2 - e^{\frac{1}{4}c_{15}t}\psi^{\frac{1}{2}} - \psi\right\}.$$

In [11] we show this implies $\|B\| \leq c_{18}e^{-\tau t}$, where τ is a positive constant. Since $F_t = F_0 + \int_0^t H$ and the mean curvature is exponentially decreasing we can conclude that $F_t(p)$ lies on a compact region of $\overline{M}$, and for any sequence $t_N \to +\infty$ we obtain a subsequence t_n such that F_{t_n} converges uniformly to a spacelike graph of a map f_∞. By (5) this map must be constant. Furthermore, in this case the limit is the same, for any sequence $t_N \to +\infty$ we take. This gives the next theorem, obtained with no need of using Bernstein results:

Theorem 3.3 ([11]). *If $Ricci_1 > 0$ everywhere and $K_1 \geq K_2$, Σ_2 not necessarily compact, all the flow converges to a unique slice.*

4. Homotopy to a constant map

We will give some applications of theorem 5. We assume in this section Σ_1 is closed and Σ_2 is complete with R_2 bounded and its derivatives. We also assume either $K_1 > 0$ everywhere, or $Ricci_1 > 0$ and $K_2 \leq -c < 0$ everywhere.

Given a constant $\rho > 0$ we consider a new metric $g_1 - g_2'$ on $\overline{M} = \Sigma_1 \times \Sigma_2$ where $g_2' = \rho^{-1}g_2$. Now if $f : \Sigma_1 \to \Sigma_2$ satisfies $f^*g_2 < \rho g_1$, means Γ_f is a timelike submanifold w.r.t. $g_1 - g_2'$. Then the curvature conditions in theorem 5 demands $K_1 \geq \rho K_2$, that can be translated in the following

Theorem 4.1 ([11]). *There exist a constant $0 \leq \rho \leq +\infty$, such that any map $f : \Sigma_1 \to \Sigma_2$ satisfying $f^*g_2 < \rho g_1$ is homotopically trivial. If $K_1 > 0$ everywhere we may take $\rho \leq \min_{\Sigma_1} K_1 / \sup_{K_2} K_2^+$. For $K_2 \leq -c$ everywhere, we may take $\rho = +\infty$.*

Note that, for $Ricci > 0$, $K_2 \leq -c < 0$ everywhere, then $\rho \geq \max_{\Sigma_1} K_1^- / \inf_{\Sigma_2} -K_2$, where $K^- = \max\{-K, 0\}$. This means we may take $\rho = +\infty$ if $K_2 \leq -c$ as in case $\sup_{K_2} K_2^+ = 0$ and $K_1 > 0$. This is the case $n = 1$. The homotopy is given by the flow, namely, since $F_t(p) = (\phi_t(p), f_t(\phi_t(p)))$, where $\phi_t : \Sigma_1 \to \Sigma_1$ is a diffeomorphism with $\phi_0 = id_{\Sigma_1}$, then $K(t, p) = f_t(\phi_t(p))$ is the homotopy. This gives a new proof of the classic Cartan-Hadmard theorem:

Corollary 4.1. *If $K_2 \leq 0$, $m \geq 2$, any map $f : \mathbb{S}^m \to \Sigma_2$ is homotopically trivial.*

The condition given in [14], $det(g_1 + f^*g_2) < 2$ implies $\sum_i \lambda_i^2 + 1 \leq \prod_i (1 + \lambda_i^2) < 2$ and so Γ_f is a spacelike submanifold. The next theorem, obtained in the Riemannian context, can be seen as a reformulated corollary of theorem 5:

Theorem 4.2 ([13,14]). *Assume both Σ_i are closed and with constant sectional curvature K_i and satisfying $K_1 \geq |K_2|$, $K_1 + K_2 > 0$.*
(1) If $det(g_1 + f^*g_2) < 2$, then Γ_f can be deformed by a family of graphs to the one of a constant map.
(2) If f is an area decreasing map, that is $\lambda_i \lambda_j < 1$ for $i \neq j$, then it is homotopically trivial.

The area decreasing condition is a slightly more general condition than spacelike graph for $n \geq 2$. In case $n = 1$ any map is area decreasing, but it is included in the case $K_2 \leq 0$. We note that in the previous theorem it is used the Riemannian structure, and in this setting K_2 cannot be given arbitrarily negative, a somehow artificial condition, that can be dropped if one uses the pseudo-Riemannian structure of the product.

Acknowledgements

The first author is partially supported by NSFC (No.10501011) and by Fundação Ciência e Tecnologia (FCT) through a FCT fellowship SFRH/BPD/26554/2006. The second author is partially supported by FCT through the Plurianual of CFIF and POCI-PPCDT/MAT/60671/2004.

References

1. A. Albujer A. and L. Alías, arXiv:math/0709.4363, 2007
2. E. Calabi, *Proc. Sympos. Pure Math.* **15**(1970), 223–230.
3. S. Cheng and S. T. Yau, *Comm. Pure Appl. Math.* **28**(1975), 333–354.
4. J. Cheeger and D.G. Ebin, *Comparison theorems in Riemannian geometry.* North-Holland Mathematical Library, Vol. 9. North-Holland Publishing Co., Amsterdam-Oxford; American Elsevier Publishing Co., Inc., New York, 1975.
5. S.S. Chern, *Enseignement Math. II. Sér* **15** (1969), 53-61.
6. K. Ecker and G. Huisken, *Comm. Math. Phys.* **135** (1991), no. 3, 595–613.
7. J. Eells and J.H. Sampson, *Amer. J. Math.* **86** (1964), 109–160.
8. M. Gromov, *Partial differential relations.* Ergebnisse der Mathematik und ihrer Grenzgebiete (3) [Results in Mathematics and Related Areas (3)], 9. Springer-Verlag, Berlin, 1986.
9. J. Jost and Y. Xin, *Results Math.* **40**(2001), 233–245.
10. G. Li and I.M.C. Salavessa, Arxiv.0801.3850.
11. G. Li and I.M.C. Salavessa, Arxiv.0804.0783.
12. I.M.C. Salavessa, *Bull. Bel. Math. Soc.* **15** (2008), 65-76.
13. M-P. Tsui, and M-T. Wang, *Comm. Pure Appl. Math.* **57** (2004), 1110–1126.
14. M-T. Wang, *Invent. Math.* **148** (2002), 525-543.
15. Y. Xin, *Chin. Ann. Math. Ser. B* **29** (2008), 121–134.
16. X-P. Zhu, *Lectures on mean curvature flows.* AMS/IP Studies in advanced mathematics, **32**, American Mathematical Society, Providence, RI; International Press, (2002).

RIEMANNIAN GEOMETRIC REALIZATIONS FOR RICCI TENSORS OF GENERALIZED ALGEBRAIC CURVATURE OPERATORS

P. Gilkey

Mathematics Department, University of Oregon
Eugene OR 97403 USA
E-mail: gilkey@uoregon.edu

S. Nikčević

Mathematical Institute, Sanu, Knez Mihailova 35, p.p. 367
11001 Belgrade, Serbia
E-mail: stanan@mi.sanu.ac.rs

D. Westerman

Mathematics Department, University of Oregon
Eugene OR 97403 USA
E-mail: dwesterm@uoregon.edu

We examine questions of geometric realizability for algebraic structures which arise naturally in affine and Riemannian geometry.

Keywords: Constant scalar curvature, geometric realization, generalized algebraic curvature operator, Ricci tensor, Ricci antisymmetric, Ricci flat, Ricci symmetric, Ricci trace free

Mathematics Subject Classification 2000: 53B20

1. Introduction

Many questions in Riemannian geometry involve constructing geometric realizations of algebraic objects where the objects in question are invariant under the action of the structure group G. We present several examples to illustrate this point. We first review previously known results. Section 1.1 deals with Riemannian algebraic curvature tensors, Section 1.2 deals with Osserman tensors, and Section 1.3 deals with generalized algebraic curvature operators.

In Section 1.4 we present the new results of this paper that deal with a mixture of affine and Riemannian geometry; this mixture has not been considered previously. The results in the real analytic context can perhaps be considered as extensions of previous results in affine geometry; the results in the C^s context are genuinely new and require additional estimates. We refer to Section 1.4 for further details. To simplify the discussion, we shall assume that the underlying dimension m is at least 3 as the 2-dimensional case is a bit exceptional. We adopt the *Einstein convention* and sum over repeated indices henceforth.

1.1. Realizing Riemannian algebraic curvature tensors

Let V be an m-dimensional real vector space and let $\mathfrak{r}(V) \subset \otimes^4 V^*$ be the set of all *Riemannian algebraic curvature tensors*; $A \in \mathfrak{r}(V)$ if and only if A has the symmetries of the Riemannian curvature tensor of the Levi-Civita connection:

$$A(x,y,z,w) = -A(y,x,z,w), \quad A(x,y,z,w) = A(z,w,x,y),$$
$$A(x,y,z,w) + A(y,z,x,w) + A(z,x,y,w) = 0. \tag{1}$$

Let $A \in \mathfrak{r}(V)$ and let $\langle \cdot, \cdot \rangle$ be a non-degenerate symmetric bilinear form on V of signature (p,q). The triple $\mathfrak{M} := (V, \langle \cdot, \cdot \rangle, A)$ is said to be a *pseudo-Riemannian algebraic curvature model*; let $\Xi(V)$ be the set of such models.

Let $\mathcal{M} := (M, g)$ be a pseudo-Riemannian manifold. Let ∇^g be the associated Levi-Civita connection and let $R_P^g \in \otimes^4 T_P^* M$ be the curvature tensor at a point P of M. Since R_P^g satisfies the symmetries of Equation (1), $\mathfrak{M}_P(\mathcal{M}) := (T_P M, g_P, R_P^g) \in \Xi(T_P M)$.

The following result shows every $\mathfrak{M} \in \Xi(V)$ is geometrically realizable; in particular, the symmetries of Equation (1) generate the universal symmetries of the curvature tensor of the Levi-Civita connection.

Theorem 1.1. *Let $\mathfrak{M} \in \Xi(V)$. There exists a pseudo-Riemannian manifold $\mathcal{M}$, a point $P \in M$, and an isomorphism ϕ from $T_P M$ to V so that $\mathfrak{M}_P(\mathcal{M}) = \phi^* \mathfrak{M}$.*

1.2. Osserman geometry

The relevant structure group which arises in this context is the orthogonal group $O(V, \langle \cdot, \cdot \rangle)$; one can ask geometric realization questions concerning any $O(V, \langle \cdot, \cdot \rangle)$ invariant subset of $\mathfrak{r}(V)$. If $\mathfrak{M} = (V, \langle \cdot, \cdot \rangle, A) \in \Xi(V)$, the *Jacobi operator* $\mathcal{J}_{\mathfrak{M}} \in \mathrm{End}(V) \otimes V^*$ is characterized by the relation:

$$\langle \mathcal{J}_{\mathfrak{M}}(x)y, z \rangle = A(y, x, x, z).$$

If $p > 0$, then $\mathfrak{M}$ is said to be *timelike Osserman* if the spectrum of $\mathcal{J}_\mathfrak{M}$ is constant on the pseudo-sphere of unit timelike vectors in V. The notion *spacelike Osserman* is defined similarly if $q > 0$. If $p > 0$ and if $q > 0$, work of N. Blažić et al.[1] and of García-Río et al.[2] shows these two notions are equivalent and thus we shall simply say $\mathfrak{M}$ is *Osserman* in this context. As this definition is invariant under the action of the structure group $O(V, \langle \cdot, \cdot \rangle)$, it extends to the geometric setting. Thus a pseudo-Riemannian manifold $\mathcal{M}$ will be said to be *Osserman* provided that the associated model $\mathfrak{M}_P(\mathcal{M})$ is Osserman for every $P \in M$.

Work of Chi[3] shows there are 4-dimensional Osserman Riemannian algebraic curvature tensors which are not geometrically realizable by Osserman manifolds. The field is a vast one and we refer to Nikolayevsky[4] for further details in the Riemannian setting and to García-Río et al.[5] for a discussion in the pseudo-Riemannian setting; it is possible to construct many examples of Osserman tensors in the algebraic context which have no corresponding geometrical analogues.

1.3. Affine geometry

Let ∇ be a torsion free connection on M. The associated curvature operator $\mathcal{R} \in T^*M \otimes T^*M \otimes \mathrm{End}(TM)$ is a $(3,1)$ tensor which has the symmetries

$$\mathcal{R}(x,y)z = -\mathcal{R}(y,x)z, \quad \mathcal{R}(x,y)z + \mathcal{R}(y,z)x + \mathcal{R}(z,x)y = 0. \quad (2)$$

As we are in the affine setting, there is no analogue of the additional curvature symmetry $A(x,y,z,w) = A(z,w,x,y)$ which appears in the pseudo-Riemannian setting. In the algebraic context, let $\mathfrak{A}(V) \subset V^* \otimes V^* \otimes \mathrm{End}(V)$ be the set of $(3,1)$ tensors satisfying the relations of Equation (2). An element $\mathcal{A} \in \mathfrak{A}(V)$ is said to be a *generalized algebraic curvature operator*.

If ∇ is a torsion free connection and if $P \in M$, then $\mathcal{R}_P^\nabla \in \mathfrak{A}(T_PM)$. The following geometric realizability result is closely related to Theorem 1.1. It shows that any universal symmetry of the curvature tensor of an affine connection is generated by the symmetries of Equation (2).

Theorem 1.2. *Let $\mathcal{A} \in \mathfrak{A}(V)$. There exists a torsion free connection ∇ on a smooth manifold M, a point $P \in M$, and an isomorphism ϕ from T_PM to V so that $\mathcal{R}_P^\nabla = \phi^* \mathcal{A}$.*

We contract indices to define the *Ricci tensor* $\rho(\mathcal{A}) \in V^* \otimes V^*$ by setting

$$\rho(\mathcal{A})(x,y) := \mathrm{Trace}\{z \to \mathcal{A}(z,x)y\}.$$

The decomposition $V^* \otimes V^* = \Lambda^2(V^*) \oplus S^2(V^*)$ sets $\rho(\mathcal{A}) = \rho_a(\mathcal{A}) + \rho_s(\mathcal{A})$ where $\rho_a(\mathcal{A})$ and $\rho_s(\mathcal{A})$ are the antisymmetric and symmetric Ricci tensors. The natural structure group for $\mathfrak{A}(V)$ is the general linear group $GL(V)$. The Ricci tensor defines a $GL(V)$ equivariant short exact sequence

$$0 \to \ker(\rho) \to \mathfrak{A}(V) \to V^* \otimes V^* \to 0.$$

Strichartz[6] showed this short exact sequence is $GL(V)$ equivariantly split and gives a $GL(V)$ equivariant decomposition

$$\mathfrak{A}(V) = \ker(\rho) \oplus \Lambda^2(V^*) \oplus S^2(V^*)$$

into irreducible $GL(V)$ modules. The *Weyl projective curvature operator* $\mathcal{P}(\mathcal{A})$ is the projection of $\mathcal{A}$ on $\ker(\rho)$; $\mathcal{A}$ is said to be *projectively flat* if $\mathcal{P}(\mathcal{A}) = 0$, $\mathcal{A}$ is said to be *Ricci symmetric* if $\rho_a(\mathcal{A}) = 0$, and $\mathcal{A}$ is said to be *Ricci antisymmetric* if $\rho_s(\mathcal{A}) = 0$. These notions for a connection are defined similarly. There are 8 natural geometric realization questions which arise in this context and whose realizability[7] may be summarized in the following table – the possibly non-zero components being indicated by $\star$:

$\ker(\rho)$	$S^2(V^*)$	$\Lambda^2(V^*)$		$\ker(\rho)$	$S^2(V^*)$	$\Lambda^2(V^*)$	
$\star$	$\star$	$\star$	yes	0	$\star$	$\star$	yes
$\star$	$\star$	0	yes	0	$\star$	0	yes
$\star$	0	$\star$	yes	0	0	$\star$	no
$\star$	0	0	yes	0	0	0	yes

Thus, for example, if $\mathcal{A}$ is projectively flat and Ricci symmetric, then $\mathcal{A}$ can be geometrically realized by a projectively flat Ricci symmetric torsion free connection. But if $\mathcal{A} \neq 0$ is projectively flat and Ricci antisymmetric, then $\mathcal{A}$ can not be geometrically realized by a projectively flat Ricci antisymmetric torsion free connection.

1.4. *Torsion free connections and Riemannian geometry*

We now combine the settings of Sections 1.1 and 1.3. Let $\langle \cdot, \cdot \rangle$ be a non-degenerate symmetric inner product on V of signature (p, q). Fix a basis $\{e_i\}$ for V and let $g_{ij} := \langle e_i, e_j \rangle$ give the components of $\langle \cdot, \cdot \rangle$. Let g^{ij} be the inverse matrix. If $\mathcal{A} \in \mathfrak{A}(V)$, expand $\mathcal{A}(e_i, e_j)e_k = \mathcal{A}_{ijk}{}^\ell e_\ell$. The scalar curvature τ and trace free Ricci tensor are then given, respectively, by

$$\tau(\mathcal{A}, \langle \cdot, \cdot \rangle) := g^{ij} \mathcal{A}_{kij}{}^k, \quad \rho_0(\mathcal{A}, \langle \cdot, \cdot \rangle) := \rho_s(\mathcal{A}) - \tfrac{\tau(\mathcal{A}, \langle \cdot, \cdot \rangle)}{m} \langle \cdot, \cdot \rangle.$$

Let $S_0^2(V^*, \langle \cdot, \cdot \rangle)$ be the space of trace free symmetric bilinear forms. One has an $O(V, \langle \cdot, \cdot \rangle)$ invariant decomposition of $V^* \otimes V^*$ into irreducible

$O(V, \langle \cdot, \cdot \rangle)$ modules

$$V^* \otimes V^* = \Lambda^2(V^*) \oplus S_0^2(V^*, \langle \cdot, \cdot \rangle) \oplus \mathbb{R}.$$

This decomposition leads to 8 geometric realization questions which are natural with respect to the structure group $O(V, \langle \cdot, \cdot \rangle)$ and which can all be solved either in the real analytic category or in the C^s category of s-times differentiability for any $s \geq 1$. The following is the main result of this paper; as our considerations are local, we take $M = V$ and $P = 0$.

Theorem 1.3. *Let g be a C^s (resp. real analytic) pseudo-Riemannian metric on V. Let $\mathcal{A} \in \mathfrak{A}(V)$. There exists a torsion free C^s (resp. real analytic) connection ∇ defined on a neighborhood of 0 in V such that:*

(a) $\mathcal{R}_0^\nabla = \mathcal{A}$.
(b) ∇ has constant scalar curvature.
(c) If $\mathcal{A}$ is Ricci symmetric, then ∇ is Ricci symmetric.
(d) If $\mathcal{A}$ is Ricci antisymmetric, then ∇ is Ricci antisymmetric.
(e) If $\mathcal{A}$ is Ricci traceless, then ∇ is Ricci traceless.

The subspace $\ker(\rho) \subset \mathfrak{A}(V)$ is not an irreducible $O(V, \langle \cdot, \cdot \rangle)$ module but decomposes as the direct sum of 5 additional irreducible factors – see Bokan.[8] This decomposition will play no role in our further discussion and studying the additional realization questions which arise from this decomposition is a topic for future investigation.

2. The proof of Theorem 1.3

We assume $s \geq 1$ and $m \geq 3$ henceforth; fix $\mathcal{A} \in \mathfrak{A}(V)$. We introduce the following notational conventions. Choose a basis $\{e_i\}$ for V to identify $M = V = \mathbb{R}^m$ and let $\{x_1, ..., x_m\}$ be the associated coordinates.

For $\delta > 0$, let $B_\delta := \{x \in \mathbb{R}^m : |x| < \delta\}$ where $|x|$ is the usual Euclidean norm on $\mathbb{R}^m$. Let C_δ^s be the set of functions on B_δ which are s-times differentiable. Let $\alpha = (\alpha_1, ..., \alpha_m)$ be a multi-index. Set

$$\partial_i := \tfrac{\partial}{\partial x_i}, \quad \partial_x^\alpha := (\partial_1)^{\alpha_1}...(\partial_m)^{\alpha_m}, \quad |\alpha| = \alpha_1 + ... + \alpha_m.$$

If $\mathfrak{z}$ is a real vector space, let $C_\delta^s(\mathfrak{z})$ be the set of C^s functions on B_δ with values in $\mathfrak{z}$. Fix a basis $\{f_\sigma\}$ for $\mathfrak{z}$ and expand $P \in C_\delta^s(\mathfrak{z})$ as $P = P^\sigma f_\sigma$ for $P^\sigma \in C_\delta^s$. Let $\nu \in \mathbb{R}$. Set

$$|P| := \sup_\sigma |P^\sigma| \in C_\delta^0 \quad \text{and} \quad ||P||_{\delta,\nu,-1} := 0.$$

For $0 \leq r \leq s$, define $\|P\|_{\delta,\nu,r} \in [0,\infty]$ by setting

$$\|P\|_{\delta,\nu,r} := \sup_{|\alpha|=r,\ |x|<\delta} |\partial_x^\alpha P(x)| \cdot |x|^{-\nu}.$$

Thus $\|P\|_{\delta,\nu,r} \leq C$ implies $|\partial_x^\alpha P(x)| \leq C|x|^\nu$ for $|\alpha| = r$ and $|x| < \delta$. Let

$$\mathfrak{G} := S^2((\mathbb{R}^m)^*) \otimes \mathbb{R}^m, \quad \mathfrak{S} := S^2((\mathbb{R}^m)^*), \quad \text{and} \quad \mathfrak{A} := \mathfrak{A}(\mathbb{R}^m).$$

We use the basis $\{e_i\}$ and the coordinate frame $\{\partial_i\}$ to determine the components of tensors of all types; if computing relative to some orthonormal frame $\{E_i\}$, we shall make this explicit. Thus, for example, if $\mathcal{S} \in \mathfrak{S}$, then $\mathcal{S}_{ij} = \mathcal{S}(e_i, e_j)$ while if $\mathcal{S} \in C^s_\delta(\mathfrak{S})$, then $\mathcal{S}_{ij} := \mathcal{S}(\partial_{x_i}, \partial_{x_j})$. If $\Gamma, \mathcal{E} \in C^s_\delta(\mathfrak{G})$, define $\mathcal{L}(\Gamma) \in C^{s-1}(\mathfrak{A})$ and $\Gamma \star \mathcal{E} \in C^s_\delta(\mathfrak{A})$ by setting

$$\begin{aligned}\mathcal{L}(\Gamma)_{ijk}{}^l &:= \partial_i \Gamma_{jk}{}^l - \partial_j \Gamma_{ik}{}^l, \\ (\Gamma \star \mathcal{E})_{ijk}{}^\ell &:= \mathcal{E}_{in}{}^\ell \Gamma_{jk}{}^n + \Gamma_{in}{}^\ell \mathcal{E}_{jk}{}^n - \mathcal{E}_{jn}{}^\ell \Gamma_{ik}{}^n - \Gamma_{jn}{}^\ell \mathcal{E}_{ik}{}^n.\end{aligned} \quad (3)$$

If $\Gamma \in C^s_\delta(\mathfrak{G})$, let $\nabla(\Gamma)$ be the C^s torsion free connection on B_δ with Christoffel symbol Γ. One has:

$$\begin{aligned}\mathcal{R}^{\nabla(\Gamma)} &= \mathcal{L}(\Gamma) + \tfrac{1}{2} \Gamma \star \Gamma, \\ \rho(\Gamma \star \Gamma)_{jk} &= 2\Gamma_{\ell n}{}^\ell \Gamma_{jk}{}^n - 2\Gamma_{jn}{}^\ell \Gamma_{\ell k}{}^n = \rho(\Gamma \star \Gamma)_{kj}, \\ \rho_a(\mathcal{R}^{\nabla(\Gamma)})_{jk} &= \rho_a(\mathcal{L}(\Gamma))_{jk} = \tfrac{1}{2}\left\{\partial_k \Gamma_{ji}{}^i - \partial_j \Gamma_{ki}{}^i\right\}.\end{aligned} \quad (4)$$

One says that Γ is *normalized* if

(1) $\Gamma(0) = 0$ and $\mathcal{R}^{\nabla(\Gamma)} = \mathcal{A} + O(|x|^2)$.
(2) $\rho_s(\mathcal{R}^{\nabla(\Gamma)})$ is C^s.
(3) $\rho_a(\mathcal{R}^{\nabla(\Gamma)})(\partial_i, \partial_j) = \rho_a(\mathcal{A})(e_i, e_j)$ on B_δ.

We remark that Assertion (2) is non-trivial as $\mathcal{R}^{\nabla(\Gamma)}$ need only be C^{s-1}. This is a technical condition used subsequently to avoid loss of smoothness.

Theorem 1.2 follows from the following observation which forms the starting point in our proof of Theorem 1.3:

Lemma 2.1. *If* $\Gamma_{uv}{}^l := \tfrac{1}{3}(\mathcal{A}_{wuv}{}^l + \mathcal{A}_{wvu}{}^l)x^w$, *then* Γ *is normalized.*

Proof. Since $\Gamma(0) = 0$, one has:

$$\begin{aligned}\mathcal{R}_0^{\nabla(\Gamma)}(\partial_i, \partial_j)\partial_k &= \left\{\partial_i \Gamma_{jk}{}^l(0) - \partial_j \Gamma_{ik}{}^l(0)\right\} \partial_\ell \\ &= \tfrac{1}{3}\left\{\mathcal{A}_{ijk}{}^l + \mathcal{A}_{ikj}{}^l - \mathcal{A}_{jik}{}^l - \mathcal{A}_{jki}{}^l\right\} \partial_\ell \\ &= \tfrac{1}{3}\left\{\mathcal{A}_{ijk}{}^l - \mathcal{A}_{kij}{}^l + \mathcal{A}_{ijk}{}^l - \mathcal{A}_{jki}{}^l\right\} \partial_\ell = \mathcal{A}_{ijk}{}^l \partial_\ell.\end{aligned}$$

By Equation (4), $\rho_a(\mathcal{R}^{\nabla(\Gamma)})_{ij} = \rho_a(\mathcal{L}(\Gamma))_{ij} = \rho_a(\mathcal{R}_0^{\nabla(\Gamma)})_{ij} = \rho_a(\mathcal{A})_{ij}$. □

We continue our analysis with the following basic solvability result:

Lemma 2.2. *If $\Theta \in C^s_\delta(\mathfrak{S})$, then there exists $\mathcal{E} \in C^s_\delta(\mathfrak{G})$ so $\rho(\mathcal{L}(\mathcal{E})) = \Theta$, so $\mathcal{E}_{ij}{}^j = 0$, and so $||\mathcal{E}||_{\delta,\nu+1,r} \leq ||\Theta||_{\delta,\nu,r} + r||\Theta||_{\delta,\nu+1,r-1}$.*

Proof. By assumption, $m \geq 3$. For each pair of indices $\{i,j\}$, not necessarily distinct, choose $k = k(i,j) = k(j,i)$ with $k \neq i$ and $k \neq j$. Set

$$\mathcal{E}_{ij}{}^\ell := \begin{cases} \int_0^{x_k} \Theta_{ij}(x_1,...,x_{k-1},u,x_{k+1},...,x_m)du & \text{if } \ell = k, \\ 0 & \text{if } \ell \neq k. \end{cases}$$

Since $k \neq j$, $\mathcal{E}_{ij}{}^j = 0$. Consequently Equation (3) yields

$$\rho(\mathcal{L}(\mathcal{E}))_{ij} = \partial_\ell \mathcal{E}_{ij}{}^\ell = \Theta_{ij}.$$

Expand $\partial_x^\alpha = \partial_k^\mu \partial_x^\beta$ where β does not involve the index k. Then:

$$\partial_x^\alpha \mathcal{E}_{ij}{}^k = \begin{cases} \int_0^{x_k} \partial_x^\alpha \Theta_{ij}(x_1,...,x_{k-1},u,x_{k+1},...,x_m)du & \text{if } \mu = 0, \\ \mu \partial_k^{\mu-1} \partial_x^\beta \Theta_{ij} & \text{if } \mu > 0. \end{cases}$$

Assume that $|\partial_x^\alpha \Theta_{ij}(x)| \leq C|x|^\nu$ for all $x \in B_\delta$ and all $|\alpha| = j$. Then

$$|\int_0^{x_k} \partial_x^\alpha \Theta_{ij}(x_1,...,x_{k-1},u,x_{k+1},...,x_m)du|$$
$$\leq |x_k| \int_0^1 |\partial_x^\alpha \Theta_{ij}(x_1,...,x_{k-1},tx_k,x_{k+1},...,x_m)|dt$$
$$\leq |x_k| \cdot C|x|^\nu \leq C|x|^{\nu+1}.$$

The estimates of the Lemma now follow. $\square$

Let g be a C^s pseudo-Riemannian metric on B_δ for $\delta < 1$, let $\{E_i\}$ be a C^s g-orthonormal frame for the tangent bundle of B_δ, and let $e_i := E_i(0)$. Let $\Gamma \in C^s_\delta(\mathfrak{G})$. Define $\Theta = \Theta(\Gamma) \in C^s_\delta(\mathfrak{S})$ by:

$$\Theta_{ij} := \rho_s(\mathcal{R}^{\nabla(\Gamma)})(E_i, E_j) - \rho_s(\mathcal{A})(e_i, e_j). \tag{5}$$

Use Lemma 2.2 to define $\mathcal{E} = \mathcal{E}(\Gamma) \in C^s_\delta(\mathfrak{G})$ so that $\rho_s(\mathcal{L}(\mathcal{E})) = -\Theta$. We use Lemma 2.1 to choose an initial Christoffel symbol $\Gamma_1 \in C^s_\delta(\mathfrak{G})$ which is normalized. Inductively, set

$$\Theta_\nu := \Theta(\Gamma_\nu), \quad \mathcal{E}_{\nu+1} := \mathcal{E}(\Gamma_\nu), \quad \Gamma_{\nu+1} := \Gamma_\nu + \mathcal{E}_{\nu+1}.$$

We will set $\Gamma_\infty := \Gamma_1 + \mathcal{E}_2 + ...$, we will establish convergence, and we will show Γ_∞ defines a connection with the desired properties. We begin by using Equation (4) to compute:

$$\Theta_{\nu+1,ij} = \rho_s(\mathcal{R}_\nu)(E_i, E_j) - \rho_s(\mathcal{A})(e_i, e_j) + \rho_s(\mathcal{L}(\mathcal{E}_{\nu+1}))_{ij}$$
$$+ \rho_s(\mathcal{L}(\mathcal{E}_{\nu+1}))(E_i, E_j) - \rho_s(\mathcal{L}(\mathcal{E}_{\nu+1}))_{ij}$$
$$+ \rho_s\{(\Gamma_\nu + \tfrac{1}{2}\mathcal{E}_{\nu+1}) \star \mathcal{E}_{\nu+1}\}(E_i, E_j).$$

As $\rho_s(\mathcal{L}(\mathcal{E}_{\nu+1})) = -\Theta_\nu$, the first line vanishes and

$$\Theta_{\nu+1,ij} = -\Theta_\nu(E_i, E_j) + \Theta_{\nu,ij} + \rho_s\{(\Gamma_\nu + \tfrac{1}{2}\mathcal{E}_{\nu+1}) \star \mathcal{E}_{\nu+1}\}(E_i, E_j). \quad (6)$$

Choose $\kappa \geq 1$ so we have the following estimates for any $x \in B_\delta$:

$$|\mathcal{S}(E_i, E_j) - \mathcal{S}_{ij}| \leq \kappa|\mathcal{S}| \cdot |x|^2 \quad \forall \ \mathcal{S} \in C_\delta^0(\mathfrak{S}),$$
$$|\rho_s(\Gamma \star \mathcal{E})(E_i, E_j)| \leq \kappa\{|\Gamma| \cdot |\mathcal{E}|\} \quad \forall \ \Gamma, \mathcal{E} \in C_\delta^0(\mathfrak{S}).$$

Lemma 2.3. *Adopt the notation established above. Then Γ_ν is normalized for all ν. Furthermore, there exists $\delta_0 > 0$ and there exist constants $C_r > 0$ for $0 \leq r \leq s$ so for $\nu = 1, 2, \ldots$ we have the estimates:*

(a) $\|\Gamma_\nu\|_{\delta_0, 1-r, r} \leq \tfrac{1}{4\kappa} C_r$.
(b) $\|\Theta_\nu\|_{\delta_0, 2\nu-r, r} \leq C_r^\nu$.
(c) $\|\mathcal{E}_{\nu+1}\|_{\delta_0, 2\nu+1-r, r} \leq C_r^\nu + r C_{r-1}^\nu$.

Proof. By assumption Γ_1 is normalized. We assume inductively Γ_ν is normalized and show $\Gamma_{\nu+1}$ is normalized. As $\mathcal{E}_{\nu+1,ij}{}^j = 0$, Equation (4) yields

$$\rho_a(\mathcal{R}_{\nu+1})_{ij} = \rho_a(\mathcal{R}_\nu)_{ij} = \rho_a(\mathcal{A})_{ij} \quad \text{on} \quad B_\delta. \quad (7)$$

Since $\Theta_\nu = \rho_s(\mathcal{R}_\nu)(E_i, E_j) - \rho_s(\mathcal{A})_{ij} = O(|x|^2)$, $\mathcal{E}_{\nu+1} = O(|x|^3)$ and

$$\mathcal{R}_{\nu+1} = \mathcal{R}_\nu + O(|x|^2) = \mathcal{A} + O(|x|^2).$$

As Γ_ν is normalized, $\Theta_\nu \in C_\delta^s(\mathfrak{S})$. Hence $\mathcal{E}_{\nu+1} \in C_\delta^s(\mathfrak{S})$ and $\Gamma_{\nu+1} \in C_\delta^s(\mathfrak{S})$. Since $\rho_s(\mathcal{L}(\mathcal{E}_{\nu+1})) = -\Theta_\nu$ is C^s, we may conclude that $\rho_s(\mathcal{R}_{\nu+1})$ is C^s even though $\mathcal{R}_{\nu+1}$ need only be C^{s-1}. Thus $\Gamma_{\nu+1}$ is normalized.

We establish the estimates by induction on r and then on ν; Assertion $(3)_{\nu,r}$ follows from Assertions $(2)_{\nu,r}$ and $(2)_{\nu,r-1}$ and from Lemma 2.2. Suppose first that $r = 0$; this is, somewhat surprisingly, the most difficult case. As Γ_1 is normalized, $\mathcal{R}_1 = \mathcal{A} + O(|x|^2)$. One has $E_i(0) = \partial_i$. Thus $\Theta_1 = O(|x|^2)$. As $\Gamma_1 = O(|x|)$, by shrinking δ, we may choose $\bar{C}_0$ so

$$|\Gamma_1|(x) \leq \bar{C}_0|x| \quad \text{and} \quad |\Theta_1|(x) \leq \bar{C}_0|x|^2 \quad \text{on} \quad B_\delta.$$

Choose C_0 and $\delta_0 < \delta < 1$ so that

$$\bar{C}_0 + 1 < \tfrac{1}{4\kappa} C_0 < C_0, \quad \kappa + \tfrac{1}{4} C_0 + \tfrac{1}{2}\kappa \leq C_0, \quad \delta_0^2 C_0 < \tfrac{1}{2}.$$

If $\nu = 1$, then Assertions (1) and (2) follow from the choices made. Assume the Assertions hold for $\mu \leq \nu$ where $\nu \geq 1$. Then

$$|\Gamma_{\nu+1}| \leq |\Gamma_1| + |\mathcal{E}_2| + \ldots + |\mathcal{E}_{\nu+1}| \leq \bar{C}_0|x| + C_0|x|^3 + C_0^2|x|^5 + \ldots$$
$$\leq \bar{C}_0|x| + |x|\tfrac{C_0|x|^2}{1-C_0|x|^2} \leq (\bar{C}_0 + 1)|x| \leq \tfrac{1}{4\kappa} C_0.$$

We use Equation (6) to complete the induction step for $r = 0$ by checking

$$|\Theta_{\nu+1}| \leq \kappa\{|x|^2 \cdot |\Theta_\nu| + (|\Gamma_\nu| + |\mathcal{E}_{\nu+1}|)|\mathcal{E}_{\nu+1}|\}$$
$$\leq \kappa\{C_0^\nu |x|^{2\nu+2} + (\tfrac{1}{4\kappa}C_0|x| + C_0^\nu|x|^{2\nu+1})C_0^\nu|x|^{2\nu+1}\}$$
$$\leq C_0^\nu|x|^{2\nu+2}\{\kappa + \tfrac{1}{4}C_0 + \tfrac{1}{2}\kappa\} \leq C_0^{\nu+1}|x|^{2\nu+2}.$$

We now suppose $r = 1$; we get 1 less power of $|x|$ in the decay estimates. We choose $\bar{C}_1$ so $|\partial_k \Gamma_1| \leq \bar{C}_1$ and $|\partial_k \Theta_1| \leq \bar{C}_1|x|$; the desired estimates then hold for $\nu = 1$ for C_1 sufficiently large. We proceed by induction on ν. We then have for sufficiently large C_1 and small δ_0 that:

$$|\partial_k \Gamma_{\nu+1}| \leq |\partial_k \Gamma_1| + |\partial_k \mathcal{E}_2| + \ldots + |\partial_k \mathcal{E}_{\nu+1}|$$
$$\leq \bar{C}_1 + \{C_1 + C_0\}|x|^2 + \{C_1^2 + C_0^2\}|x|^4 + \ldots \leq \tfrac{1}{4\kappa}C_1.$$

We differentiate Equation (6) to obtain

$$\partial_k \Theta_{\nu+1,ij} = -(\partial_k \Theta_\nu)(E_i, E_j) + \partial_k \Theta_{\nu,ij} - \Theta_\nu(\partial_k E_i, E_j) - \Theta_\nu(E_i, \partial_k E_j)$$
$$+ \rho_s\{(\partial_k \Gamma_\nu + \tfrac{1}{2}\partial_k \mathcal{E}_{\nu+1}) \star \mathcal{E}_{\nu+1} + (\Gamma_\nu + \tfrac{1}{2}\mathcal{E}_{\nu+1}) \star \partial_k \mathcal{E}_{\nu+1}\}(E_i, E_j)$$
$$+ \rho_s\{(\Gamma_\nu + \tfrac{1}{2}\mathcal{E}_{\nu+1}) \star \mathcal{E}_{\nu+1}\}\{(\partial_k E_i, E_j) + (E_i, \partial_k E_j)\}.$$

Thus for a suitably chosen constant $\kappa_1 = \kappa_1(\mathcal{A}, E, \Gamma_1)$ which is independent of ν and for suitably chosen $C_1 > C_0$, we have

$$|\partial_k \Theta_{\nu+1}| \leq \kappa_1\{|x|^2|\partial_k \Theta_\nu| + |x| \cdot |\Theta_\nu| + (|\partial_k \Gamma_\nu| + |\partial_k \mathcal{E}_{\nu+1}|)|\mathcal{E}_{\nu+1}|$$
$$+ (|\Gamma_\nu| + |\mathcal{E}_{\nu+1}|) \cdot |\partial_k \mathcal{E}_{\nu+1}| + (|\Gamma_\nu| + |\mathcal{E}_{\nu+1}|) \cdot |\mathcal{E}_{\nu+1}|\}$$
$$\leq \kappa_1 |x|^{2\nu+1}\{C_1^\nu + C_0^\nu + (\tfrac{1}{4\kappa}C_1 + (C_1^\nu + C_0^\nu)|x|^{2\nu})C_0^\nu$$
$$+ (\tfrac{1}{4\kappa}C_0 + C_0^\nu|x|^{2\nu})(C_1^\nu + C_0^\nu) + (\tfrac{1}{4\kappa}C_0|x| + C_0^\nu|x|^{2\nu+1})C_0^\nu\}.$$

A crucial point is that there are no $C_1^{\nu+1}$ terms present. The desired estimate now follows for C_1 sufficiently large and δ_0 sufficiently small. This completes the proof of the case $r = 1$; the higher order derivatives are estimated similarly. □

Since $|\partial_x^\alpha \mathcal{E}_{\nu+1}| \leq C_r^\nu |x|^{2\nu+1-r}$, the series $\mathcal{E}_2 + \mathcal{E}_3 + \ldots$ converges geometrically for small x and thus the sequence Γ_ν converges in the C^r topology to a limit Γ_∞. Note that as we have to shrink δ at each stage, we do not get convergence in the C^∞ topology even if the initial metric is smooth. We use Equation (5) to see that for small x we have:

$$\rho_a(\mathcal{R}_\infty(x))_{ij} = \lim_{\nu \to \infty} \rho_a(\mathcal{R}_\nu(x))_{ij} = \rho_a(\mathcal{A})_{ij}. \tag{8}$$

This controls the antisymmetric part of the Ricci tensor. To control the symmetric part of the Ricci tensor, we use the g-orthonormal frame $\{E_i\}$. We compute, using Equation (7), that:

$$\rho_s(\mathcal{R}_\infty(x))(E_i, E_j) = \lim_{\nu \to \infty} \rho_s(\mathcal{R}_\nu(x))(E_i, E_j) \\ = \lim_{\nu \to \infty} \Theta_{\nu, ij}(x) + \rho_s(\mathcal{A})_{ij} = \rho_s(\mathcal{A})_{ij}. \quad (9)$$

The frame $\{E_i\}$ is g-orthonormal. Thus $\mathcal{R}_\infty$ has constant scalar curvature. By Equation (8) if $\mathcal{A}$ is Ricci symmetric, then so is $\mathcal{R}_\infty$. By Equation (9), if $\mathcal{A}$ is Ricci antisymmetric or is Ricci tracefree, so is $\mathcal{R}_\infty$. This completes the proof of Theorem 1.3 in the C^s category.

In the real analytic category, we complexify and consider the complex ball of radius δ in $\mathbb{C}^m$. Since C^0 convergence of holomorphic functions gives convergence in the holomorphic setting, Theorem 1.3 follows in the real analytic context as well.

Acknowledgments

Research of P. Gilkey supported by Project MTM2006-01432 (Spain) and PIP 6303-2006-2008 Conicet (Argentina). Research of S. Nikčević supported by Project 144032 (Srbija). Research of D. Westerman supported by the University of Oregon.

References

1. N. Blažić, N. Bokan, and P. Gilkey, *Bull. London Math. Soc.* **29**, 227 (1997).
2. E. García-Río, D. N. Kupeli, M. E. Vázquez-Abal, *Diff. Geom. and Appl.* **7**, 85 (1997).
3. Q. S. Chi, *J. Diff. Geom.* **28**, 187 (1988).
4. Y. Nikolayevsky, *Mat. Annalen* **331**, 505 (2005).
5. E. García-Río, D. N. Kupeli, R. Vázquez-Lorenzo; *Osserman manifolds in semi-Riemannian geometry*, Lect. Notes Math. **1777**, Springer-Verlag, Berlin (2002).
6. R. Strichartz, *Can. J. Math.* **XL**, 1105 (1988).
7. P. Gilkey, S. Nikčević, and D. Westerman, arXiv:0811.3180.
8. N. Bokan, *Rend. Circ. Mat. Palermo* **XXIX**, 331 (1990).

CONFORMALLY OSSERMAN MULTIPLY WARPED PRODUCT STRUCTURES IN THE RIEMANNIAN SETTING

M. Brozos-Vázquez

Mathematics Department, University of A Coruña
Ferrol, E.U. Politécnica 15405, Spain
E-mail: mbrozos@udc.es

M. E. Vázquez-Abal* and R. Vázquez-Lorenzo**

Faculty of Mathematics, University of Santiago de Compostela
Santiago de Compostela, 15782, Spain
* E-mail: elena.vazquez.abal@usc.es
** E-mail: ravazlor@edu.xunta.es

We survey some known relations between Osserman and conformally Osserman properties and the algebraic structure of a warped metric tensor, extending them to multiply warped product structures.

Keywords: Riemannian manifold, Osserman manifold, conformally Osserman manifold, multiply warped product

1. Introduction

The group of isometries of any Riemannian two-point-homogeneous space acts transitively on the unit sphere bundle. This implies that any Riemannian two-point-homogeneous manifold is Osserman, i.e., the eigenvalues of the Jacobi operator are constant on the unit sphere bundle. Moreover, the converse was conjectured by Osserman and proved to be true in dimension different from 16.[1–3] Later, the Osserman condition was extended to the conformal setting by means of the conformal Jacobi operator. Remarkably, in dimensions greater than 5 and different from 16, both the Osserman and conformally Osserman conditions are equivalent for Einstein manifolds. Moreover it has been recently shown that out of those dimensions a Riemannian manifold is conformally Osserman if and only if it is in the conformal class of an Osserman manifold;[4] however, this equivalence fails to be true in the higher signature setting.[5] We refer to Ref. 4–8 for basic results on conformally Osserman manifolds.

The aim of this paper is to survey some known relations between geometric properties such as being Osserman or conformally Osserman and the algebraic structure of the metric tensor, and extend them to multiply warped product structures. More specifically, for each of the sections of this paper, we recall the necessary results on warped products to classify multiply warped products with the analogous condition. Thus, after reviewing some basic definitions in Section 2, we impose the conformally Osserman condition on multiply warped products in Section 3, showing that they are necessarily locally conformally flat. This motivates the analysis of local conformal flatness carried out in Section 4 for these structures. Sections 5 and 6 are devoted to multiply warped products which are Osserman and of constant curvature, respectively. Note that, on the one hand, the Osserman condition is equivalent to the conformal Osserman and the Einstein conditions,[8] and, on the other hand, a manifold has constant sectional curvature if and only if it is Einstein and locally conformally flat. Therefore, the results in Sections 5 and 6 can be derived from the corresponding results in Sections 3 and 4 by imposing the Einstein condition.

Detailed information about the results included in this survey and other related material can be found in Ref. 9–14.

2. Preliminaries

Let (M, g) be a Riemannian manifold of dimension $n \geq 4$ and let R be the associated Riemann curvature tensor. R decomposes as $R = C \bullet g + W$, where C is the Schouten tensor, W is the Weyl tensor and $\bullet$ denotes de Kulkarni–Nomizu product. Let

$$\mathcal{J}(X)Y = R(X,Y)X, \quad \text{and} \quad \mathcal{J}_W(X)Y = W(X,Y)X$$

be the usual *Jacobi operator* and the *conformal Jacobi operator*, respectively. (M, g) is said to be *pointwise Osserman* (respectively, *conformally Osserman*) if for every $p \in M$ the Jacobi operator $\mathcal{J}_p$ (respectively, the conformal Jacobi operator $\mathcal{J}_W$) has constant eigenvalues on the unit tangent sphere $S_p = \{x \in T_pM : g_p(x,x) = 1\}$, where T_pM is the space tangent to M on p.

Let (B, g_B), (F_1, g_1), ..., (F_k, g_k) be Riemannian manifolds. The product manifold $M = B \times F_1 \times \cdots \times F_k$ equipped with the metric

$$g = g_B \oplus f_1^2 g_1 \oplus \cdots \oplus f_k^2 g_k$$

where $f_1, \ldots, f_k : B \longrightarrow \mathbb{R}$ are positive functions is called a *multiply warped product*. B is the *base*, $F_1, \ldots, F_k$ are the *fibers* and $f_1, \ldots, f_k$ are referred to as the *warping functions*. In what follows we will denote a multiply warped product manifold as above by $M = B \times_{f_1} F_1 \times \cdots \times_{f_k} F_k$. (See Ref. 15, 16 and the references therein for more information on multiply warped products). The special case of a multiply warped product with one fiber is referred to as a *warped product*.

The possible order of the fibers is irrelevant for our purposes and, in order to avoid ambiguity, we assume henceforth that warping functions are nonconstant and fibers with the same warping function, up to a scale factor, are joined in one fiber.

3. Conformally Osserman multiply warped products

The aim of this section is to classify multiply warped products which are conformally Osserman. In order to attain that objective, as a first step we investigate direct products. Then we study the warped product structure, as the simplest case of a multiply warped product, to later obtain the complete classification in the general context. We refer to Ref. 9 for detailed proofs of the following lemmas.

Lemma 3.1. *Let (M, g) be a conformally Osserman manifold which decomposes as a direct product $B \times F$. Then (M, g) is locally conformally flat.*

Given an arbitrary warped product manifold $B \times_f F$ with metric $g_B \oplus f^2 g_F$ one considers the conformal change given by $1/f^2$ to see that it belongs to the conformal class of a direct product metric. Therefore, since the conformal Osserman condition is invariant under conformal transformations,[8] one gets the following result for warped products as a consequence of Lemma 3.1.

Lemma 3.2. *A warped product $B \times_f F$ is conformally Osserman if and only if it is locally conformally flat.*

Remark 3.1. Although this paper deals with positive definite signature, we digress slightly in this remark to point out that Lemma 3.2 cannot be extended with full generality to higher signature, see Ref. 9 for details; however it remains true in dimension 4 for any possible signature.

Theorem 3.1. *A multiply warped product $B \times_{f_1} F_1 \times \cdots \times_{f_k} F_k$ is conformally Osserman if and only if it is locally conformally flat.*

Proof. A multiply warped product $B \times_{f_1} F_1 \times \cdots \times_{f_k} F_k$ can be thought of as a warped product with base $B \times_{f_1} F_1 \times \cdots \times_{f_{k-1}} F_{k-1}$ and fiber F_k. Hence Lemma 3.2 directly applies. □

4. Locally conformally flat multiply warped products

Motivated by the results in Section 3, it is natural to proceed studying multiply warped products which are locally conformally flat. As we did in the previous section, we begin by analyzing the simplest case which is that of a warped product. See Ref. 10 and 11 for the proofs of the results in this section.

Theorem 4.1. Let $M = B \times_f F$ be a warped product. Then the following holds:

(i) If $\dim B = 1$, then $M = B \times_f F$ is locally conformally flat if and only if (F, g_F) is a space of constant curvature.

(ii) If $\dim B > 1$ and $\dim F > 1$, then $M = B \times_f F$ is locally conformally flat if and only if

(ii.a) (F, g_F) is a space of constant curvature c_F.

(ii.b) The function $f : B \to \mathbb{R}^+$ defines a conformal change on B such that $(B, \frac{1}{f^2} g_B)$ is a space of constant curvature $\tilde{c}_B = -c_F$.

(iii) If $\dim F = 1$, then $M = B \times_f F$ is locally conformally flat if and only if the function $f : B \to \mathbb{R}^+$ defines a conformal change on B such that $(B, \frac{1}{f^2} g_B)$ is a space of constant curvature.

Before giving a detailed description of locally conformally flat multiply warped products, we provide an upper bound on the number of different fibers.

Theorem 4.2. Let $M = B \times_{f_1} F_1 \times \cdots \times_{f_k} F_k$ be a multiply warped product. If M is locally conformally flat then $k \leq 3$ and the fibers (F_i, g_{F_i}) are spaces of constant sectional curvature (provided that $\dim F_i \geq 2$), for all $i = 1, \ldots, k$.

We divide the subsequent analysis in terms of the dimension of the base. Thus, the following result provides a precise description of all locally conformally flat manifolds $M = I \times_{f_1} F_1 \times \cdots \times_{f_k} F_k$, where I is a real interval.

Theorem 4.3. Let $M = I \times_{f_1} F_1 \times \cdots \times_{f_k} F_k$ be a multiply warped product. M is locally conformally flat if and only if one of the following holds:

(i) $M = I \times_f F$ is a warped product with fiber F of constant sectional curvature (provided dim $F \geq 2$) and any (positive) warping function f.

(ii) $M = I \times_{f_1} F_1 \times_{f_2} F_2$ is a multiply warped product with two fibers of constant sectional curvature (provided that dim $F_i \geq 2$) and warping functions
$$f_1 = (\xi \circ f)\frac{1}{f'}, \qquad f_2 = \frac{1}{f'}$$
where f is a strictly increasing function and ξ is a warping function making $I \times_\xi F_1$ of constant sectional curvature (cf. Theorem 6.1) and $(\xi \circ f) > 0$.

(iii) $M = I \times_{f_1} F_1 \times_{f_2} F_2 \times_{f_3} F_3$ is a multiply warped product with three fibers of constant sectional curvature (provided that dim $F_i \geq 2$) and warping functions
$$f_1 = (\xi_1 \circ f)\frac{1}{f'}, \qquad f_2 = (\xi_2 \circ f)\frac{1}{f'}, \qquad f_3 = \frac{1}{f'}$$
where f is a strictly increasing function and ξ_i are warping functions making $I \times_{\xi_1} F_1 \times_{\xi_2} F_2$ of constant sectional curvature (as in Theorem 6.1) such that $(\xi_i \circ f) > 0$, $i = 1, 2$.

For a general multiply warped product with base of dimension greater than one, we do not have a classification like in the theorem above. However we summarize next some interesting results for this case. We start with the following theorem.

Theorem 4.4. *Let $M = \mathfrak{U}^s \times_{f_1} F_1 \times \cdots \times_{f_k} F_k$ be a multiply warped product with $\mathfrak{U}^s \subset \mathbb{R}^s$, $s \geq 2$. Then M is locally conformally flat if and only if the warping functions satisfy*
$$f_i(\vec{\mathbf{x}}) = a_i\|\vec{\mathbf{x}}\|^2 + \langle \vec{\mathbf{b}}_i, \vec{\mathbf{x}} \rangle + c_i \tag{1}$$
for all $\vec{\mathbf{x}} \in \mathfrak{U}^s$, where $a_i > 0$, $c_i \in \mathbb{R}$ and $\vec{\mathbf{b}}_i \in \mathbb{R}^s$ and moreover the warping functions are compatible in the sense that
$$\langle \vec{\mathbf{b}}_i, \vec{\mathbf{b}}_j \rangle = 2(a_i c_j + a_j c_i), \qquad i \neq j \tag{2}$$
and the sectional curvature of each fiber of $\dim F_i \geq 2$ is given by
$$K^{F_i} = \|\vec{\mathbf{b}}_i\|^2 - 4a_i c_i, \qquad i,j = 1, \ldots, k. \tag{3}$$

Remark 4.1. Note that the previous theorem can be extended for not necessarily flat locally conformally flat bases (B, g_B) to get a local description of locally conformally flat multiply warped spaces. Since (B, g_B) is locally conformally flat, there exist local coordinates such that $g_B = \Psi^2 g_{\mathfrak{U}^s}$. Hence, when using such coordinates the multiply warped metric satisfies

$$g_B \oplus f_1^2 g_1 \oplus \cdots \oplus f_k^2 g_k = \Psi^2 \left(g_{\mathfrak{U}^s} \oplus (\frac{f_1}{\Psi})^2 g_1 \oplus \cdots \oplus (\frac{f_k}{\Psi})^2 g_k \right).$$

Therefore the multiply warped product $g_B \oplus f_1^2 g_1 \oplus \cdots \oplus f_k^2 g_k$ is locally conformally flat if and only if so is $g_{\mathfrak{U}^s} \oplus (\frac{f_1}{\Psi})^2 g_1 \oplus \cdots \oplus (\frac{f_k}{\Psi})^2 g_k$. Hence the warping functions are determined locally up to a conformal factor Ψ, since the warping functions, in local coordinates where $g_B = \Psi^2 g_{\mathfrak{U}^s}$, are given by $f_i(\mathbf{x}) = (a_i \|\vec{\mathbf{x}}\|^2 + \langle \vec{\mathbf{b}}_i, \vec{\mathbf{x}} \rangle + c_i) \Psi$ for all $i = 1, \ldots, k$.

Remark 4.2. Locally conformally flat multiply warped spaces can now be easily constructed as follows. Since any warping function of a locally conformally flat multiply warped space $M = \mathfrak{U}^s \times_{f_1} F_1 \times \cdots \times_{f_k} F_k$ is completely determined by scalars $a_i, c_i \in \mathbb{R}$ and vectors $\vec{\mathbf{b}}_i = (b_{i1}, \ldots, b_{is}) \in \mathbb{R}^s$, consider the vectors $\vec{\xi}_i = (b_{i1}, \ldots, b_{is}, a_i, c_i)$ in $\mathbb{R}^{s+2}$. Next, define a Lorentzian inner product in $\mathbb{R}^{s+2}$ by

$$\begin{pmatrix} 1 & & & & \\ & \ddots^s & & & \\ & & 1 & & \\ \hline & & & 0 & -2 \\ & & & -2 & 0 \end{pmatrix}$$

and note that equations (2) and (3) at Theorem 4.4 are interpreted in terms of the orthogonality $\vec{\xi}_i \perp \vec{\xi}_j$ (for all $i \neq j$) and $K^{F_i} = \|\vec{\xi}_i\|^2$ (whenever $\dim F_i \geq 2$), respectively. Hence it follows from Remark 4.1 that

(i) *A locally conformally flat space $M = B \times_{f_1} F_1 \times \cdots \times_{f_k} F_k$ has, at most, $(s+2)$ different fibers, where $s = \dim B$.*

(ii) *Let $s = \dim B$. The sectional curvature of the fibers (F_i, g_i) of a locally conformally flat multiply warped space is as follows*

  (ii.1) *There are at most $(s+1)$ fibers of positive curvature.*
  (ii.2) *There are at most one fiber of nonpositive curvature.*

  whenever $\dim F_i \geq 2$.

(iii) *For any locally conformally flat manifold (B^s, g_B), there exist $(s+2)$ locally defined warping functions $f_i : \mathfrak{U} \subset B \to \mathbb{R}^+$ and (F_i, g_i) spaces of constant curvature such that $M = \mathfrak{U} \times_{f_1} F_1 \times \cdots \times_{f_{s+2}} F_{s+2}$ is locally conformally flat.*

5. Osserman multiply warped products

Since the Osserman condition is equivalent to both the Einstein and the conformal Osserman conditions, we take advantage of the results in Section 3 to study the Osserman condition. Thus, the following result which was proved in Ref. 9 is a consequence of Theorem 3.1.

Theorem 5.1. *Let $M = B \times_{f_1} F_1 \times \cdots \times_{f_k} F_k$ be a manifold with local structure of a multiply warped product. Then M is pointwise Osserman if and only if it is a space of constant sectional curvature.*

Every two-point-homogeneous space is Osserman, in particular $\mathbb{C}P^n$, $\mathbb{Q}P^n$, $\mathbb{O}P^n$, $\mathbb{C}H^n$, $\mathbb{Q}H^n$ and $\mathbb{O}H^n$ are Osserman. Furthermore, they are the only Osserman manifolds of dimensions n not equal to 4 and 16 which do not have constant sectional curvature. Dimensions 4 and 16 are excluded because for $n = 4$ there exist pointwise Osserman manifolds which are not globally Osserman and for $n = 16$ the Cayley plane is Osserman. This relation between Osserman manifolds and two-point-homogeneous manifolds gives the following consequence of Theorem 5.1.

Corollary 5.1. *Neither $\mathbb{C}P^n$, $\mathbb{Q}P^n$, $\mathbb{O}P^n$ nor their negative curvature duals may be decomposed as a multiply warped product.*

6. Multiply warped products of constant curvature

As we did in Section 4, we first consider multiply warped products with 1-dimensional base, to then extend the classification results to the higher dimensional setting.

The following theorem has been first obtained by Mignemi and Schmidt;[15] we refer to Ref. 11 for a proof following the line of our discussion here. As a consequence of Theorem 4.2, if $M = I \times_{f_1} F_1 \times \cdots \times_{f_k} F_k$ is a space of constant sectional curvature, then each fiber F_i must be of constant sectional curvature.

Theorem 6.1. *Let $M = I \times_{f_1} F_1 \times \cdots \times_{f_k} F_k$ be a multiply warped product. Then M is of constant sectional curvature K if and only if $k \leq 2$ and, moreover, one of the following holds:*

(i) *If $K = 0$, then $M = I \times_{\alpha_1} F_1$ or $M = I \times_{\alpha_1} F_1 \times_{\alpha_2} F_2$, with warping functions given by*

$$\alpha_i(t) = a_i\, t + b_i, \qquad i = 1, 2.$$

Moreover, the fibers (F_i, g_i) are necessarily of constant sectional curvature $K^{F_i} = a_i^2$, provided that $\dim F_i \geq 2$ ($i = 1, 2$), and the warping functions satisfy the compatibility condition $a_1 a_2 = 0$ in the case of two fibers.

(ii) If $K = c^2$, then $M = I \times_{\beta_1} F_1$ or $M = I \times_{\beta_1} F_1 \times_{\beta_2} F_2$, with warping functions given by

$$\beta_i(t) = a_i \sin ct + b_i \cos ct, \qquad i = 1, 2.$$

Moreover, the fibers (F_i, g_i) are necessarily of constant sectional curvature $K^{F_i} = c^2(a_i^2 + b_i^2)$, provided that $\dim F_i \geq 2$ ($i = 1, 2$), and the warping functions satisfy the compatibility condition $a_1 a_2 + b_1 b_2 = 0$ in the case of two fibers.

(iii) If $K = -c^2$, then $M = I \times_{\gamma_1} F_1$ or $M = I \times_{\gamma_1} F_1 \times_{\gamma_2} F_2$, with warping functions given by

$$\gamma_i(t) = a_i \sinh ct + b_i \cosh ct, \qquad i = 1, 2.$$

Moreover, the fibers (F_i, g_i) are necessarily of constant sectional curvature $K^{F_i} = c^2(a_i^2 - b_i^2)$, provided that $\dim F_i \geq 2$ ($i = 1, 2$), and the warping functions satisfy the compatibility condition $a_1 a_2 - b_1 b_2 = 0$ in the case of two fibers.

Remark 6.1. Two immediate consequences follow from previous theorem on the multiply warped structure of a space of constant curvature. (Compare with the bounds given in Theorem 4.2).

(a) No more than two fibers are admissible for a space $I \times_{f_1} F_1 \times \cdots \times_{f_k} F_k$ to be of constant sectional curvature.
(b) The (constant) sectional curvatures of the fibers are subject to some restrictions.

We turn back our attention to the case of multiply warped products with base of dimension $s \geq 2$. See Ref. 14 for a complete proof of the results henceforth. Note from Theorem 4.2 that if $M = B \times_{f_1} F_1 \times \cdots \times_{f_k} F_k$ is a space of constant curvature κ, then each fiber F_i must also have constant sectional curvature. Moreover since the base inherits the geometry of the whole manifold in a multiply warped product, the base B has constant sectional curvature equal to κ.

Theorem 6.2. Let $M = B \times_{f_1} F_1 \times \cdots \times_{f_k} F_k$ be a multiply warped product with $B, F_1, \ldots, F_k$ of constant sectional curvature, $s = \dim B \geq 2$ and

$K^B = \kappa$. Then M is a space of constant sectional curvature ($K = \kappa$) if and only if the warping functions have locally the expression

$$f_i(\vec{\mathbf{x}}) = \frac{\frac{-\kappa c_i}{4}\langle \vec{\mathbf{x}}, \vec{\mathbf{x}} \rangle + \langle \vec{\mathbf{b_i}}, \vec{\mathbf{x}} \rangle + c_i}{1 + \frac{\kappa}{4}\langle \vec{\mathbf{x}}, \vec{\mathbf{x}} \rangle}, \quad c_i \in \mathbb{R}, \quad \vec{\mathbf{b_i}} \in \mathbb{R}^s,$$

with

(i) $\langle \vec{\mathbf{b_i}}, \vec{\mathbf{b_j}} \rangle + \kappa\, c_i\, c_j = 0, \quad i \neq j$,
(ii) $\langle \vec{\mathbf{b_i}}, \vec{\mathbf{b_i}} \rangle + \kappa\, c_i^2 = K^{F_i}, \quad \text{if } \dim F_i \geq 2$,

for all $i, j = 1, \ldots, k$, where $\langle \cdot, \cdot \rangle$ denotes the scalar product in the Euclidean space $\mathbb{R}^s$.

Remark 6.2. The local description of multiply warped product spaces of constant curvature in the previous theorem leads to some restrictions on the number and the geometry of the fibers. Consider vectors $\vec{\xi_i} = (\vec{\mathbf{b_i}}, c_i) = (b_{i1}, \ldots, b_{ic}, c_i)$ in $\mathbb{R}^{s+1}$ endowed with the symmetric bilinear form given by $\mathrm{diag}[1, \ldots, 1, \kappa]$. Then, conditions (i) and (ii) in Theorem 6.2 mean that the vectors $\vec{\xi_i}$, $i = 1, \ldots, k$, must be orthogonal to each other and, if $\dim F_i \geq 2$, then the associated vector $\vec{\xi_i}$ must satisfy $\langle \vec{\xi_i}, \vec{\xi_i} \rangle = K^{F_i}$. Moreover, it follows that the above bilinear form is a scalar product in $\mathbb{R}^{s+1}$ of index $(1, \ldots, 1, \frac{\kappa}{|\kappa|})$ if $\kappa \neq 0$ and $(1, \ldots, 1, 0)$ if $\kappa = 0$. Now, if $\kappa = 0$, Theorem 6.2 implies that the scaling functions are polynomials of degree one, $f_i(\vec{\mathbf{x}}) = \langle \vec{\mathbf{b_i}}, \vec{\mathbf{x}} \rangle + c_i$, and the corresponding vectors $\vec{\xi_i}$ are orthogonal if and only if so are the vectors $\vec{\mathbf{b_i}}$. This shows that no more than s fibers may exist if $\kappa = 0$. (Recall that warping functions are assumed to be nonconstant, and thus the case $\vec{\xi} = (\vec{\mathbf{0}}, c)$ is excluded). Therefore, proceeding as in Remark 4.2, we conclude:

(i) A multiply warped product $M = B \times_{f_1} F_1 \times \cdots \times_{f_k} F_k$ of constant sectional curvature κ, with a $(s \geq 2)$-dimensional base has, at most, s fibers if $\kappa = 0$, and at most $s + 1$ if $\kappa \neq 0$.
(ii) The sectional curvatures of the fibers F_i of a multiply warped product of constant curvature and with s-dimensional base are as follows:

  (ii.1) If $\kappa > 0$, then no more than $s + 1$ fibers of dimension ≥ 2 and positive sectional curvature are admissible and none fibers of dimension ≥ 2 and negative sectional curvature may occur.
  (ii.2) If $\kappa < 0$, then no more than s fibers of dimension ≥ 2 and positive sectional curvature are admissible and none fibers of dimension ≥ 2 and negative sectional curvature may occur.

(ii.3) If $\kappa = 0$, then no more than s fibers of dimension ≥ 2 and positive sectional curvature are admissible and no more than 1 fiber of dimension ≥ 2 and negative sectional curvature may occur.

(iii) For any $(s \geq 2)$-dimensional base (B, g_B) of constant curvature κ, there exist $s + 1$ (if $\kappa \neq 0$) or s (if $\kappa = 0$) locally defined scaling functions $f_i : \mathfrak{U} \subset B \to \mathbb{R}^+$ and (F_i, g_{F_i}) spaces of constant curvature so that $M = \mathfrak{U} \times_{f_1} M_1 \times \cdots \times_{f_\nu} M_\nu$ has constant sectional curvature.

Acknowledgements

The authors would like to express their gratitude to the referee for his useful suggestions and to Prof. E. García-Río who is a coauthor in many of the results exposed in this paper. Research supported by projects MTM2006-01432 and PGIDIT06PXIB207054PR (Spain).

References

1. Q.-S. Chi, *Pacific J. Math.* **150**, 31–42 (1991).
2. Y. Nikolayevsky, *Manuscr. Math.* **115**, 31–53 (2004).
3. Y. Nikolayevsky, *Math. Ann.* **331**, 505–522 (2005).
4. Y. Nikolayevsky, arXiv:0810.5621 [math.DG].
5. M. Brozos-Vázquez, E. García–Río, R. Vázquez-Lorenzo, *Proc. Royal Society A* **462**, 1425–1441 (2006).
6. N. Blažić, P. Gilkey, *Int. J. Geom. Methods Mod. Phys.* **1**, 97–106 (2004).
7. N. Blažić, P. Gilkey, *Proceedings DGA Conf., Czech Republic 2004*, 15–18 (2005).
8. N. Blažić, P. Gilkey, S. Nikčević, U. Simon, *PDEs, submanifolds and affine differential geometry*, Banach Center Publ. **69**, Polish Acad. Sci., Warsaw, 195–203 (2005).
9. M. Brozos-Vázquez, E. García–Río, R. Vázquez-Lorenzo, *Result. Math.* **52**, 211-221 (2008).
10. M. Brozos-Vázquez, E. García–Río, R. Vázquez-Lorenzo, *Pacific J. Math.* **226**, 201–219 (2006).
11. M. Brozos-Vázquez, E. García-Río, R. Vázquez-Lorenzo, *Mat. Contemp.* **28**, 91-110 (2005).
12. M. Brozos-Vázquez, E. García-Río, R. Vázquez-Lorenzo, *J. Math. Phys.* **46**, 11pp. (2005).
13. M. Brozos-Vázquez, E. García-Río, R. Vázquez-Lorenzo, *J. Cosmol. Astropart. Phys.* **12**, 13pp. (2004).
14. M. Brozos-Vázquez, E. García-Río, R. Vázquez-Lorenzo, *Class. Quantum Grav.* **22**, 3119-3133 (2005).
15. S. Mignemi, H. J. Schmidt, *J. Math. Phys.* **39**, 998-1010 (1998).
16. R. Tojeiro, *Houston J. Math.* **32**, 725–743 (2006).

RIEMANNIAN Γ-SYMMETRIC SPACES

Michel Goze* and Elisabeth Remm**

*Université de Haute Alsace, Laboratoire de Mathématiques
4, rue des Frères Lumière, 68 093 Mulhouse cedex, France*
* E-mail: m.goze@uha.fr
** E-mail: e.remm@uha.fr

A Γ-symmetric space is a reductive homogeneous space $M = G/H$ provided in each of its points with a finite abelian group of "symmetries" isomorphic to Γ. In case of $\Gamma = \mathbb{Z}_2^p$, the Lie algebra $\mathfrak{g}$ of G is graded by Γ and this grading permits to construct again the symmetries of M. An adapted Riemannian metric will be a tensor metric for which the symmetries are isometries. We give in case of $p = 2$ the classification of compact $\mathbb{Z}_2^2$-symmetric spaces corresponding to G compact simple. In particular we find again the oriented flag manifold. For this class of nonsymmetric spaces and for the space $SO(2m)/Sp(m)$ we describe all these metrics. In particular we prove that in the definite positive case, these metrics are not, in general, naturally reductive. We class also the Lorentzian $\mathbb{Z}_2^2$-symmetric metrics.

Keywords: Γ-symmetric spaces, adapted Riemannian metrics, graded Lie algebras

1. Riemannian reductive homogeneous spaces

Let $M = G/H$ be a homogeneous space where G is a connected Lie group which acts effectively on M. It is a reductive space if the Lie algebra $\mathfrak{g}$ of G can be decomposed into a direct sum of vector spaces of the Lie algebra $\mathfrak{h}$ of H and an $ad(H)$-invariant subspace $\mathfrak{m}$, i.e., $\mathfrak{g} = \mathfrak{h} + \mathfrak{m}$, $ad(H)\mathfrak{m} \subset \mathfrak{m}$. If H is connected, and we will assume it in all the following, the second condition is equivalent to $[\mathfrak{h}, \mathfrak{m}] \subset \mathfrak{m}$.

We denote by ∇_M (or simply ∇) the canonical G-invariant connection on M. Recall that its torsion tensor T_{∇_M} and curvature tensor R_{∇_M} satisfy:

$$\begin{cases} T_{\nabla_M}(X,Y)_0 = [X,Y]_\mathfrak{m}, \quad X,Y \in \mathfrak{m} \quad (0 \text{ denotes the class of } 1_G \text{ in } G/H), \\ (R_{\nabla_M}(X,Y)Z)_0 = -[[X,Y],Z], \quad X,Y,Z \in \mathfrak{m}, \\ \nabla_M T_{\nabla_M} = \nabla_M R_{\nabla_M} = 0. \end{cases}$$

We denote by $\tilde{\nabla}_M$ the natural (complete) torsion-free G-invariant connection on M. It admits the same geodesics as ∇_M and it is defined by

$$\tilde{\nabla}_M(X)(Y) = \frac{1}{2}[X,Y]_{\mathfrak{m}}, \quad X,Y \in \mathfrak{m}.$$

Let g be a G-invariant indefinite Riemannian metric on the reductive homogeneous space $M = G/H$. It is completely determined by an $ad(H)$-invariant non degenerate symmetric bilinear form B on $\mathfrak{m}$, the correspondence is given by $B(X,Y) = g(X,Y)_0$, for all $X,Y \in \mathfrak{m}$. Recall that g is positive definite if and only if B is positive definite. We denote by ∇_g the corresponding Riemannian connection. It coincides with the natural connection $\tilde{\nabla}_M$ if and only if we have $B(X,[Z,Y]_{\mathfrak{m}}) + B([Z,X]_{\mathfrak{m}},Y) = 0$, $\forall X,Y,Z \in \mathfrak{m}$. In this case the Riemannian reductive homogeneous space M is said to be naturally reductive.

For example, if M is a symmetric space, that is the decomposition of $\mathfrak{g}$ satisfies $\mathfrak{g} = \mathfrak{h} + \mathfrak{m}$ with $[\mathfrak{h},\mathfrak{h}] \subset \mathfrak{h}$, $[\mathfrak{h},\mathfrak{m}] \subset \mathfrak{m}$, $[\mathfrak{m},\mathfrak{m}] \subset \mathfrak{h}$ then the canonical and the natural connections coincide. Moreover if ∇_g is a Riemannian symmetric connection, that is the symmetries of M are isometries, we have $\nabla_M = \tilde{\nabla}_M = \nabla_g$ and M is naturally reductive.

2. Riemannian Γ-symmetric spaces

2.1. Γ-symmetric spaces

Let Γ be a finite abelian group. A Γ-symmetric space is a triple (G, H, Γ_G) where G is a connected Lie group, H a closed subgroup of G and Γ_G an abelian finite subgroup of the group of automorphisms of G isomorphic to Γ: $\Gamma_G = \{\rho_\gamma \in Aut(G), \gamma \in \Gamma\}$ such that H lies between G_Γ, the closed subgroup of G consisting of all elements left fixed by the automorphisms of Γ_G, and the identity component of G_Γ. The elements of Γ_G satisfy:

$$\begin{cases} \rho_{\gamma_1} \circ \rho_{\gamma_2} = \rho_{\gamma_1\gamma_2}, & \forall \gamma_1, \gamma_2 \in \Gamma \\ \rho_e = Id & \text{where } e \text{ is the unit of } \Gamma \\ (\rho_\gamma(g) = g \ \forall \gamma \in \Gamma) \iff g \in H. \end{cases}$$

We also suppose that H does not contain any proper normal subgroup of G.

Given a Γ-symmetric space (G, H, Γ_G), we construct for each point x of $M = G/H$ a subgroup Γ_x of $\textit{Diff}(M)$, the group of diffeomorphisms of M, isomorphic to Γ and which has x as an isolated fixed point. We denote by $\bar{g}$ the class of $g \in G$ in M and by e the identity of G. We consider

$$\Gamma_{\bar{e}} = \{s_{(\gamma,\bar{e})} \in \textit{Diff}(M), \gamma \in \Gamma\},$$

with $s_{(\gamma,\bar{e})}(\bar{g}) = \overline{\rho_\gamma(g)}$. For another point $x = \overline{g_0}$ of M we have

$$\Gamma_x = \{s_{(\gamma,x)} \in \mathit{Diff}(M),\ \gamma \in \Gamma\},$$

with $s_{(\gamma,\bar{g_0})}(y) = g_0(s_{(\gamma,\bar{e})})(g_0^{-1}y)$. All these subgroups Γ_x of $\mathit{Diff}(M)$ are isomorphic to Γ. The elements of Γ_x will be called the symmetries of M at the point x or more generally the symmetries of M.

Remark. If (G, H, Γ_G) is a Γ-symmetric space, we will also say that the homogeneous space $M = G/H$ is Γ-symmetric.

2.2. Γ-grading of the Lie algebra $\mathfrak{g}$ of G

Let (G, H, Γ_G) be a Γ-symmetric space. Let $\mathfrak{g}$ (resp. $\mathfrak{h}$) be the Lie algebra of G (resp. H). Each automorphism ρ_γ of G induces an automorphism τ_γ of $\mathfrak{g}$ and the set $\Gamma_\mathfrak{g} = \{\tau_\gamma \in \mathit{Aut}(\mathfrak{g}), \gamma \in \Gamma\}$ is a finite group isomorphic to Γ such that $(\forall \gamma \in \Gamma,\ \tau_\gamma(X) = X) \Leftrightarrow X \in \mathfrak{h}$. As each of the linear morphism τ_γ is diagonalizable, the relation $\tau_\gamma \circ \tau_{\gamma'} = \tau_{\gamma'} \circ \tau_\gamma$ implies that $\mathfrak{g}$ is a vectorial direct sum of root spaces $\mathfrak{g}_\chi, \chi \in \Gamma^*$ where Γ^* denotes the dual group of Γ. As Γ is abelian, Γ^* is isomorphic to Γ and we can identify Γ and Γ^*. We deduce that $\mathfrak{g}$ is Γ-graded, $\mathfrak{g} = \underset{\gamma \in \Gamma}{\oplus} \mathfrak{g}_\gamma$ with $\mathfrak{g}_e = \mathfrak{h}$ where e is the identity of Γ.

Conversely, we prove in[1] that every Γ-graded Lie algebra $\mathfrak{g} = \underset{\gamma \in \Gamma}{\oplus} \mathfrak{g}_\gamma$ determines a Γ-symmetric space (G, H, Γ) where G (resp. H) is a connected Lie group associated to $\mathfrak{g}$ (resp. $\mathfrak{g}_e$).

Theorem 2.1. *Every Γ-symmetric space is a reductive homogeneous space.*

In fact if $M = G/H$ is a Γ-symmetric space, the Lie algebra $\mathfrak{g}$ of G is Γ-graded that is $\mathfrak{g} = \underset{\gamma \in \Gamma}{\oplus} \mathfrak{g}_\gamma$. Then we put $\mathfrak{h} = \mathfrak{g}_e$ and $\mathfrak{m} = \underset{\gamma \in \Gamma, \gamma \neq e}{\oplus} \mathfrak{g}_\gamma$. We have $[\mathfrak{h}, \mathfrak{m}] \subset \mathfrak{m}$ and $[\mathfrak{m}, \mathfrak{m}] \subset \mathfrak{h}$. If we assume that H is connected, this implies that $M = G/H$ is a reductive space

2.3. Riemannian and Indefinite Riemannian Γ-symmetric spaces

Let $M = G/H$ be a Γ-symmetric space. Recall that for all $x \in M, \Gamma_x = \{s(\gamma, x), \gamma \in \Gamma\}$ is a subgroup of $\mathit{Diff}(M)$ isomorphic to Γ. Let $S(\gamma, x)$ be the tangent map $T_x s(\gamma, x)$. Thus $S(\gamma, x) \in GL(T_x M)$ and $\{S(\gamma, x), \gamma \in \Gamma\}$ is a finite subgroup of $GL(T_x M)$ isomorphic to Γ.

Definition 2.1. A Riemannian (resp. Indefinite Riemannian) metric on M is called a Γ-symmetric Riemannian (resp. Indefinite Riemannian) metric if for all $x \in M$, the linear symmetries $S(\gamma, x)$, $\gamma \in \Gamma$ are isometries.

Remark. If $\Gamma = \mathbb{Z}_2 = \mathbb{Z}/2\mathbb{Z}$, we find the classical notion of Riemannian symmetric space again. When the Riemannian metric g is positive definite, the natural and canonical connections coincide, and they are also equal to the Riemannian connection associated to g. All these spaces are naturally reductive. In general, this property is false for the Riemannian Γ-symmetric spaces, when Γ is not isomorphic to $\mathbb{Z}_2$. We will see in the following section examples of $\mathbb{Z}_2^2$-Riemannian spaces which are not naturally reductive.

2.4. Irreducible Riemannian Γ-symmetric spaces

Let (G, H, Γ_G) be a Γ-symmetric space. Since G/H is a reducible homogeneous space with an $ad(H)$-invariant decomposition $\mathfrak{g} = \mathfrak{g}_e \oplus \mathfrak{m}$, the Lie algebra of the holonomy group of ∇ is spanned by the endomorphisms of $\mathfrak{m}$ given by $R(X,Y)_0$ for all $X, Y \in \mathfrak{m}$. Recall that $(R(X,Y)Z)_0 = -[[X,Y]_\mathfrak{h}, Z]$ for all $X, Y, Z \in \mathfrak{m}$. In particular we have $R(X,Y)_0 = 0$ as soon as $X \in \mathfrak{g}_\gamma, Y \in \mathfrak{g}_{\gamma'}$ with $\gamma, \gamma' \neq e$. For example, if $\Gamma = \mathbb{Z}_2^2$ then $\mathfrak{g} = \mathfrak{g}_e \oplus \mathfrak{g}_a \oplus \mathfrak{g}_b \oplus \mathfrak{g}_c$ and $R(\mathfrak{g}_a, \mathfrak{g}_b)_0 = R(\mathfrak{g}_a, \mathfrak{g}_c)_0 = R(\mathfrak{g}_b, \mathfrak{g}_c)_0 = 0$.

Lemma 2.1. *Let $\mathfrak{g}$ be a simple Lie algebra $\mathbb{Z}_2^2$-graded. Then*

$$[\mathfrak{g}_a, \mathfrak{g}_a] \oplus [\mathfrak{g}_b, \mathfrak{g}_b] \oplus [\mathfrak{g}_c, \mathfrak{g}_c] = \mathfrak{g}_e.$$

Proof. Let U denote $[\mathfrak{g}_a, \mathfrak{g}_a] \oplus [\mathfrak{g}_b, \mathfrak{g}_b] \oplus [\mathfrak{g}_c, \mathfrak{g}_c]$. Then $I = U \oplus \mathfrak{g}_a \oplus \mathfrak{g}_b \oplus \mathfrak{g}_c$ is an ideal of $\mathfrak{g}$. In fact, $X \in I$ is decomposed as $X_U + X_a + X_b + X_c$. The main point is to prove that $[X_U, Y]$ is in I for any $Y \in \mathfrak{g}_e$. But X_U is decomposed as $[X_a, Y_a] + [X_b, Y_b] + [X_c, Y_c]$. The Jacobi identity shows that $[[X_a, Y_a], Y] \in [\mathfrak{g}_a, \mathfrak{g}_a]$ and it is similar for the other components. Then I is an ideal of $\mathfrak{g}$ which is simple, and hence $U = \mathfrak{g}_e$. □

One should note that in any case, as soon as Γ is not $\mathbb{Z}_2$, the representation $ad\,\mathfrak{g}_e$ is not irreducible on $\mathfrak{m}$. In fact, each component $\mathfrak{g}_\gamma$ is an invariant subspace of $\mathfrak{m}$.

Definition 2.2. The representation $ad\,\mathfrak{g}_e$ on $\mathfrak{m}$ is called Γ-*irreducible* if $\mathfrak{m}$ cannot be written $\mathfrak{m} = \mathfrak{m}_1 \oplus \mathfrak{m}_2$ with $\mathfrak{g}_e \oplus \mathfrak{m}_1$ and $\mathfrak{g}_e \oplus \mathfrak{m}_2$ Γ-graded Lie algebras.

Example. Let $\mathfrak{g}_1$ be a simple Lie algebra and $\mathfrak{g} = \mathfrak{g}_1 \oplus \mathfrak{g}_1 \oplus \mathfrak{g}_1 \oplus \mathfrak{g}_1$. Let $\sigma_1, \sigma_2, \sigma_3$ be the automorphisms of $\mathfrak{g}$ given by

$$\begin{cases} \sigma_1(X_1, X_2, X_3, X_4) = (X_2, X_1, X_3, X_4), \\ \sigma_2(X_1, X_2, X_3, X_4) = (X_1, X_2, X_4, X_3), \\ \sigma_3 = \sigma_1 \circ \sigma_2. \end{cases}$$

They define a $(\mathbb{Z}_2^2)$-graduation on $\mathfrak{g}$ and we have $\mathfrak{g}_e = \{(X,X,Y,Y)\}, \mathfrak{g}_a = \{(0,0,Y,-Y)\}, \mathfrak{g}_b = \{(X,-X,0,0)\}$ and $\mathfrak{g}_c = \{(0,0,0,0)\}$ with $X,Y \in \mathfrak{g}_1$. In particular, $\mathfrak{g}_a$ is isomorphic to $\mathfrak{g}_1$, hence $[\mathfrak{g}_e, \mathfrak{g}_a] = \mathfrak{g}_a$ and, since $\mathfrak{g}_1$ is simple, we cannot have $\mathfrak{g}_a = \mathfrak{g}_a^1 + \mathfrak{g}_a^2$ with $[\mathfrak{g}_e, \mathfrak{g}_a^i] = \mathfrak{g}_a^i$ for $i = 1, 2$. Then $\mathfrak{g}$ is $(\mathbb{Z}_2^2)$-graded and this decomposition is $(\mathbb{Z}_2^2)$-irreducible.

Suppose now that $\mathfrak{g}$ is a simple Lie algebra. Let K be the Killing-Cartan form of $\mathfrak{g}$. It is invariant by all automorphisms of $\mathfrak{g}$. In particular

$$K(\tau_\gamma X, \tau_\gamma Y) = K(X, Y)$$

for any $\tau_\gamma \in \check{\Gamma}$. If $X \in \mathfrak{g}_\alpha$ and $Y \in \mathfrak{g}_\beta$, $\alpha \neq \beta$ there exists $\gamma \in \Gamma$ such that $\tau_\gamma X = \lambda(\alpha, \gamma) X$ and $\tau_\gamma Y = \lambda(\beta, \gamma) Y$ with $\lambda(\alpha, \gamma)\lambda(\beta, \gamma) \neq 1$. Thus $K(X,Y) = 0$ and the homogeneous components $\mathfrak{g}_\gamma$ are pairewise orthogonal with respect to K. Moreover $K_\gamma = K|_{\mathfrak{g}_\gamma}$ is a nondegenerate bilinear form. Since $\mathfrak{g}$ is a simple Lie algebra, there exists an $ad\,\mathfrak{g}_e$-invariant inner product $\tilde{B}$ on $\mathfrak{g}$ such that the restriction $B = \tilde{B}|_\mathfrak{m}$ to $\mathfrak{m}$ defines a Riemannian Γ-symmetric structure on G/H. This means that $\tilde{B}(\mathfrak{g}_\gamma, \mathfrak{g}_{\gamma'}) = 0$ for $\gamma \neq \gamma' \in \Gamma$. We consider an orthogonal basis of $\tilde{B}$. For each $X \in \mathfrak{g}_e$, $ad\,X$ is expressed by a skew-symmetric matrix $(a_{ij}(X))$ and $K(X,X) = \sum_{i,j} a_{ij}(X) a_{ji}(X) < 0$. This implies that K is negative-definite on $\mathfrak{g}_e$.

Let K_γ and B_γ be the restrictions of K and B to the homogeneous component $\mathfrak{g}_\gamma$. Let $\beta \in \mathfrak{m}^*$ be such that $K_\gamma(X,Y) = B_\gamma(\beta_\gamma(X), Y)$ for all $X, Y \in \mathfrak{g}_\gamma$ and $\beta_\gamma = \beta|_{\mathfrak{g}_\gamma}$. Since B_γ is nondegenerate on $\mathfrak{g}_\gamma$, the eigenvalues of β_γ are real and non-zero. The eigenspaces $\mathfrak{g}_\gamma^1, ..., \mathfrak{g}_\gamma^p$ of β_γ are pairwise orthogonal with respect to B_γ and K_γ. But for every $Z \in \mathfrak{g}_e$ we have

$$K_\gamma([Z,X], Y) = K_\gamma(X, [Z,Y]) = B_\gamma(\beta_\gamma(X), [Z,Y])$$

and hence $B_\gamma(\beta_\gamma[Z,X], Y) = B_\gamma([Z, \beta_\gamma(X)], Y)$ for every $Y \in \mathfrak{g}_\gamma$ and $\beta_\gamma[Z,X] = [Z, \beta_\gamma(X)]$, implying that $\beta_\gamma \circ ad\,Z = ad\,Z \circ \beta_\gamma$ for any $Z \in \mathfrak{g}_e$. This yields $[\mathfrak{g}_e, \mathfrak{g}_\gamma^i] \subset \mathfrak{g}_\gamma^i$.

Now we examine the particular case of $\Gamma = \mathbb{Z}_2^2$. The eigenvalues of the involutive automorphisms τ_γ being real, the Lie algebra $\mathfrak{g}$ admits a real Γ-decomposition $\mathfrak{g} = \sum_{\gamma \in \mathbb{Z}_2^2} \mathfrak{g}_\gamma$. Then we can assume that $\mathfrak{g}$ is a real Lie algebra. Now if $i \neq j$ we have

$$K_\gamma([\mathfrak{g}_\gamma^i, \mathfrak{g}_\gamma^j], [\mathfrak{g}_\gamma^i, \mathfrak{g}_\gamma^j]) \subset K([\mathfrak{g}_\gamma^i, \mathfrak{g}_\gamma^j], \mathfrak{g}_e) \subset (\mathfrak{g}_\gamma^i, \mathfrak{g}_\gamma^j) = 0$$

and we have, for $i \neq j$:

$$[\mathfrak{g}_\gamma^i, \mathfrak{g}_\gamma^j] = \{0\} \ .$$

Let $\{e, a, b, c\}$ be the elements of $\mathbb{Z}_2^2$ with $a^2 = b^2 = c^2 = e$ and $ab = c$. Each component $\mathfrak{g}_\gamma$, $\gamma \neq e$, satisfies $[\mathfrak{g}_\gamma, \mathfrak{g}_\gamma] \subset \mathfrak{g}_e$ and $\mathfrak{g}_e \oplus \mathfrak{g}_\gamma$ is a symmetric Lie algebra. Endowed with the inner product $\tilde{B}$, the Lie algebra $\mathfrak{g}_e \oplus \mathfrak{g}_\gamma$ is an orthogonal symmetric Lie algebra. The Killing-Cartan form is not degenerate on $\mathfrak{g}_e \oplus \mathfrak{g}_\gamma$. Then $\mathfrak{g}_e \oplus \mathfrak{g}_\gamma$ is semi-simple. It is a direct sum of orthogonal symmetric Lie algebras of the following two kinds:

$i)$ $\mathfrak{g} = \mathfrak{g}' + \mathfrak{g}'$ with $\mathfrak{g}'$ simple
$ii)$ $\mathfrak{g}$ is simple.

The first case has been studied above and the representation is $(\mathbb{Z}_2^2)$-irreducible. In the second case $ad\,[\mathfrak{g}_\gamma, \mathfrak{g}_\gamma]$ is irreducible in $\mathfrak{g}_\gamma$ and the representation is $(\mathbb{Z}_2^2)$-irreducible on $\mathfrak{m}$.

3. Classification of compact simple $\mathbb{Z}_2^2$ symmetric spaces

The paper[1] is, in a large part, devoted to the classification of $\mathbb{Z}_2^2$-symmetric spaces G/H in case of $\mathfrak{g}$ is simple of classical type. Recently, in,[5] the exceptional case has been developped. Combining both results, we have a complete classification of $\mathbb{Z}_2^2$-symmetric space G/H when $\mathfrak{g}$ is simple complex. From these classifications, we can easily deduce the classification of compact $\mathbb{Z}_2^2$-symmetric spaces when $\mathfrak{g}$ is compact and real simple. Following the terminology of[3] and,[1] if $(G, H, \mathbb{Z}_2^2)$ is a $\mathbb{Z}_2^2$-symmetric space, the corresponding pair $(\mathfrak{g}, \mathfrak{h})$ of Lie algebras is called local symmetric space. From the local classification, it is very easy to exhibit the classification of the $\mathbb{Z}_2^2$-symmetric spaces when G and H are connected and G simply connected. In the following table, we give the list of local $\mathbb{Z}_2^2$-symmetric spces when $\mathfrak{g}$ is real simple of compact type and non exceptional.

$(\mathfrak{g},\mathfrak{h})$
$(su(2n), su(n))$
$(su(k_1+k_2), su(k_1) \oplus su(k_2) \oplus \mathbb{C})$
$(su(k_1+k_2+k_3), su(k_1) \oplus su(k_2) \oplus su(k_3) \oplus \mathbb{C}^2)$
$(su(k_1+k_2+k_3+k_4), su(k_1) \oplus su(k_2) \oplus su(k_3) \oplus su(k_4) \oplus \mathbb{C}^3)$
$(su(n), so(n))$
$(su(2m), sp(m))$
$(su(k_1+k_2), so(k_1) \oplus so(k_2))$
$(su2(k_1+k_2), sp(2k_1) \oplus sp(2k_2))$
$(so(k_1+k_2+k_3), so(k_1) \oplus so(k_2) \oplus so(k_3)$
$(so(k_1+k_2+k_3+k_4), so(k_1) \oplus so(k_2) \oplus so(k_3) \oplus so(k_4))$
$(so(4m), sp(2m))$
$(so(2m), so(m))$
$(so(8), su(3) \oplus su(1))$
$(sp(k_1+k_2+k_3+k_4), sp(k_1) \oplus sp(k_2) \oplus sp(k_3) \oplus sp(k_4))$
$(sp(4m), sp(2m))$
$(sp(2m), so(m))$

4. On the classification of Riemannian compact $\mathbb{Z}_2^2$-symmetric spaces

Let $(M = G/H, g)$ be a Riemannian $\mathbb{Z}_2^2$-symmetric space. We assume that the local pair is in the previous table. In this case, M is compact. To each local pair, we have to classify, up to an isometry, the corresponding $ad(H)$-invariant bilinear form B. With regard to the symmetric case, the computation is more complicated because we obtain, not only the Killing Cartan metric, but also a large class of definite positive or only non degenerate invariant bilinear forms. In the work, we are interested by two classes

- The $\mathbb{Z}_2^2$-symmetric flag manifolds. These spaces have been very well studied and our approach permits to look these metrics with the symmetric point of view.
- The homogeneous space $SO(4m)/Sp(2m)$ because it has no equivalent in the symmetric case.

4.1. $\mathbb{Z}_2^2$-symmetric metrics on flag manifolds

Let $M = SO(2l+1)/SO(r_1) \times SO(r_2) \times SO(r_3) \times SO(r_4)$ be an oriented flag manifold (with $r_1 r_2 r_3 \neq 0$. This manifold is a $\mathbb{Z}_2^2$-symmetric space and the grading of the Lie algebra $so(2l+1)$ is given by

$$\mathfrak{g}_e = \begin{pmatrix} X_1 & 0 & 0 & 0 \\ 0 & X_2 & 0 & 0 \\ 0 & 0 & X_3 & 0 \\ 0 & 0 & 0 & X_4 \end{pmatrix}, \mathfrak{g}_a = \begin{pmatrix} 0 & A_1 & 0 & 0 \\ -{}^tA_1 & 0 & 0 & 0 \\ 0 & 0 & 0 & A_2 \\ 0 & 0 & -{}^tA_2 & 0 \end{pmatrix}$$

$$\mathfrak{g}_b = \begin{pmatrix} 0 & 0 & B_1 & 0 \\ 0 & 0 & 0 & B_2 \\ -{}^tB_1 & 0 & 0 & 0 \\ 0 & -{}^tB_2 & 0 & 0 \end{pmatrix}, \mathfrak{g}_c = \begin{pmatrix} 0 & 0 & 0 & C_1 \\ 0 & 0 & C_2 & 0 \\ 0 & -{}^tC_2 & 0 & 0 \\ -{}^tC_1 & 0 & 0 & 0 \end{pmatrix}$$

where A_1 (resp. B_1, C_1, C_2, A_2, X_i) is a matrix of order (r_1, r_2) (resp. $(r_1, r_3), (r_1, r_4), (r_2, r_3), (r_2, r_4), (r_3, r_4)$ and (r_i, r_i)). Let B be a $\mathfrak{g}_e$-invariant inner product on $\mathfrak{g}$. By hypothesis $B(\mathfrak{g}_\alpha, \mathfrak{g}_\beta) = 0$ as soon as $\alpha \neq \beta$ in $\mathbb{Z}_2^2$. This shows that B is written as $B = B_{\mathfrak{g}_e} + B_{\mathfrak{g}_a} + B_{\mathfrak{g}_b} + B_{\mathfrak{g}_c}$ where $B_{\mathfrak{g}_x}$ is an inner product on $\mathfrak{g}_x$. We denote by $\{\alpha_{ij}^1, \alpha_{ij}^2, \beta_{ij}^1, \beta_{ij}^2, \gamma_{ij}^1, \gamma_{ij}^2\}$ the dual basis of the elemetary basis of $A_1 \oplus A_2 \oplus B_1 \oplus B_2 \oplus C_1 \oplus C_2$ given by the elementary matrices. A direct computation conduces to the following result:

Proposition 4.1. *Every $ad(\mathfrak{g}_e)$-invariant inner product on $\mathfrak{m} = \mathfrak{g}_a \oplus \mathfrak{g}_b \oplus \mathfrak{g}_c$ is given by the formula*

$$B = t_{A_1}\Sigma(\alpha_{ij}^1)^2 + t_{A_2}\Sigma(\alpha_{ij}^2)^2 + t_{B_1}\Sigma(\beta_{ij}^1)^2 + t_{B_2}\Sigma(\beta_{ij}^2)^2 + t_{C_1}\Sigma(\gamma_{ij}^1)^2 + t_{C_2}\Sigma(\gamma_{ij}^2)^2$$

with $t_{A_1}, t_{A_2}, t_{B_1}, t_{B_2}, t_{C_1}, t_{C_2}$ not zero.

Consequences.
1. Such a bilinear form defines a naturally reductive structure on M if and only if

$$t_{A_1} = t_{A_2} = t_{B_1} = t_{B_2} = t_{C_1} = t_{C_2} = \lambda > 0.$$

2. On Lorentzian $\mathbb{Z}_2^2$-symmetric structure.

Definition 4.1. Let (G, H, Γ_G) be a Γ-symmetric space, g a semi-Riemannian metric of signature $(1, n-1)$ where $n = \dim M$ and B the corresponding $ad\mathfrak{g}_e$-invariant symmetric bilinear form on $\mathfrak{m}$. Then $M = G/H$ is called a Γ-*symmetric Lorentzian space* if the homogeneous components of $\mathfrak{m}$ are pairwise orthogonal with respect to B.

From the classification of $ad\mathfrak{g}_e$-invariant forms on $so(2l+1)$ given in Proposition 4.1, the $(\mathbb{Z}_2^2)$-symmetric space $SO(2l+1)/SO(r_1) \times ... \times SO(r_4)$ is Lorentzian if and only if there exists one homogeneous component of $\mathfrak{m}$ of dimension 1. For example if we consider the $(\mathbb{Z}_2^2)$-symmetric space $SO(5)/SO(2) \times SO(2) \times SO(1)$ the homogeneous components are of dimension 2 and every semi-Riemannian metric is of signature $(2p, 8-2p)$ and cannot be a Lorentzian metric. So $SO(5)/SO(2) \times SO(2) \times SO(1)$ cannot be Lorentzian (as a $\mathbb{Z}_2^2$-symmetric space). Nevertheless one may consider the grading of $so(5)$ given by

$$\begin{pmatrix} 0 & a_1 & b_1 & b_2 & b_3 \\ -a_1 & 0 & c_1 & c_2 & c_3 \\ -b_1 & -c_1 & 0 & x_1 & x_2 \\ -b_2 & -c_2 & -x_1 & 0 & x_3 \\ -b_3 & -c_3 & -x_2 & -x_3 & 0 \end{pmatrix}$$

where $\mathfrak{g}_e$ is parametrized by x_1, x_2, x_3, $\mathfrak{g}_a$ by a_1, $\mathfrak{g}_b$ by b_1, b_2, b_3 and $\mathfrak{g}_c$ by c_1, c_2, c_3. Let us denote by $\{X_1, X_2, X_3, A_1, B_1, B_2, B_3, C_1, C_2, C_3\}$ the corresponding graded basis. Here $\mathfrak{g}_e$ is isomorphic to $so(3) \oplus so(1) \oplus so(1)$ and we obtain the $\mathbb{Z}_2^2$-symmetric homogeneous space

$$SO(5)/SO(3) \times SO(1) \times SO(1) = SO(5)/SO(3).$$

Every nondegenerate symmetric bilinear form on $so(5)$ invariant by $g_e = so(3)$ is written

$$q = t(\omega_1^2 + \omega_2^2 + \omega_3^2) + u\alpha_1^2 + v(\beta_1^2 + \beta_2^2 + \beta_3^2) + w(\gamma_1^2 + \gamma_2^2 + \gamma_3^2)$$

where $\{\omega_i, \alpha_1, \beta_i, \gamma_i\}$ is the dual basis of the basis $\{X_i, A_1, B_i, C_i\}$. In particular, we obtain:

4.2. The $\mathbb{Z}_2^2$-Riemannian symmetric space $SO(2m)/Sp(m)$

We first define the grading of $so(2m)$. Consider the matrices

$$S_m = \begin{pmatrix} 0 & I_m \\ -I_n & 0 \end{pmatrix}, X_a = \begin{pmatrix} -1 & 0 \\ 0 & 1 \end{pmatrix}, X_b = \begin{pmatrix} 0 & 1 \\ 1 & 0 \end{pmatrix}, X_c = \begin{pmatrix} 0 & -1 \\ 1 & 0 \end{pmatrix}.$$

The linear maps on $so(2m)$ given by $\tau_\alpha(M) = J_\alpha^{-1} M J_\alpha$, $\alpha = a, b, c$ where $J_a = S_m \otimes X_a$, $J_b = S_m \otimes X_b$, $J_c = S_m \otimes X_c$ are involutive automorphisms of $so(2m)$ which pairwise commute. Thus $\{Id, \tau_a, \tau_b, \tau_c\}$ is a finite subgroup of $Aut(so(2m))$ isomorphic to $\mathbb{Z}_2^2$. We deduce the $\mathbb{Z}_2^2$-grading

$$so(2m) = \mathfrak{g}_e \oplus \mathfrak{g}_a \oplus \mathfrak{g}_b \oplus \mathfrak{g}_c$$

where

$$\mathfrak{g}_e = \left\{ \begin{pmatrix} A_1 & B_1 & A_2 & B_2 \\ -B_1 & A_1 & B_2 & -A_2 \\ \hline -{}^tA_2 & -{}^tB_2 & A_1 & B_1 \\ -{}^tB_2 & {}^tA_2 & -B_1 & A_1 \end{pmatrix} \text{ with } \begin{matrix} {}^tA_1 = -A_1, & {}^tB_1 = B_1 \\ {}^tA_2 = A_2, & {}^tB_2 = B_2 \end{matrix} \right\}$$

$$\mathfrak{g}_a = \left\{ \begin{pmatrix} X_a & Y_a & Z_a & T_a \\ Y_a & -X_a & -T_a & Z_a \\ \hline -{}^tZ_a & {}^tT_a & -X_a & -Y_a \\ -{}^tT_a & -{}^tZ_a & -Y_a & X_a \end{pmatrix} \text{ with } \begin{matrix} {}^tX_a = -X_a, & {}^tY_a = -Y_a \\ {}^tZ_a = -Z_a, & {}^tT_a = T_a \end{matrix} \right\}$$

$$\mathfrak{g}_b = \left\{ \begin{pmatrix} X_b & Y_b & Z_b & T_b \\ -Y_b & X_b & T_b & Z_b \\ \hline -{}^tZ_b & -{}^tT_b & -X_b & -Y_b \\ -{}^tT_b & -{}^tZ_b & Y_b & -X_b \end{pmatrix} \text{ with } \begin{matrix} {}^tX_b = -X_b, & {}^tY_b = Y_b \\ {}^tZ_b = -Z_b, & {}^tT_b = -T_b \end{matrix} \right\}$$

$$\mathfrak{g}_c = \left\{ \begin{pmatrix} X_c & Y_c & Z_c & T_c \\ Y_c & -X_c & -T_c & Z_c \\ \hline -{}^tZ_c & {}^tT_c & X_c & Y_c \\ -{}^tT_c & -{}^tZ_c & Y_c & -X_c \end{pmatrix} \text{ with } \begin{matrix} {}^tX_c = -X_c, & {}^tY_c = -Y_c \\ {}^tZ_c = Z_c, & {}^tT_c = -T_c \end{matrix} \right\}.$$

The subalgebra $\mathfrak{g}_e$ is isomorphic to $sp(m)$ and from[1], every $\mathbb{Z}_2^2$-grading of $so(2m)$ such that $\mathfrak{g}_e$ is isomorphic to $sp(m)$ is equivalent to the previous one.'The symmetries of the $\mathbb{Z}_2^2$-symmetric space $SO(2m)/Sp(m)$ at any point x can be described as soon as we know the expression of the symmetries at the point $\bar{1}$, the class on the quotient $SO(2m)/Sp(m)$ of 1, the unit of the group $SO(2m)$. Moreover we have $s_{\gamma,\bar{1}}(\overline{A}) = \overline{(\rho_\gamma(A))}$ with $\rho_\alpha(A) = J_\alpha^{-1} A J_\alpha$, $\alpha = a, b, c$. Let g be a Riemannian $\mathbb{Z}_2^2$-symmetric metric on $SO(2m)/Sp(m)$. The corresponding bilinear form B on $\mathfrak{g}_a \oplus \mathfrak{g}_b \oplus \mathfrak{g}_c$ is $\mathfrak{g}_e$-invariant and the linear spaces $\mathfrak{g}_a, \mathfrak{g}_b, \mathfrak{g}_c$ are orthogonal. Then B writes $B = B_a + B_b + B_c$ where B_γ is a nondegenerate bilinear form on $\mathfrak{g}_\gamma$ for $\gamma = a, b, c$ such that the kernel contains $\oplus_{\gamma' \neq \gamma} \mathfrak{g}_{\gamma'}$. Let $\{X_{\gamma,ij}, Y_{\gamma,ij}, Z_{\gamma,ij}, T_{\gamma,ij}\}$ the basis of $\mathfrak{g}_\gamma$ given by elementary matrices which generate $(X_\gamma, Y_\gamma, Z_\gamma, T_\gamma)$ and $\{\alpha_{\gamma,ij}, \beta_{\gamma,ij}, \gamma_{\gamma,ij}, \delta_{\gamma,ij}\}$ its dual basis.

Proposition 4.2. *Every Riemannian (indefinite) $\mathbb{Z}_2^2$-symmetric on*

$SO(2m)/Sp(m)$ is defined from the bilinear form B whose quadratic form is written $q_B = q_{\mathfrak{g}_a} + q_{\mathfrak{g}_b} + q_{\mathfrak{g}_b}$ with

$$\begin{cases} q_{\mathfrak{g}_a} = \lambda_1^a \left(\sum(\alpha_{a,ij}^2 + \beta_{a,ij}^2 + \gamma_{a,ij}^2) + \sum_{i \neq j} \delta_{a,ij}^2\right) + \lambda_2^a(\delta_{a,ii}^2) \\ \quad + (\lambda_2^a - \frac{\lambda_1^a}{2})(\sum_{i<j}(\delta_{a,ii}\delta_{a,jj})) \\ q_{\mathfrak{g}_b} = \lambda_1^b(\sum(\alpha_{b,ij}^2 + \gamma_{ij}^2) + \delta_{b,ij}^2 + \sum_{i \neq j} \beta_{b,ij}^2) + \lambda_2^b(\beta_{b,ii}^2) \\ \quad + (\lambda_2^b - \frac{\lambda_1^b}{2})(\sum_{i<j}(\beta_{b,ii}\beta_{b,jj})) \\ q_{\mathfrak{g}_c} = \lambda_1^c(\sum(\beta_{c,ij}^2 + \gamma_{c,ij}^2) + \delta_{c,ij}^2 + \sum_{i \neq j} \alpha_{c,ij}^2) + \lambda_2^c(\alpha_{c,ii}^2) \\ \quad + (\lambda_2^c - \frac{\lambda_1^c}{2})(\sum_{i<j}(\alpha_{c,ii}\alpha_{c,jj})). \end{cases}$$

Let $\gamma \in \{a,b,c\}$. The eigenvalues of $q_{\mathfrak{g}_\gamma}$ are

$$\mu_{1,\gamma} = \lambda_1^\gamma, \quad \mu_{2,\gamma} = \lambda_2^\gamma/2 + \lambda_1^\gamma/4, \quad \mu_{3,\gamma} = \lambda_2^\gamma \frac{r+1}{2} - \lambda_1^\gamma \frac{r-1}{4},$$

where $r = \frac{m^2+m-2}{m^2+m+2}$ is the order of symmetric matrices T_a, Y_b, Z_c. These roots are respectively of multiplicities $\dim \mathfrak{g}_\gamma - r, r-1, 1$. We deduce

Theorem 4.1. *Every definite positive Riemannian $\mathbb{Z}_2^2$-symmetric metric on $SO(2m)/Sp(m)$ is given from the bilinear form B whose quadratic associated form*

$$q_B = q_{\mathfrak{g}_a}(\lambda_1^a, \lambda_2^a) + q_{\mathfrak{g}_b}(\lambda_1^b, \lambda_2^b) + q_{\mathfrak{g}_b}(\lambda_1^b, \lambda_2^b)$$

satisfies $\lambda_1^\gamma > 0$ and $\lambda_2^\gamma > \lambda_1^\gamma \frac{m^2+m-2}{2(m^2+m+2)})$ for all $\gamma \in \{a,b,c\}$. Such a metric is naturally reductive if and only if

$$\lambda_1^a = \lambda_1^b = \lambda_1^c = 2\lambda_2^a = 2\lambda_2^b = 2\lambda_2^c.$$

For the lorentzian case, we have

Theorem 4.2. *Every lorentzian $\mathbb{Z}_2^2$-symmetric metric on $SO(2m)/Sp(m)$ is given from the bilinear form B whose quadratic associated form*

$$q_B = q_{\mathfrak{g}_a}(\lambda_1^a, \lambda_2^a) + q_{\mathfrak{g}_b}(\lambda_1^b, \lambda_2^b) + q_{\mathfrak{g}_b}(\lambda_1^b, \lambda_2^b)$$

satisfies

$$\begin{cases} \forall \gamma \in \{a,b,c\}, \ \lambda_1^\gamma > 0, \\ \exists \gamma_0 \in \{a,b,c\} \text{ such that } -\lambda_1^{\gamma_0}/2 < \lambda_2^{\gamma_0} < \lambda_1^{\gamma_0} \frac{r-1}{2(r+1)}, \\ \forall \gamma \neq \gamma_0, \ \lambda_2^\gamma > \lambda_1^\gamma \frac{r-1}{2(r+1)}. \end{cases}$$

References

1. Y. Bahturin and M. Goze, *Pacific J. Math.* **236** (2008), 1–21.
2. A. Bouyakoub, M. Goze, and E. Remm, On Riemannian nonsymmetric spaces and flag manifolds. Preprint Mulhouse (2006).
3. M. Berger, *Ann. Sci. École Norm. Sup. (3)* **74** (1957), 85–177.
4. S. Kobayashi and K. Nomizu, *Foundations of differential geometry, Vol. II.* Interscience Tracts in Pure and Applied Mathematics, No. 15 Vol. II Interscience Publishers John Wiley & Sons, Inc., New York-London-Sydney 1969
5. Kollross, Andreas. arXiv:0808.0306.

METHODS FOR SOLVING THE JACOBI EQUATION. CONSTANT OSCULATING RANK VS. CONSTANT JACOBI OSCULATING RANK

Teresa Arias-Marco

Departamento de Matemáticas, Universidad de Extremadura
Badajoz, 06071, Spain
E-mail: ariasmarco@unex.es

This article shows the relation between the known methods for solving the Jacobi equation. As example, we work on the H-type group given by A. Kaplan.[6] Here, the method for solving the Jacobi equation based in the used of the constant Jacobi osculating rank of the g.o. space is applied explicitly for the first time using the constant osculating rank of the Jacobi operator. Moreover, we show an explicit case where the constant Jacobi osculating rank and the constant osculating rank are not the same.

Keywords: Jacobi equation, H-type groups, g.o. spaces, constant osculating rank of the Jacobi operator, Jacobi osculating rank of a g.o. space

1. Introduction and preliminaries

Let M be a Riemannian manifold and let ∇ denote the Levi-Civita connection. For an arbitrary vector $x \in T_p M$, we denote by γ_t or γ the geodesic through $p \in M$ whose tangent vector at $\gamma_0 = p$ is $\dot{\gamma}_0 = x$. The *Jacobi operator* along γ is defined by $\mathcal{J}_t(\cdot) = R(\cdot, \dot{\gamma}_t)\dot{\gamma}_t$ where R denotes the curvature tensor. Let $\{E_i\}$ be an orthonormal basis of $T_p M$. We denote by $\{Q_i\}$ the orthonormal frame field obtained by $\nabla_{\dot{\gamma}}$-parallel translation of the basis $\{E_i\}$ along γ. Moreover, let Q_t be the vector formed by the elements of $\{Q_i\}$. A vector field Y_t along a geodesic γ of M is called *Jacobi vector field* if it satisfies the *Jacobi equation* along γ: $Y_t'' + \mathcal{J}_t(Y_t) = 0$ with initial conditions $Y_0 = 0$, $Y_0' = Q_0$.

The resolution of the Jacobi equation on a Riemannian manifold can be quite difficult. However, in a symmetric space the problem is reduced to a system of differential equations with constant coefficients. I. Chavel[4,5] solved this problem for some particular directions of the geodesic on the naturally reductive spaces $Sp(2)/SU(2)$ and $SU(5)/(Sp(2) \times S^1)$. Nonetheless, the method used by I. Chavel does not seem to solve in a simple way the Jacobi equation along a geodesic with arbitrary direction. For naturally reductive compact homogeneous spaces, W. Ziller[9] solved the Jacobi equation but the solution can be considered of qualitative type (it does not allow us to obtain in an easy way the Jacobi vectors fields neither for any particular example nor for an arbitrary geodesic). On the other hand, J. Berndt, F. Tricerry and L. Vanhecke[3] solved it along an arbitrary geodesic on any H-type group. The methods used by all these authors are special cases of a more general procedure based on the use of the canonical connection. This procedure is valid, in particular, on any g.o. space and any generalized Heisenberg group. From now on, we will refer to it as *standard method* and it will be presented in Section 3.2.

A g.o. space is a Riemannian homogeneous space on which every geodesic is an orbit of a one-parameter group of isometries. The first counter-example of a g.o. space which is not naturally reductive is Kaplan's example. This is a six-dimensional H-type group. On the other hand, K. Tsukada[8] pointed out on naturally reductive spaces and, the author and A. M. Naveira[2] showed on g.o. spaces that $\mathcal{J}_t$ has *constant osculating rank* i.e., for every fix geodesic γ there is a natural number r_γ and there are $\beta_1, ..., \beta_{r_\gamma}$ constant such that $\beta_1 \mathcal{J}_t^{1)} + \cdots + \beta_{r_\gamma} \mathcal{J}_t^{r_\gamma)} + \mathcal{J}_t^{r_\gamma+1)} = 0$. In Ref. 2, the authors settled the concept of *Jacobi osculating rank of a g.o. space* as the number $\mathbf{r} = max\{r_\gamma :$ for all γ of the g.o. space$\}$ and they established that a g.o. space has *constant Jacobi osculating rank* if there is also a relation $\beta_1 \mathcal{J}_t^{1)} + \cdots + \beta_r \mathcal{J}_t^{r)} + \mathcal{J}_t^{r+1)} = 0$ that it is the same for all γ of the g.o. space. It is clear that $r_\gamma \leq \mathbf{r}$ for all γ of a g.o. space. Moreover, $r_\gamma = \mathbf{r} = 0$ for all γ of a symmetric space (i.e. $\nabla R = 0$). Therefore, the following question is natural : *is $r_\gamma = \mathbf{r}$ for all γ of a g.o. space with constant Jacobi osculating rank?* This question was not answer in Ref. 2 where the authors proved that Kaplan's example has $\mathbf{r} = 4$ and constant Jacobi osculating rank due to $\frac{1}{4}|x|^4 \mathcal{J}_t^{1)} + \frac{5}{4}|x|^2 \mathcal{J}_t^{3)} + \mathcal{J}_t^{5)} = 0$. In Section 3.1, we will answer the question providing a family of geodesics on Kaplan's example with $r_\gamma = 2$.

Recently, constant Jacobi osculating rank was used by A. M. Naveira and A. Tarrío[7] to develop a recursive method for solving the Jacobi equation. They worked on the space $Sp(2)/SU(2)$ although they did not applied it explicitly. In Ref. 1, the author and S. Bartoll presented the non-recursive version of the previous method. Moreover, they applied explicitly the non-recursive method working on a geodesic of the naturally reductive space $U(3)/U(1) \times U(1) \times U(1)$ with $r_\gamma = \mathbf{r} = 4$. In Section 3.2, we will recall the recursive method namely, *the constant osculating rank method* and we will applied it on the family of geodesics presented in Section 3.1. Moreover, we will propose a simplification of this method based in the use of r_γ instead of r when we want to solve the Jacobi equation on geodesics with $r_\gamma \neq r$.

Finally, we will devote Section 3.3 to compare the results obtained by standard and constant osculating rank methods in Section 3.2.

2. Preliminaries about H-type groups

Let $\mathfrak{n}$ be a 2-step nilpotent Lie algebra with an inner product $\langle \, , \, \rangle$. Let $\mathfrak{z}$ be the center of $\mathfrak{n}$ and let $\mathfrak{v}$ be its orthogonal complement. For each vector $A \in \mathfrak{z}$, the operator $j(A) : \mathfrak{v} \to \mathfrak{v}$ is defined by the relation

$$\langle j(A)X, Y \rangle = \langle A, [X,Y] \rangle \quad \text{for all} \quad X, Y \in \mathfrak{v}. \tag{1}$$

The algebra $\mathfrak{n}$ is called *H-type algebra* if, for each $A \in \mathfrak{z}$, the operator $j(A)$ satisfies the identity $j(A)^2 = -|A|^2 I_\mathfrak{v}$ where $|\ |^2$ denotes the quadratic form of the inner product $\langle \, , \, \rangle$. A connected, simply connected Lie group whose Lie algebra is an H-type algebra is diffeomorphic to $\mathbb{R}^n$ and it is called *H-type group*. It is endowed with a left-invariant metric.

In particular, the Lie algebra structure on $\mathfrak{n}$ is defined by extending the skew-symmetric bilinear map $[\, ,\,] : \mathfrak{v} \times \mathfrak{v} \to \mathfrak{z}$ to a bracket $[A+X, B+Y] = [X,Y]$ where $A, B \in \mathfrak{z}$ and $X, Y \in \mathfrak{v}$.

Moreover, a geodesic γ_t through the origin p of an H-type group N is described by means of two vector-valued functions $t \to X_t \in \mathfrak{v}$, $t \to A_t \in \mathfrak{z}$ as follows: $\gamma_t = \exp(X_t + A_t)$ such that $X_0 = 0$, $A_0 = 0$ and $\dot\gamma_0 = \dot X_0 + \dot A_0$. (See Refs. 3 and 6 for more information about H-type groups).

Finally, we recall a recursive expression for the nth covariant derivative of the Jacobi operator at p of an H-type group given in Ref. 2. Let $\zeta_{(n,A)} : \mathfrak{z} \to \mathfrak{z}$, $\nu_{(n,A)} : \mathfrak{z} \to \mathfrak{v}$, $\zeta_{(n,X)} : \mathfrak{v} \to \mathfrak{z}$ and $\nu_{(n,X)} : \mathfrak{v} \to \mathfrak{v}$ be the mappings defined by the following recurrent formulas where $B \in \mathfrak{z}$ and $Y \in \mathfrak{v}$.

$$\zeta_{(0,A)}(B) = \tfrac{1}{4}|\dot X_0|^2 B, \quad \nu_{(0,A)}(B) = \tfrac{1}{2} j(B) j(\dot A_0) \dot X_0 - \tfrac{1}{4} j(\dot A_0) j(B) \dot X_0,$$

$$\zeta_{(0,X)}(Y) = \tfrac{1}{4}[\dot X_0, j(\dot A_0)Y] - \tfrac{1}{2}[Y, j(\dot A_0)\dot X_0],$$

$$\nu_{(0,X)}(Y) = \tfrac{1}{4}|\dot A_0|^2 Y + \tfrac{3}{4} j([Y, \dot X_0]) \dot X_0.$$

$$\zeta_{(n,A)}(B) = \tfrac{1}{2}([\dot X_0, \nu_{(n-1,X)}(B)] + \zeta_{(n-1,X)}(j(B)\dot X_0)),$$

$$\nu_{(n,A)}(B) = \tfrac{1}{2}(\nu_{(n-1,X)}(j(B)\dot X_0) - j(\dot A_0)\nu_{(n-1,A)}(B) - j(\zeta_{(n-1,A)}(B))\dot X_0),$$

$$\zeta_{(n,X)}(Y) = \tfrac{1}{2}([\dot X_0, \nu_{(n-1,X)}(Y)] + \zeta_{(n-1,X)}(j(\dot A_0)Y) - \zeta_{(n-1,A)}([\dot X_0, Y])),$$

$$\nu_{(n,X)}(Y) = \tfrac{1}{2}(\nu_{(n-1,X)}(j(\dot A_0)Y) - j(\dot A_0)\nu_{(n-1,X)}(Y)$$
$$- j(\zeta_{(n-1,X)}(Y))\dot X_0 - \nu_{(n-1,A)}([\dot X_0, Y])). \tag{2}$$

Proposition 2.1. *The nth covariant derivative of the Jacobi operator at the origin of an H-type group for any $B \in \mathfrak{z}$, $Y \in \mathfrak{v}$ is given by*

$$\mathcal{J}_0^{n)}(B) = \zeta_{(n,A)}(B) + \nu_{(n,A)}(B), \quad \mathcal{J}_0^{n)}(Y) = \zeta_{(n,X)}(Y) + \nu_{(n,X)}(Y).$$

3. Kaplan's example

3.1. *Constant osculating rank of the Jacobi operator along a special family of geodesics*

Let $\mathfrak{n}$ be a vector space of dimension 6 equipped with a scalar product and let $\{E_i\}_{i=1}^6$ form an orthonormal basis. The elements E_5 and E_6 span the center $\mathfrak{z}$ of the Lie algebra $\mathfrak{n}$. The structure of a Lie algebra on $\mathfrak{n}$ is given by the following relations:

$$[E_1, E_2] = [E_3, E_4] = [E_5, E_6] = [E_k, E_5] = [E_k, E_6] = 0, \ k = 1,...,4,$$
$$[E_1, E_3] = E_5, \quad [E_1, E_4] = E_6, \quad [E_2, E_3] = E_6, \quad [E_2, E_4] = -E_5. \tag{3}$$

Moreover, from (1) and (3) we easily obtain that

$$j(E_5)E_1 = E_3, \ j(E_5)E_2 = -E_4, \ j(E_5)E_3 = -E_1, \ j(E_5)E_4 = E_2,$$
$$j(E_6)E_1 = E_4, \ j(E_6)E_2 = E_3, \ j(E_6)E_3 = -E_2, \ j(E_6)E_4 = -E_1. \tag{4}$$

The H-type group corresponding to $\mathfrak{n}$ is named *Kaplan's example* and from now on we will denote it by N. Moreover, in the following we will always suppose that $x \in \mathfrak{n}$ is an arbitrary unit vector with null-center; i.e., $x = \dot{X}_0 + \dot{A}_0$, $\dot{X}_0 = \sum_{i=1}^4 x_i E_i$, $\dot{A}_0 = 0$ and $|x|^2 = 1$. In addition, γ will denote the family of geodesics on N with $\gamma_0 = p$ and $\dot{\gamma}_0 = x$.

Now, we want to calculate the constant osculating rank of the Jacobi operator along γ, i.e. the number r_γ and the constants such that $\beta_1 \mathcal{J}_t^{1)} + \cdots + \beta_{r_\gamma} \mathcal{J}_t^{r_\gamma)} + \mathcal{J}_t^{r_\gamma+1)} = 0$.

Let us start with the following technical lemmas on N. Their proofs are direct considering the linearity of all operators and using (2), (3) and (4).

Lemma 3.1. *The operator $[\dot{X}_0, Y] \in \mathfrak{z}$, $Y \in \mathfrak{v}$, is given by*

$$[\dot{X}_0, E_1] = -x_3 E_5 - x_4 E_6, \quad [\dot{X}_0, E_2] = x_4 E_5 - x_3 E_6,$$
$$[\dot{X}_0, E_3] = x_1 E_5 + x_2 E_6, \quad [\dot{X}_0, E_4] = -x_2 E_5 + x_1 E_6. \tag{5}$$

$j(\dot{A}_0) : \mathfrak{v} \to \mathfrak{v}$ *is the null operator and* $j(\cdot)(\dot{X}_0) : \mathfrak{z} \to \mathfrak{v}$ *is given by*

$$j(E_5)(\dot{X}_0) = -x_3 E_1 + x_4 E_2 + x_1 E_3 - x_2 E_4,$$
$$j(E_6)(\dot{X}_0) = -x_4 E_1 - x_3 E_2 + x_2 E_3 + x_1 E_4. \tag{6}$$

Lemma 3.2. *The mappings* $\zeta_{(0,X)}$, $\nu_{(0,X)}$, $\zeta_{(0,A)}$, $\nu_{(0,A)}$ *are defined by*

$\nu_{(0,X)}(E_1) = \frac{-3}{4}((x_3^2 + x_4^2)E_1 - (x_1x_3 + x_2x_4)E_3 + (x_2x_3 - x_1x_4)E_4)$,

$\nu_{(0,X)}(E_2) = \frac{-3}{4}((x_3^2 + x_4^2)E_2 + (x_1x_4 - x_2x_3)E_3 - (x_1x_3 + x_2x_4)E_4)$,

$\nu_{(0,X)}(E_3) = \frac{3}{4}((x_1x_3 + x_2x_4)E_1 + (x_2x_3 - x_1x_4)E_2 - (x_1^2 + x_2^2)E_3)$,

$\nu_{(0,X)}(E_4) = \frac{3}{4}((x_1x_4 - x_2x_3)E_1 + (x_1x_3 + x_2x_4)E_2 - (x_1^2 + x_2^2)E_4)$,

$\zeta_{(0,X)}(E_i) = 0$, $i = 1,...,4$, $\nu_{(0,A)}(E_\alpha) = 0$, $\zeta_{(0,A)}(E_\alpha) = \frac{|x|^2}{4}E_\alpha$, $\alpha = 5, 6$.

Lemma 3.3. *The mappings* $\zeta_{(1,X)}$, $\nu_{(1,X)}$, $\zeta_{(1,A)}$, $\nu_{(1,A)}$ *are defined by*

$\zeta_{(1,X)}(E_1) = \frac{|x|^2}{2}(x_3E_5 + x_4E_6)$, $\zeta_{(1,X)}(E_2) = \frac{|x|^2}{2}(-x_4E_5 + x_3E_6)$,

$\zeta_{(1,X)}(E_3) = \frac{|x|^2}{2}(-x_1E_5 - x_2E_6)$, $\zeta_{(1,X)}(E_4) = \frac{|x|^2}{2}(x_2E_5 - x_1E_6)$,

$\nu_{(1,X)}(E_i) = 0$, $i = 1,...,4$, $\nu_{(1,A)}(E_5) = \frac{|x|^2}{2}(x_3E_1 - x_4E_2 - x_1E_3 + x_2E_4)$,

$\nu_{(1,A)}(E_6) = \frac{|x|^2}{2}(x_4E_1 + x_3E_2 - x_2E_3 - x_1E_4)$, $\zeta_{(1,A)}(E_\alpha) = 0$, $\alpha = 5, 6$.

Lemma 3.4. *The mappings* $\zeta_{(2,X)}$, $\nu_{(2,X)}$, $\zeta_{(2,A)}$, $\nu_{(2,A)}$ *are defined by*

$\nu_{(2,X)}(E_1) = \frac{|x|^2}{2}((x_3^2 + x_4^2)E_1 - (x_1x_3 + x_2x_4)E_3 + (x_2x_3 - x_1x_4)E_4)$,

$\nu_{(2,X)}(E_2) = \frac{|x|^2}{2}((x_3^2 + x_4^2)E_2 + (x_1x_4 - x_2x_3)E_3 - (x_1x_3 + x_2x_4)E_4)$,

$\nu_{(2,X)}(E_3) = \frac{|x|^2}{2}(-(x_1x_3 + x_2x_4)E_1 + (x_1x_4 - x_2x_3)E_2 + (x_1^2 + x_2^2)E_3)$,

$\nu_{(2,X)}(E_4) = \frac{|x|^2}{2}((x_2x_3 - x_1x_4)E_1 - (x_1x_3 + x_2x_4)E_2 + (x_1^2 + x_2^2)E_4)$,

$\zeta_{(2,X)}(E_i) = 0$, $i = 1,...,4$, $\nu_{(2,A)}(E_\alpha) = 0$, $\zeta_{(2,A)}(E_\alpha) = \frac{-|x|^4}{2}E_\alpha$, $\alpha = 5,6$.

Lemma 3.5. *The mappings* $\zeta_{(3,X)}$, $\nu_{(3,X)}$, $\zeta_{(3,A)}$, $\nu_{(3,A)}$ *are defined by*

$\zeta_{(3,X)}(E_1) = \frac{|x|^4}{2}(-x_3E_5 - x_4E_6)$, $\zeta_{(3,X)}(E_2) = \frac{|x|^4}{2}(x_4E_5 - x_3E_6)$,

$\zeta_{(3,X)}(E_3) = \frac{|x|^4}{2}(x_1E_5 + x_2E_6)$, $\zeta_{(3,X)}(E_4) = \frac{|x|^4}{2}(-x_2E_5 + x_1E_6)$,

$\nu_{(3,X)}(E_i) = 0, i = 1,...,4, \nu_{(3,A)}(E_5) = \frac{|x|^4}{2}(-x_3E_1 + x_4E_2 + x_1E_3 - x_2E_4)$,

$\nu_{(3,A)}(E_6) = \frac{|x|^4}{2}(-x_4E_1 - x_3E_2 + x_2E_3 + x_1E_4)$, $\zeta_{(3,A)}(E_\alpha) = 0$, $\alpha = 5, 6$.

Moreover, by a lengthy but elementary calculation using Proposition 2.1, the lemmas 3.2, 3.3, 3.4, 3.5 and the linearity of all operators, we get the nth covariant derivative of the Jacobi operator at p of N. It is given by the symmetric matrix $\mathcal{J}_0^{n)} = (\mathcal{J}_{ij}^{n)})$, $i,j = 1,...,6$, where $\mathcal{J}_{\alpha\beta}^n = \langle \zeta_{(n,A)}(Q_\alpha), Q_\beta \rangle(0) = \langle \zeta_{(n,A)}(E_\alpha), E_\beta \rangle$, $\mathcal{J}_{\alpha j}^n = \langle \nu_{(n,A)}(E_\alpha), E_j \rangle$,

$\mathcal{J}_{i\beta}^n = \langle \zeta_{(n,X)}(E_i), E_\beta \rangle$, $\mathcal{J}_{ij}^n = \langle \nu_{(n,X)}(E_i), E_j \rangle$ for $i,j = 1,...,4$, $\alpha, \beta = 5, 6$.
The non-zero elements of $\mathcal{J}_0^{n)}$, $n = 0,...,3$ are the following:

$$\mathcal{J}_{11}^0 = \mathcal{J}_{22}^0 = \frac{-3}{4}(x_3^2 + x_4^2), \; \mathcal{J}_{33}^0 = \mathcal{J}_{44}^0 = \frac{-3}{4}(x_1^2 + x_2^2), \; \mathcal{J}_{55}^0 = \mathcal{J}_{66}^0 = \frac{|x|^2}{4},$$
$$\mathcal{J}_{13}^0 = \mathcal{J}_{24}^0 = \frac{3}{4}(x_1 x_3 + x_2 x_4), \; \mathcal{J}_{14}^0 = -\mathcal{J}_{23}^0 = \frac{3}{4}(-x_2 x_3 + x_1 x_4).$$
(7)

$$\mathcal{J}_{15}^1 = \mathcal{J}_{26}^1 = \frac{|x|^2}{2} x_3, \; \mathcal{J}_{16}^1 = -\mathcal{J}_{25}^1 = \frac{|x|^2}{2} x_4,$$
$$\mathcal{J}_{35}^1 = \mathcal{J}_{46}^1 = \frac{-|x|^2}{2} x_1, \; \mathcal{J}_{36}^1 = -\mathcal{J}_{45}^1 = \frac{-|x|^2}{2} x_2.$$
(8)

$$\mathcal{J}_{11}^2 = \mathcal{J}_{22}^2 = \frac{|x|^2}{2}(x_3^2 + x_4^2), \; \mathcal{J}_{33}^2 = \mathcal{J}_{44}^2 = \frac{|x|^2}{2}(x_1^2 + x_2^2), \; \mathcal{J}_{55}^2 = \mathcal{J}_{66}^2 = \frac{-|x|^4}{2},$$
$$\mathcal{J}_{14}^2 = -\mathcal{J}_{23}^2 = \frac{|x|^2}{2}(x_2 x_3 - x_1 x_4), \; \mathcal{J}_{13}^2 = \mathcal{J}_{24}^2 = \frac{-|x|^2}{2}(x_1 x_3 + x_2 x_4).$$
(9)

$$\mathcal{J}_{15}^3 = \mathcal{J}_{26}^3 = \frac{-|x|^4}{2} x_3, \; \mathcal{J}_{16}^3 = -\mathcal{J}_{25}^3 = \frac{-|x|^4}{2} x_4,$$
$$\mathcal{J}_{35}^3 = \mathcal{J}_{46}^3 = \frac{|x|^4}{2} x_1, \; \mathcal{J}_{36}^3 = -\mathcal{J}_{45}^3 = \frac{|x|^4}{2} x_2.$$
(10)

Proposition 3.1. *The constant osculating rank of the Jacobi operator along the family of geodesics γ on N is given by $r_\gamma = 2$, $|x|^2 \mathcal{J}_t^{1)} + \mathcal{J}_t^{3)} = 0$.*

Proof. Let us consider the linear homogeneous system of equations form by $A\mathcal{J}_{11}^1 + B\mathcal{J}_{11}^2 = 0$ and $A\mathcal{J}_{15}^1 + B\mathcal{J}_{15}^2 = 0$. Using (8) and (9) we conclude that $A = B = 0$. Therefore, $\mathcal{J}_0^{1)}$ and $\mathcal{J}_0^{2)}$ are linear independent.

On the other hand, we easily check from (8), (9) and (10) that

$$|x|^2 \mathcal{J}_0^{1)} + \mathcal{J}_0^{3)} = 0.$$
(11)

Finally, we use (11) in the expansion in Taylor's series of $|x|^2 \mathcal{J}_t^{1)} + \mathcal{J}_t^{3)}$. □

Note that $\beta_1 = |x|^2$, $\beta_2 = 0$ and $\beta_3 = 1$. Moreover, the value of these constants is always the same on every member of the family of geodesics γ on N. Therefore, we can use the constant osculating rank method for solving the Jacobi equation along γ. The following relation will be the key in the next section. It is obtained using (11) and the induction method.

$$\mathcal{J}_0^{2l+m)} = (-1)^l |x|^{2l} \mathcal{J}_0^{m)}, \quad m = 1, 2, \quad l = 1, 2, 3, ...$$
(12)

3.2. Resolution of the Jacobi equation

Here, we use the constant osculating rank method and the standard method to calculate the first Jacobi vector field along the previous family of geodesics with $\dot{A}_0 = 0$ on Kaplan's example N.

The constant osculating rank method.[7] Let $Y_t = D_t Q_t$ be the Jacobi vector field along a geodesic γ. Then, the Jacobi equation $Y_t'' + \mathcal{J}_t(Y_t) = 0$ with initial conditions $Y_0 = 0$, $Y_0' = Q_0$, can be rewritten as $(D_t Q_t)'' + \mathcal{J}_t(D_t Q_t) = (D_t)'' Q_t + D_t \mathcal{J}_t(Q_t) = 0$ and, consequently, as $D_t^{2)} + D_t \mathcal{J}_t = 0$ with initial values $D_0 = 0$, $D_0^{1)} = I$ where we consider the covariant differentiation with respect to $\dot{\gamma}$ and I is the identity transformation of $T_p M$. Therefore, to obtain the expression of the Jacobi fields it is enough to know the development in Taylor's series of D_t. Thus, we successively derive $D_t^{2)} = -D_t \mathcal{J}_t$ and we apply the initial conditions. Finally, D_t along the geodesic γ on N is given by $D_t = \sum_{k=0}^{\infty} D_0^{k)} \frac{t^k}{k!}$ where $D_t^{0)} = C_t^0 = C_t^1 = 0$, $D_t^{1)} = I$ and, for $k \geq 2$, $D_t^{k)} = C_t^{k-1} + \left(D_t^{k-1)}\right)'$, $C_t^k = \left(C_t^{k-1}\right)' - \mathcal{J}_t D_t^{k-1)}$. Moreover, if we work on a g.o. space with constant Jacobi osculating rank, $D_0^{k)}$ only depend of $\mathcal{J}_0, \mathcal{J}_0^{1)}, ..., \mathcal{J}_0^{r)}$.

Now, we use the previous methodology assuming that γ denotes the family of geodesics with $\dot{A}_0 = 0$ on Kaplan's example N. In Ref. 2, the authors proved that N has constant Jacobi osculating rank with $r = 4$. However, we use that $r_\gamma = 2$ (see Proposition 3.1) and the relation (12) to obtain the Jacobi vector fields instead of the information provided by the constant Jacobi osculating rank. Thus, $D_0^{k)}$ only depend of $\mathcal{J}_0, \mathcal{J}_0^{1)}$ and $\mathcal{J}_0^{2)}$. Note that we simplified the method proposed in Ref. 7. More explicitly,

Lemma 3.6. *Each Jacobi vector field along the family of geodesics with $\dot{A}_0 = 0$ on N is given by $(Y_t)_i = \sum_{j=1}^{6} (D_t)_{ij} Q_j$, $i = 1, ..., 6$ where*

$$(D_t)_{ij} = I_{ij} t - (\mathcal{J}_0)_{ij} \frac{t^3}{3!} - 2(\mathcal{J}_0^{1)})_{ij} \frac{t^4}{4!} + ((\mathcal{J}_0 \mathcal{J}_0)_{ij} - 3(\mathcal{J}_0^{2)})_{ij}) \frac{t^5}{5!} + (4|x|^2$$
$$(\mathcal{J}_0^{1)})_{ij} + 2(\mathcal{J}_0^{1)} \mathcal{J}_0)_{ij} + 4(\mathcal{J}_0 \mathcal{J}_0^{1)})_{ij}) \frac{t^6}{6!} + (5|x|^2 (\mathcal{J}_0^{2)})_{ij} - (\mathcal{J}_0 \mathcal{J}_0 \mathcal{J}_0)_{ij}$$
$$+ 3(\mathcal{J}_0^{2)} \mathcal{J}_0)_{ij} + 10(\mathcal{J}_0^{1)} \mathcal{J}_0^{1)})_{ij} + 10(\mathcal{J}_0 \mathcal{J}_0^{2)})_{ij}) \frac{t^7}{7!} + (18(\mathcal{J}_0^{2)} \mathcal{J}_0^{1)})_{ij}$$
$$+ 30(\mathcal{J}_0^{1)} \mathcal{J}_0^{2)})_{ij} - 2(\mathcal{J}_0^{1)} \mathcal{J}_0 \mathcal{J}_0)_{ij} - 4(\mathcal{J}_0 \mathcal{J}_0^{1)} \mathcal{J}_0)_{ij} - 4|x|^2 (\mathcal{J}_0^{1)} \mathcal{J}_0)_{ij}$$
$$- 6(\mathcal{J}_0 \mathcal{J}_0 \mathcal{J}_0^{1)})_{ij} - 20|x|^2 (\mathcal{J}_0 \mathcal{J}_0^{1)})_{ij} - 6|x|^4 (\mathcal{J}_0^{1)})_{ij}) \frac{t^8}{8!} + O(t^9).$$

Finally, using (7), (8) and (9) in Lemma 3.6 we obtain that

$$(Y_t)_1 = tQ_1 + S(t)((x_3^2 + x_4^2)Q_1 - (x_1 x_3 + x_2 x_4)Q_3 \qquad (13)$$
$$+ (x_2 x_3 - x_1 x_4)Q_4) + T(t)(\tfrac{1}{2} t^2 (x_3 Q_5 + x_4 Q_6)),$$

where $S(t) = \frac{1}{8} t^3 - \frac{1}{128} t^5 + \frac{1}{9216} t^7 + O(t^9) = \sum_{n=0}^{\infty} \frac{(-1)^n}{(n+1)(2n)!} \left(\frac{t}{2}\right)^{(2n+3)} = \frac{1}{2} t(-2 + 2\cos(t/2) + t\sin(t/2))$ and $T(t) = -\frac{1}{12} t^2 + \frac{1}{480} t^4 - \frac{1}{53760} t^6 + O(t^7) = \sum_{n=1}^{\infty} \frac{(-1)^n}{(2n-1)(2n+1)!} \left(\frac{t}{2}\right)^{(2n)} = \cos(t/2) - \frac{2}{t} \sin(t/2)$.

The standard method.[3] This method is based in the transformation of the Jacobi equation using a new covariant derivative. Let M be a Riemannian manifold with Levi-Civita connection ∇ and γ_t a geodesic in M parametrized by arc length. Suppose there exists a $\nabla_{\dot\gamma}$-parallel skew-symmetric tensor field T_t along γ such that the Jacobi operator $\mathcal{J}_t$ along γ satisfies $\nabla_{\dot\gamma}\mathcal{J}_t := [\mathcal{J}_t, T_t]$. Thus, we define a new covariant derivative $\bar\nabla_{\dot\gamma} := \nabla_{\dot\gamma} + T_t$, and put $\bar{\mathcal{J}}_t := \mathcal{J}_t + T_t^2$. Then $\mathcal{J}_t$, $\bar{\mathcal{J}}_t$, T_t are $\bar\nabla_{\dot\gamma}$-parallel along γ and the Jacobi equation becames $\bar\nabla_{\dot\gamma}\bar\nabla_{\dot\gamma}B_t - 2T_t\bar\nabla_{\dot\gamma}B_t + \bar{\mathcal{J}}_tB_t = 0$ where B_t is a vector field along γ.

Now, let us denote by $\{P_i\}_{i=1}^6$ the orthonormal frame field obtained by $\bar\nabla_{\dot\gamma}$-parallel translation of the basis $\{E_i\}_{i=1}^6$ along γ. The standard method was used in p. 52 of Ref. 3 for solving the Jacobi equation on any H-type group. In such case, $\bar\nabla$ is the canonical connection. Now, using their result we conclude that the first Jacobi field along the geodesic γ with $\dot{A}_0 = 0$ is

$$(B_t)_1 = tP_1 + \tfrac{1}{2}t^2[P_1, \dot\gamma_0] \stackrel{(3)}{=} tP_1 + \tfrac{1}{2}t^2(x_3P_5 + x_4P_6). \qquad (14)$$

3.3. *Relation between both methods*

It is well-known that on any n-dimensional homogeneous space the canonical and the Levi-Civita connection have the same geodesics. Moreover, Lemma 5 of Ref. 5 establishes the relation between the parallel orthonormal frame fields $\{Q_i\}$ and $\{P_i\}$ along an arbitrary geodesic γ. This is

$$Q_i(t) = \sum_{j=1}^n a_{ij}(t)P_j(t), \quad i = 1,...,n \qquad (15)$$

where $(a_{ij}(0)) = I$ and $a'_{ij}(t) = \sum_{k=1}^n a_{ik}(t)(T_\gamma P_k(t))_j$ for $i,j = 1,...,n$.

In this section, we will use (15) to relate (13) with (14). From now on, if there is no confusion we will denote the matrix $(a_{ij}(t))$ by (a_{ij}).

Lemma 3.7. (a_{ij}), $i,j = 1,...,6$ *along the family of geodesics with* $\dot{A}_0 = 0$ *on Kaplan's example is given by*

$$a_{11} = a_{22} = x_1^2 + x_2^2 + \cos(t/2)(x_3^2 + x_4^2), \ a_{16} = -a_{25} = a_{52} = \sin(t/2)x_4,$$

$$a_{13} = a_{24} = a_{31} = a_{42} = (1 - \cos(t/2))(x_1x_3 + x_2x_4),$$

$$a_{12} = a_{21} = a_{34} = a_{43} = a_{56} = a_{65} = 0, \quad a_{55} = a_{66} = \cos(t/2),$$

$$a_{14} = -a_{23} = -a_{32} = a_{41} = (-1 + \cos(t/2))(x_2x_3 - x_1x_4),$$

$$a_{33} = a_{44} = \cos(t/2)(x_1^2 + x_2^2) + x_3^2 + x_4^2, \ a_{15} = a_{26} = -a_{51} = \sin(t/2)x_3,$$

$$a_{35} = a_{46} = -a_{53} = -\sin(t/2)x_1, \quad a_{36} = -a_{45} = a_{54} = -\sin(t/2)x_2.$$
$$\qquad (16)$$

Proof. we know from p. 45 of Ref. 3 that $T_\gamma P_i = \frac{1}{2}[P_i, x]$, $i = 1, 2, 3, 4$, $T_\gamma P_\alpha = \frac{1}{2}j(P_i)(x)$, $\alpha = 5, 6$. More specifically, using (3) and (4) we have

$$T_\gamma P_1 = \tfrac{1}{2}(x_3 P_5 + x_4 P_6), \quad T_\gamma P_2 = \tfrac{1}{2}(x_3 P_6 - x_4 P_5),$$
$$T_\gamma P_3 = \tfrac{-1}{2}(x_1 P_5 + x_2 P_6), \; T_\gamma P_5 = \tfrac{1}{2}(x_1 P_3 - x_2 P_4 - x_3 P_1 + x_4 P_2), \quad (17)$$
$$T_\gamma P_4 = \tfrac{1}{2}(x_2 P_5 - x_1 P_6), \; T_\gamma P_6 = \tfrac{1}{2}(x_1 P_4 + x_2 P_3 - x_3 P_2 - x_4 P_1).$$

Thus, using (17) in (15), we obtain an homogeneous systems of ordinary differential equations, $a_i' = M \cdot a_i$, with initial condition $a_{ii}(0) = 1$, $a_{ij}(0) = 0$, $j \neq i$ for each $i \in \{1, 2, 3, 4, 5, 6\}$ where a_i and a_i' denote the vectors (a_{ij}) and (a_{ij}'), $j = 1, ..., 6$, respectively, and M is the antisymmetric matrix whose non-zero elements are $M_{15} = M_{26} = -\frac{1}{2}x_3$, $M_{16} = -M_{25} = -\frac{1}{2}x_4$, $M_{35} = M_{46} = \frac{1}{2}x_1$, $M_{36} = -M_{45} = \frac{1}{2}x_2$.

Finally, we will solve the system following the general theory of ordinary differential equations. The eigenvalues associated to M are $\{0, -\frac{1}{2}\imath, \frac{1}{2}\imath\}$, all of them with multiplicity 2. A basis $\{v_1, v_2\}$ of $\ker M^2$ and a basis $\{w_1, w_2\}$ of $\ker(M - \frac{\imath}{2}Id)^2$ are given by $v_1 = (\frac{x_1 x_3 + x_2 x_4}{x_3^2 + x_4^2}, \frac{x_2 x_3 - x_1 x_4}{x_3^2 + x_4^2}, 1, 0, 0, 0)$, $v_2 = (\frac{x_1 x_4 - x_2 x_3}{x_3^2 + x_4^2}, \frac{x_1 x_3 + x_2 x_4}{x_3^2 + x_4^2}, 0, 1, 0, 0)$, $w_1 = (\frac{x_3}{x_2}, \frac{-x_4}{x_2}, \frac{-x_1}{x_2}, 1, \frac{-\imath}{x_2}, 0)$, $w_2 = (\frac{\imath(x_1 x_3 + x_2 x_4)}{x_2}, \frac{\imath(x_2 x_3 - x_1 x_4)}{x_2}, \frac{-\imath(x_1^2 + x_2^2)}{x_2}, 0, \frac{x_1}{x_2}, 1)$. Moreover, $e^{tM} v_1 = v_1 + tM \cdot v_1 = v_1$, $e^{tM} v_2 = v_2 + tM \cdot v_2 = v_2$, $e^{tM} w_1 = e^{\frac{\imath t}{2}}(w_1 + t(M - \frac{\imath}{2}Id) \cdot w_1) = e^{\frac{\imath t}{2}} w_1$, $e^{tM} w_2 = e^{\frac{\imath t}{2}}(w_2 + t(M - \frac{\imath}{2}Id) \cdot w_2) = e^{\frac{\imath t}{2}} w_2$. Thus, the general solution of the system $a_i' = M \cdot a_i$ is $a_i = C_{i1} e^{tM} v_1 + C_{i2} e^{tM} v_2 + C_{i3} \text{Re}(e^{tM} w_1) + C_{i4} \text{Im}(e^{tM} w_1) + C_{i5} \text{Re}(e^{tM} w_2) + C_{i6} \text{Im}(e^{tM} w_2)$ or, more explicitly, it is

$$a_{i1}(t) = \tfrac{x_1 x_3 + x_2 x_4}{x_3^2 + x_4^2} C_{i1} + \tfrac{x_1 x_4 - x_2 x_3}{x_3^2 + x_4^2} C_{i2} + \tfrac{x_3}{x_2}(\cos(t/2) C_{i3} + \sin(t/2) C_{i4})$$
$$+ \tfrac{x_1 x_3 + x_2 x_4}{x_2}(\cos(t/2) C_{i6} - \sin(t/2) C_{i5}),$$
$$a_{i2}(t) = \tfrac{x_2 x_3 - x_1 x_4}{x_3^2 + x_4^2} C_{i1} + \tfrac{x_1 x_3 + x_2 x_4}{x_3^2 + x_4^2} C_{i2} - \tfrac{x_4}{x_2}(\cos(t/2) C_{i3} + \sin(t/2) C_{i4})$$
$$+ \tfrac{x_2 x_3 - x_1 x_4}{x_2}(\cos(t/2) C_{i6} - \sin(t/2) C_{i5}),$$
$$a_{i3}(t) = C_{i1} + \tfrac{-x_1}{x_2}(\cos(t/2) C_{i3} + \sin(t/2) C_{i4}) \qquad (18)$$
$$+ \tfrac{(x_1^2 - x_2^2)}{x_2}(\sin(t/2) C_{i5} - \cos(t/2) C_{i6}),$$
$$a_{i4}(t) = C_{i2} + \cos(t/2) C_{i3} + \sin(t/2) C_{i4},$$
$$a_{i5}(t) = \tfrac{1}{x_2}(\sin(t/2) C_{i3} - \cos(t/2) C_{i4}) + \tfrac{x_1}{x_2}(\cos(t/2) C_{i5} + \sin(t/2) C_{i6}),$$
$$a_{i6}(t) = \cos(t/2) C_{i5} + \sin(t/2) C_{i6},$$

where $C_{i1}, C_{i2}, C_{i3}, C_{i4}, C_{i5}, C_{i6}$ are arbitrary parameters that must be calculated for finishing the proof. We obtain their values solving the system $a_{ii}(0) = 1$, $a_{ij}(0) = 0$, $j \neq i$, for each $i \in \{1, .., 6\}$. □

Finally using (16), we substitute Q_i by $\sum_{j=1}^{6} a_{ij} P_j$, $i = 1, ..., 6$, in (13).

We obtain

$$(Y_t)_1 = tP_1 - f(t)((x_3^2 + x_4^2)P_1 - (x_1x_3 + x_2x_4)P_3 \\ + (x_2x_3 - x_1x_4)P_4) + \tfrac{1}{2}t^2 g(t)(x_3P_5 + x_4P_6), \quad (19)$$

where $f(t) = t(1 - \cos(t/2)) - \cos(t/2)S(t) + \frac{t^2}{2}\sin(t/2)T(t)$ and $g(t) = \frac{2}{t}\sin(t/2) + \cos(t/2)T(t) + \frac{2}{t^2}\sin(t/2)S(t)$. Thus, (14) is equal to (19) if and only if $f(t) = 0$, $g(t) = 1$ or, equivalently, if and only if $S(t) = \frac{1}{2}t(-2 + 2\cos(t/2) + t\sin(t/2))$, $T(t) = \cos(t/2) - \frac{2}{t}\sin(t/2)$. Therefore, (13) and (14) are obviously equivalent.

Remark 3.1. Although the standard method can be used on any Riemannian manifold it does not always give general explicit results as on H-type groups (recall the results obtained by I. Chavel[4,5] and W. Ziller[9]). Nonetheless, the constant osculating rank method can be always applied on g.o. spaces with easy and straightforward computations.

Acknowledgements

This work is partially supported by D.G.I. (Spain) and FEDER Project MTM2007-65852, the network MTM2006-27480-E/, a grant ACOMP07/088 from A.V.C.T. and by DFG Sonderforschungsbereich 647.

References

1. T. Arias-Marco, S. Bartoll: *Constant Jacobi osculating rank of U(3)/(U(1)× U(1)× U(1))*. Preprint.
2. T. Arias-Marco, A. M. Naveira, *Constant Jacobi osculating rank of a g.o. space. A method to obtain explicitly the Jacobi operator*. Preprint.
3. J. Berndt, F. Tricerri, L. Vanhecke, *Lecture Notes in Mathematics*, **1598**. Springer-Verlag, Berlin, 1995
4. I. Chavel, *Bull. Amer. Math. Soc.* **73** (1967), 477–481.
5. I. Chavel, *Comment. Math. Helvetici* **42** (1967), 237–248.
6. A. Kaplan, *Bull. London Math. Soc.* **15** (1983), 35–42.
7. A. M. Naveira, A. Tarrío, A method for the resolution of the Jacobi equation $Y'' + RY' = 0$ on the manifold $Sp(2)/SU(2)$, to appear in *Monasth. Math.*
8. K. Tsukada, *Kodai Math. J.* **19** (1996), 395–437.
9. W. Ziller, *Comment. Math. Helvetici* **52** (1977), 573–590.

ON THE REPARAMETRIZATION OF AFFINE HOMOGENEOUS GEODESICS

Zdeněk Dušek

Department of Algebra and Geometry, Palacky University
Tomkova 40, 779 00 Olomouc, Czech Republic
E-mail: dusek@prfnw.upol.cz

In this paper, the reparametrization of affine homogeneous geodesics is studied on the examples in dimension 2. In particular, the examples of affine g.o. manifolds such that almost all homogeneous geodesics require a reparametrization to obtain the affine parameter are presented.

Keywords: Homogeneous affine connection, affine Killing vector field, homogeneous geodesic, homogeneous manifold, g.o. manifold

1. Introduction

A geodesic in a homogeneous pseudo-Riemannian or affine manifold is said to be homogeneous if it is an orbit of an one-parameter group of isometries, or, of affine diffeomorphisms, respectively. A homogeneous pseudo-Riemannian or affine manifold is a g.o. manifold if every geodesic is homogeneous. On pseudo-Riemannian manifolds, the parameter of the group of isometries may be different from the affine parameter of the geodesic only for null homogeneous geodesics. Nevertheless, for all known examples of pseudo-Riemannian g.o. manifolds, these parameters are the same for all homogeneous geodesics. In the affine case, there are g.o. manifolds whose almost all geodesics must be reparametrized.

In Section 2, we recall the essential features of pseudo-Riemannian g.o. manifolds. The details can be found in Refs. 3,4,7,8. In Section 3, we present the new approach for the study of homogeneous geodesics in affine homogeneous manifold. This method was developed and used for homogeneous affine manifolds in dimension 2 in Ref. 6. In Section 4, we recall the classification of 2-dimensional homogeneous affine connection given by B. Opozda for the torsion-free case in Ref. 9 and later refined and generalized to arbitrary torsion by T. Arias-Marco and O. Kowalski in Ref. 1. In Sections 5,

6 and 7, we investigate in more details some special 2-dimensional homogeneous affine manifolds, which were studied in Ref. 6, and we also present a new example in Section 5.

2. Homogeneous geodesics in pseudo-Riemannian manifolds

Let M be a pseudo-Riemannian manifold. If there is a connected Lie group $G \subset I_0(M)$ which acts transitively on M as a group of isometries, then M is called a *homogeneous pseudo-Riemannian manifold*. Let $p \in M$ be a fixed point. If we denote by H the isotropy group at p, then M can be identified with the *homogeneous space* G/H. In general, there may exist more than one such group $G \subset I_0(M)$. The pseudo-Riemannian metric g on M can be considered as a G-invariant metric on G/H. The pair $(G/H, g)$ is then called a *pseudo-Riemannian homogeneous space*.

If the metric g is positive definite, then $(G/H, g)$ is always a *reductive homogeneous space*: We denote by $\mathfrak{g}$ and $\mathfrak{h}$ the Lie algebras of G and H respectively and consider the adjoint representation Ad: $H \times \mathfrak{g} \to \mathfrak{g}$ of H on $\mathfrak{g}$. There exists a direct sum decomposition (*reductive decomposition*) of the form $\mathfrak{g} = \mathfrak{m} + \mathfrak{h}$ where $\mathfrak{m} \subset \mathfrak{g}$ is a vector subspace such that $\mathrm{Ad}(H)(\mathfrak{m}) \subset \mathfrak{m}$. If the metric g is indefinite, the reductive decomposition may not exist. For a fixed reductive decomposition $\mathfrak{g} = \mathfrak{m} + \mathfrak{h}$ there is a natural identification of $\mathfrak{m} \subset \mathfrak{g} = T_e G$ with the tangent space $T_p M$ via the projection $\pi\colon G \to G/H = M$. Using this natural identification and the scalar product g_p on $T_p M$ we obtain a scalar product $\langle , \rangle$ on $\mathfrak{m}$. This scalar product is obviously $\mathrm{Ad}(H)$-invariant.

Definition 2.1. A geodesic $\gamma(s)$ through the point p defined in an open interval J (where s is an affine parameter) is said to be *homogeneous* if there exists
1) a diffeomorphism $s = \varphi(t)$ between the real line and the open interval J;
2) a vector $X \in \mathfrak{g}$ such that $\gamma(\varphi(t)) = \exp(tX)(p)$ for all $t \in (-\infty, +\infty)$.
The vector X is then called a *geodesic vector*.

Lemma 2.1 (see Ref. 5). *Let $X \in \mathfrak{g}$. Then the curve $\gamma(t) = \exp(tX)(p)$ is a geodesic curve with respect to some parameter s if and only if*

$$\langle [X, Z]_\mathfrak{m}, X_\mathfrak{m} \rangle = k \langle X_\mathfrak{m}, Z \rangle \tag{1}$$

for all $Z \in \mathfrak{m}$ and for some constant $k \in \mathbb{R}$.

Further, if $k = 0$, then t is an affine parameter for this geodesic. If $k \neq 0$, then $s = e^{-kt}$ is an affine parameter for the geodesic. The second case can occur only if the curve $\gamma(t)$ is a null curve in a (properly) pseudo-Riemannian manifold.

In Ref. 2, based on the examples in dimensions 6 and 7, the following conjecture was formulated:

Conjecture 2.1. *Let G/H be a pseudo-Riemannian g.o. manifold. For all null homogeneous geodesics one has $k = 0$ in Lemma 2.1.*

3. Homogeneous geodesics in affine manifolds

For the study of homogeneous geodesics in affine manifolds, we cannot use the algebraic tools as in the pseudo-Riemannian case, because we have no reductive decomposition and no scalar product here. We briefly describe the new method which is based on Killing vector fields.

Definition 3.1. A vector field X on an affine manifold (M, ∇) is called an *affine Killing vector field* if the Lie derivative $\mathcal{L}_X \nabla$ vanishes, or, equivalently, if X satisfies the equation

$$[X, \nabla_Y Z] - \nabla_Y [X, Z] - \nabla_{[X,Y]} Z = 0, \qquad (2)$$

for all vector fields Y, Z.

Proposition 3.1. *An affine manifold (M, ∇) is homogeneous if it admits at least $n = \dim M$ complete affine Killing vector fields which are linearly independent at each point.*

Definition 3.2. Let (M, ∇) be a homogeneous affine manifold. A *homogeneous geodesic* is a geodesic which is an orbit of an one-parameter group of affine diffeomorphisms. (Here the canonical parameter of the group need not be the affine parameter of the geodesic). An *affine g.o. space* is a homogeneous affine manifold (M, ∇) such that each geodesic is homogeneous.

Proposition 3.2. *Let $M = G/H$ be a homogeneous space with a left-invariant affine connection ∇. Then each regular orbit of a 1-parameter subgroup $g_t \subset G$ on M is an integral curve of an affine Killing vector field on M.*

Definition 3.3. A nonvanishing smooth vector field Z on M is said to be *geodesic along its regular integral curve γ* if the curve $\gamma(t)$ is geodesic up to a possible reparametrization. If all regular integral curves of Z are geodesics up to a reparametrization, then the vector field Z is called *a geodesic vector field*.

Proposition 3.3 (see Ref. 6). *Let Z be a nonvanishing Killing vector field on $M = (G/H, \nabla)$.*
1) Z is geodesic along its integral curve γ if and only if

$$\nabla_{Z_{\gamma(t)}} Z = k_\gamma \cdot Z_{\gamma(t)} \tag{3}$$

holds along γ, where $k_\gamma \in \mathbb{R}$ is a constant. If $k_\gamma = 0$, then t is the affine parameter of geodesic γ. If $k_\gamma \neq 0$, then the affine parameter of this geodesic is $s = e^{k_\gamma t}$.
2) Z is a geodesic vector field if and only if

$$\nabla_Z Z = k \cdot Z \tag{4}$$

holds on M. Here k is a smooth function on M, which is constant along integral curves of the vector field Z.

4. Locally homogeneous connections in dimension two

Theorem 4.1 (see Refs. 1,9). *Let ∇ be a locally homogeneous affine connection without torsion on a 2-dimensional manifold M. Then, in a neighborhood $\mathcal{U}$ of each point $m \in M$, either ∇ is locally a Levi-Civita connection of the unit sphere or, there is a system (u, v) of local coordinates and constants A, B, C, D, G, H such that ∇ is expressed in $\mathcal{U}$ by one of the following formulas:*

$$\text{type} A: \quad \nabla_{\partial_u} \partial_u = A \, \partial_u + B \, \partial_v, \ \nabla_{\partial_u} \partial_v = C \, \partial_u + D \, \partial_v,$$
$$\nabla_{\partial_v} \partial_u = C \, \partial_u + D \, \partial_v, \ \nabla_{\partial_v} \partial_v = G \, \partial_u + H \, \partial_v,$$

$$\text{type} B: \quad \nabla_{\partial_u} \partial_u = \tfrac{A}{u} \partial_u + \tfrac{B}{u} \partial_v, \ \nabla_{\partial_u} \partial_v = \tfrac{C}{u} \partial_u + \tfrac{D}{u} \partial_v,$$
$$\nabla_{\partial_v} \partial_u = \tfrac{C}{u} \partial_u + \tfrac{D}{u} \partial_v, \ \nabla_{\partial_v} \partial_v = \tfrac{G}{u} \partial_u + \tfrac{H}{u} \partial_v,$$

where not all A, B, C, D, G, H are zero.

5. G.o. manifolds of type A

Let us have a connection ∇ with constant Christoffel symbols $A, \ldots, H$ in $\mathbb{R}^2(u, v)$. It is easy to verify that the vector fields ∂_u and ∂_v satisfy the system of PDEs corresponding to the equation (2) and they are complete Killing vector fields. Hence, $(\mathbb{R}^2, \nabla)$ is a globally homogeneous manifold.

Let us consider a Killing vector field $X = x \, \partial_u + y \, \partial_v$ (where x, y are constants and $(x, y) \neq (0, 0)$) and its integral curves

$$\gamma(t) = \big(u(t), v(t)\big) = \big(xt + c_1, yt + c_2\big). \tag{5}$$

We calculate the covariant derivative of the tangent vector
$$\gamma'(t) = \big(u'(t), v'(t)\big) = \big(x, y\big) \tag{6}$$
along $\gamma(t)$. We obtain
$$\nabla_{\gamma'(t)}\gamma'(t) = \big(Ax^2 + 2C\,xy + G\,y^2\big)\partial_u + \big(B\,x^2 + 2D\,xy + H\,y^2\big)\partial_v. \tag{7}$$
We see that $\nabla_{\gamma'(t)}\gamma'(t)$ does not depend on the initial conditions of the integral curve and the formula (7) is the same for all integral curves of the vector field X. The condition
$$\nabla_{\gamma'(t)}\gamma'(t) = k\gamma'(t) \tag{8}$$
is satisfied for all x, y with $k = 0$ if and only if all the Christoffel symbols are equal to zero. In this case, $(\mathbb{R}^2, \nabla)$ is a g.o. space, every Killing field $X = x\,\partial_u + y\,\partial_v$ is geodesic and its integral curves do not require a reparametrization. For $k \neq 0$, the equality (8) is satisfied if and only if
$$Ax^2 + 2C\,xy + G\,y^2 = kx, \qquad B\,x^2 + 2D\,xy + H\,y^2 = ky, \tag{9}$$
which gives, by the elimination of the factor k, the equation
$$B\,x^3 + (2D - A)\,x^2y + (H - 2C)\,xy^2 - G\,y^3 = 0. \tag{10}$$

Example 5.1. Let us consider the affine connection whose Christoffel symbols satisfy the conditions
$$A = 2D, \qquad H = 2C, \qquad B = G = 0. \tag{11}$$
The equation (10) is obviously satisfied for all x and y. From the equations (9), we obtain
$$k = 2(Dx + Cy) \tag{12}$$
and we see that any integral curve of the vector field $X = x\partial_u + y\partial_v$ requires a reparametrization, unless $Dx + Cy = 0$. According to Proposition 3.3, the affine parameter of this geodesic is
$$s = e^{2(Dx+Cy)t}. \tag{13}$$
Now we are going to express these geodesics through the affine parameter. From the relation (13), we obtain easily
$$t = \frac{\log(s)}{2(Dx + Cy)} \tag{14}$$
and the geodesics and their tangent vectors are given by the formulas
$$\gamma(s) = \big(u(s), v(s)\big) = \Big(\frac{x\log(s)}{2(Dx + Cy)} + c_1, \frac{y\log(s)}{2(Dx + Cy)} + c_2\Big),$$

$$\gamma'(s) = \big(u'(s), v'(s)\big) = \left(\frac{x}{2(Dx+Cy)s}, \frac{y}{2(Dx+Cy)s}\right). \tag{15}$$

After a longer but straightforward computation we verify that

$$\nabla_{\gamma'(s)}\gamma'(s) = 0. \tag{16}$$

It is worth mentioning here that all integral curves of the Killing field X are reparametrized by the same diffeomorphism (13). With the affine parameter, geodesics are defined on the interval $(0, \infty)$.

By the direct solution of the system of PDEs corresponding to the equation (2), we can check that these affine connections admit another Killing vector fields, namely

$$(au + bv)(C\partial_u - D\partial_v), \tag{17}$$

for arbitrary a, b. The lines in the direction $(x, y) = (-C, D)$ are the only integral curves which do not need reparametrization when considered as the integral curves of a Killing field of the type $X = x\partial_u + y\partial_v$. As the orbits of the Killing field of the type (17), they obviously require a reparametrization.

On the other hand, if $C \neq 0$ or $D \neq 0$, integral curves of a Killing vector field $X = x\partial_u + y\partial_v$ such that $(x, y) \neq (-C, D)$ are not integral curves of any other Killing vector field. The reparametrization of these homogeneous geodesics cannot be avoided.

Example 5.2. Let us now consider the connection with the Christoffel symbols

$$A \neq 0, \quad C \neq 0, \quad B = D = H = G = 0. \tag{18}$$

If we consider the Killing vector fields of the type $X = x\partial_u + y\partial_v$, the equations (9) give us

$$Ax^2y + 2Cxy^2 = kx, \quad ky = 0. \tag{19}$$

We see that only the integral curves corresponding to $x = 0$, $y = 0$ or $Ax + 2Cy = 0$ are geodesics.

When we solve the system of PDEs corresponding to the equation (2), we find that the general Killing vector fields for these connections are

$$X = c_1 e^{-Au}(e^{-2Cv} + c_2)\partial_u + c_3\partial_v, \quad \text{for} \quad c_i \in \mathbb{R}. \tag{20}$$

Let us now consider the Killing vector field

$$X = xe^{-(Au+2Cv)}\partial_u + y\partial_v \tag{21}$$

for any $x, y \in \mathbb{R}$ and calculate the covariant derivative. We obtain

$$\nabla_X X = xe^{-(Au+2Cv)}\nabla_{\partial_u}\Big(xe^{-(Au+2Cv)}\partial_u + y\partial_v\Big)$$

$$+ y \nabla_{\partial_v}\Big(xe^{-(Au+2Cv)}\partial_u + y\partial_v\Big)$$
$$= xe^{-(Au+2Cv)}\Big(-Axe^{-(Au+2Cv)} + xe^{-(Au+2Cv)}A + yC\Big)\partial_u$$
$$+ y\Big(-2Cxe^{-(Au+2Cv)} + xe^{-(Au+2Cv)}C\Big)\partial_u = 0. \qquad (22)$$

We see that the Killing vector field X given by the formula (21) is geodesic. All its integral curves are geodesics and they do not need a reparametrization.

6. G.o. manifolds of type B

Let now ∇ be the affine connection on the manifold $\mathsf{H}_+^2 = \{\mathbb{R}^2(u,v) \mid u > 0\}$ whose Christoffel symbols are $\frac{A}{u}, \frac{B}{u}, \ldots \frac{H}{u}$, where $A, \ldots, H$ are constants. By the direct check we can verify that the vector fields ∂_v and $u\partial_u + v\partial_v$ satisfy the system of PDEs corresponding to the equation (2). These are complete affine Killing vector fields and (H_+^2, ∇) is a globally homogeneous manifold.

Let us consider the Killing vector field $X = x\,\partial_v + y\,(u\partial_u + v\partial_v)$, where x, y are arbitrary parameters and express its integral curves. We obtain

$$\gamma'(t) = \Big(u'(t), v'(t)\Big) = \Big(y\,u(t), x + y\,v(t)\Big) \qquad (23)$$

and by the integration in the case $y \neq 0$ we have

$$\gamma(t) = \Big(u(t), v(t)\Big) = \Big(c_1\,e^{yt}, c_2\,e^{yt} - x/y\Big) \qquad (24)$$

and in the case $y = 0$ we have

$$\gamma(t) = \Big(u(t), v(t)\Big) = \Big(c_1, xt + c_2\Big), \qquad (25)$$

where $c_1 > 0$ and $c_2 \in \mathbb{R}$ are integration constants. We calculate the covariant derivative of the tangent vector $\gamma'(t)$ along $\gamma(t)$ for $y \neq 0$:

$$\begin{aligned}
\nabla_{\gamma'(t)}\gamma'(t) &= c_1 y^2 e^{yt}\partial_u + c_2 y^2 e^{yt}\partial_v + c_1^2 y^2 e^{2yt}\nabla_{\partial_u}\partial_u \\
&= +2c_1 c_2 y^2 e^{2yt}\nabla_{\partial_u}\partial_v + c_2^2 y^2 e^{2yt}\nabla_{\partial_v}\partial_v \\
&= c_1 y^2 e^{yt}\partial_u + c_2 y^2 e^{yt}\partial_v + c_1^2 y^2 e^{2yt}\big(\tfrac{A}{c_1}e^{-yt}\partial_u + \tfrac{B}{c_1}e^{-yt}\partial_v\big) \\
&\quad + 2c_1 c_2 y^2 e^{2yt}\big(\tfrac{C}{c_1}e^{-yt}\partial_u + \tfrac{D}{c_1}e^{-yt}\partial_v\big) \\
&\quad + c_2^2 y^2 e^{2yt}\big(\tfrac{G}{c_1}e^{-yt}\partial_u + \tfrac{H}{c_1}e^{-yt}\partial_v\big) \\
&= \big((1+A)c_1 + G\tfrac{c_2^2}{c_1} + 2Cc_2\big)y^2 e^{yt}\partial_u \\
&\quad + \big((1+2D)c_2 + Bc_1 + H\tfrac{c_2^2}{c_1}\big)y^2 e^{yt}\partial_v.
\end{aligned} \qquad (26)$$

For $y = 0$ we obtain
$$\nabla_{\gamma'(t)}\gamma'(t) = \frac{x^2}{c_1}(G\partial_u + H\partial_v). \tag{27}$$

Now let us consider various cases of the affine connections of type B.

Example 6.1. Let the constants in the Christoffel symbols of the affine connections satisfy
$$A = -1, \quad D = -1/2, \quad B = C = H = G = 0. \tag{28}$$

Then in the equations (26) and (27), we obtain for any c_1, c_2
$$\nabla_{\gamma'(t)}\gamma'(t) = 0. \tag{29}$$

Obviously, any Killing field $X = x\,\partial_v + y\,(u\partial_u + v\partial_v)$ is geodesic and any integral curve is geodesic parametrized by the affine parameter.

Example 6.2. Let us consider the connections, whose Christoffel symbols satisfy
$$A = 2D, \quad H = 2C, \quad B = G = 0 \tag{30}$$

and $A \neq -1$ or $C \neq 0$. For $y \neq 0$, the relation (26) gives
$$\nabla_{\gamma'(t)}\gamma'(t) = \left((2D+1) + 2C\frac{c_2}{c_1}\right)y \cdot \gamma'(t). \tag{31}$$

If $C \neq 0$, the integral curve corresponding to $(2D+1)c_1 + 2Cc_2 = 0$ does not need a reparametrization. According to Proposition (3.3), other integral curves have to be reparametrized by the diffeomorphism
$$s = \exp\left(\left((2D+1) + 2C\frac{c_2}{c_1}\right)yt\right). \tag{32}$$

From the equality (32), we obtain
$$e^{yt} = s^{a_1}, \quad \text{where} \quad a_1 = \frac{c_1}{(2D+1)c_1 + 2Cc_2}. \tag{33}$$

For geodesics, we obtain
$$\gamma(s) = \big(u(s), v(s)\big) = \left(c_1 s^{a_1}, c_2 s^{a_1} - \frac{x}{y}\right),$$
$$\gamma'(s) = \big(u'(s), v'(s)\big) = \left(c_1 a_1 s^{a_1 - 1}, c_2 a_1 s^{a_1 - 1}\right). \tag{34}$$

After a longer, but straightforward calculation we can verify that
$$\nabla_{\gamma'(s)}\gamma'(s) = 0. \tag{35}$$

For $y = 0$, the equation (27) gives

$$\nabla_{\gamma'(t)}\gamma'(t) = \frac{2Cx}{c_1} \cdot \gamma'(t). \tag{36}$$

For $C \neq 0$, we use the reparametrization $s = \exp\left(\frac{2Cx}{c_1}t\right)$ and we obtain

$$\gamma(s) = \big(u(s), v(s)\big) = \left(c_1, \frac{c_1}{2C}\log(s) + c_2\right),$$
$$\gamma'(s) = \big(u'(s), v'(s)\big) = \left(0, \frac{c_1}{2C}\frac{1}{s}\right) \tag{37}$$

and we verify easily $\nabla_{\gamma'(s)}\gamma'(s) = 0$.

Because γ was arbitrary integral curve, we see again that any Killing vector field $X = x\,\partial_v + y\,(u\partial_u + v\partial_v)$ is geodesic, but the reparametrization depends on the initial conditions (c_1, c_2) and hence it is different for different integral curves of the given Killing vector field X.

We observe again, that the integral curves of the Killing vector field $X = x\,\partial_v + y\,(u\partial_u + v\partial_v)$ were defined on the interval $(-\infty, \infty)$, but the affine parameter of geodesics is from the interval $(0, \infty)$.

7. General connection of type B

Example 7.1. Let us now consider the connections, whose Christoffel symbols do not satisfy the relations (30), let us choose for example

$$A = -1, \quad B = C = 0, \quad D = 1/2, \quad G = -1, \quad H = -3. \tag{38}$$

From the equation (26), we obtain

$$\nabla_{\gamma'(t)}\gamma'(t) = -\frac{c_2^2}{c_1}y^2 e^{yt}\partial_u + \left(2c_2 - 3\frac{c_2^2}{c_1}\right)y^2 e^{yt}\partial_v. \tag{39}$$

Now, for the simplicity, we can always "normalize" the Killing vector field $X = x\,\partial_v + y\,(u\partial_u + v\partial_v)$ in a way that $y = 1$. We can also asume that the initial conditions (c_1, c_2) of the integral curve of this vector field sasisfy $c_1 = 1$, because any integral curve intersect the line $u = 1$. Hence, the tangent vector $\gamma'(t)$ is

$$\gamma'(t) = e^t \partial_u + c_2 e^t \tag{40}$$

and the equation (39) is

$$\nabla_{\gamma'(t)}\gamma'(t) = -c_2^2 e^t \partial_u + c_2(2 - 3c_2)e^t \partial_v. \tag{41}$$

We see easily that the equality $\nabla_{\gamma'(t)}\gamma'(t) = k\gamma'(t)$ is satisfied if and only if c_2 satisfies the equation

$$c_2(c_2^2 - 3c_2 + 2) = 0, \tag{42}$$

with the solutions

$$c_2 = 0 \ (k = 0), \qquad c_2 = 1 \ (k = -1), \qquad c_2 = 2 \ (k = -4). \tag{43}$$

We see that any Killing vector field $X = x\,\partial_v + y\,(u\partial_u + v\partial_v)$ is geodesic along exactly 3 integral curves corresponding to $c_1 = 1$ and c_2 from the above solutions.

We also see that there are three homogeneous geodesics through any point $p \in \mathsf{H}^2_+$ and two of them require a reparametrization. If $p = (1, p_v)$, then the components x, y of the Killing vector fields $X = x\,\partial_v + (u\partial_u + v\partial_v)$ corresponding to these geodesics can be calculated using the relation (24), because we suppose $\gamma(0) = p$, and we obtain

$$x = c_2 - p_v. \tag{44}$$

We see that the Killing vector fields corresponding to these geodesics depend on the point p.

Acknowledgments

This research was supported by the grant MSM 6198959214 of the Czech Ministry MŠMT

References

1. T. Arias-Marco, and O. Kowalski, *Monaths. Math.* **153** (2008), 1–18.
2. Z. Dušek, *Adv. Geom.* **9** (2009), 99–110.
3. Z. Dušek and O. Kowalski, Examples of pseudo-Riemannian g.o. manifolds, Geometry, Integrability and Quantization, I. Mladenov and M. de Leon (Eds.), Softex (2007), 144–155.
4. Z. Dušek and O. Kowalski, *Math. Nachr.* **254-255** (2003), 87–96.
5. Z. Dušek and O. Kowalski, *Publ. Math. Debrecen* **71** (2007), 245–252.
6. Z. Dušek, O. Kowalski and Z. Vlášek, Homogeneous geodesics in homogeneous affine manifolds, preprint.
7. Z. Dušek, O. Kowalski and S. Nikčević, *Differential Geom. Appl.* **21** (2004), 65–78.
8. O. Kowalski and S. Nikčević, *Archiv der Math.* **73** (1999), 223–234; Appendix: *Archiv der Math.* **79** (2002), 158–160.
9. B. Opozda, *Differential Geom. Appl.* **21** (2004), 173–198.

CONJUGATE CONNECTIONS AND DIFFERENTIAL EQUATIONS ON INFINITE DIMENSIONAL MANIFOLDS

M. Aghasi

Department of Mathematics
Isfahan University of Technology, Isfahan, Iran
E-mail: m.aghasi@cc.iut.ac.ir

C. T. J. Dodson

School of Mathematics
Manchester University, Manchester, M13 9PL, UK
E-mail: ctdodson@manchester.ac.uk

G. N. Galanis

Section of Mathematics
Naval Academy of Greece, Xatzikyriakion, Piraeus 185 39, Greece
E-mail: ggalanis@snd.edu.gr

A. Suri

Department of Mathematics
Isfahan University of Technology, Isfahan, Iran
E-mail: a.suri@math.iut.ac.ir

On a smooth manifold M, the vector bundle structures of the second order tangent bundle, T^2M bijectively correspond to linear connections. In this paper we classify such structures for those Fréchet manifolds which can be considered as projective limits of Banach manifolds. We investigate also the relation between ordinary differential equations on Fréchet spaces and the linear connections on their trivial bundle; the methodology extends to solve differential equations on those Fréchet manifolds which are obtained as projective limits of Banach manifolds. Such equations arise in theoretical physics. We indicate an extension to the Fréchet case for the Earle and Eells foliation theorem.

Keywords: Second order tangent bundle, connection, Banach manifold, Fréchet manifold, foliation

Mathematics Subject Classification 2000: Primary 58B25; Secondary 58A05

1. Introduction

For a smooth finite dimensional manifold M the structure of T^2M, the bundle of accelerations, was studied by Dodson and Radivoiovici.[4] They proved that T^2M admits a vector bundle structure over M if and only if M is endowed with a linear connection. Dodson and Galanis[2] have established the structure of T^2M for Banach manifolds and also for those Fréchet manifolds which are projective limits of Banach manifolds. They proved that existence of a vector bundle structure on T^2M is equivalent to the existence on M of a linear connection in the sense of Vilms.[16] By this means, vector bundle structures of T^2M were classified by Dodson, Galanis and Vassiliou[3] for the Banach case.

In this paper we extend that classification to a large class of Fréchet manifolds. Also, we investigate some relations between connections and ordinary differential equations on Fréchet spaces which generalize a result of Vassiliou[15] in the Banach case. As Galanis and Vassiliou[8] have pointed out, there is no specific method to solve a given differential equation on Fréchet spaces. Here we introduce a method for solving such problems and we give also a relation between these equations and the induced connections. This method can solve a wide class of ordinary differential equation on any Fréchet space because every Fréchet space can be considered as a projective limit of Banach spaces. Furthermore, it extends to solve differential equations on those Fréchet manifolds which are obtained as projective limits of Banach manifolds. We indicate how the methodology may be applied by suggesting an approach to generalize to a large class of Fréchet manifolds the Earle and Eells[5] foliation theorem.

There has been recent interest in the Fréchet case for various models in theoretical physics and stochastic calculus; we mention some examples. Blair[1] studied the space $\mathcal{M}$ of all C^∞-Riemannian metrics on a manifold M as an infinite-dimensional Fréchet manifold with C^∞-topology and provided certain of its geometrical properties. Sergeev[14] suggested a new realization of the homogeneous factor-space $Diff(S^1)/S^1$, which is a Fréchet manifold. He interpreted it as the space of those complex structures on the loop space ΩG of a compact Lie group G (regarded as a Fréchet-Kähler manifold equipped with a canonical action of the group $Diff(S^1)$ by symplectomorphisms) that are compatible with the symplectic structure. Minic and Tze[13] proposed a generalization of quantum mechanics in which the projective Hilbert space of quantum events is replaced by a 'nonlinear Grassmannian' $Gr(\mathbb{C}^{n+1})$ of codimension-2 compact submanifolds of $\mathbb{C}^{n+1}$, which is a Fréchet manifold. Then it admits a symplectic structure (being a coadjoint

orbit of the group of volume-preserving diffeomorphisms of $\mathbb{C}^{n+1}$) and a
(non-integrable) almost complex structure, which make it into an almost
Kähler manifold. Kinateder and McDonald[9] discussed the stochastic flow of
certain diffeomorphisms via the Fréchet manifold $\mathcal{D}$ of smoothly bounded
domains in $\mathbb{R}^n$ with compact closure; they included also a review of the relevant Fréchet geometry, developed the stochastic analysis and gave a number
of examples and applications. See also McDonald[12] who studied Brownian
motion in a complete Riemannian manifold M where, for each $v > 0$, $\mathcal{M}_v$
is the Fréchet manifold of all relatively compact smooth domains D in M
of volume v; he obtained various results when M has constant curvature.

2. Preliminaries

Let M be a smooth manifold modelled on the Banach space $\mathbb{E}$ with the
corresponding atlas $\{(U_\alpha, \psi_\alpha)\}_{\alpha \in I}$. For each $x \in M$ we define $C_x = \{f : (-\epsilon, \epsilon) \longrightarrow M$; f is smooth and f(0)=x$\}$. For $f, g \in C_x$, we define $f \sim_x g$
iff $f'(0) = g'(0)$, so $T_x M = C_x/\sim_x$ and $TM = \bigcup_{x \in M} T_x M$. It is easy to
check that TM is a smooth Banach manifold modelled on $\mathbb{E} \times \mathbb{E}$. Moreover it
is a vector bundle over M by the projection $\pi_M : TM \longrightarrow M$. Consider the
trivialization $\{(\pi_M^{-1}(U_\alpha), \Psi_\alpha)\}_{\alpha \in I}$ for TM and similarly the trivialization
$\{(\pi_{TM}^{-1}(\pi_M^{-1}(U_\alpha)), \tilde{\Psi}_\alpha)\}_{\alpha \in I}$ for T(TM).

Following e.g. Vilms,[16] a connection on M is a vector bundle morphism
$\nabla : T(TM) \longrightarrow TM$ with the local forms $\omega_\alpha : \psi_\alpha(U_\alpha) \times \mathbb{E} \longrightarrow L(\mathbb{E}, \mathbb{E})$.
Local representation of ∇ is as follows:

$$\nabla_\alpha : \psi_\alpha(U_\alpha) \times \mathbb{E} \times \mathbb{E} \times \mathbb{E} \longrightarrow \psi_\alpha(U_\alpha) \times \mathbb{E}$$

with $\nabla_\alpha = \Psi_\alpha \circ \nabla \circ \tilde{\Psi}_\alpha^{-1}$ for $\alpha \in I$, and the relation $\nabla_\alpha(y, u, v, w) = (y, w + \omega_\alpha(y, u).v)$ is satisfied. Furthermore ∇ is a linear connection iff
$\{\omega_\alpha\}_{\alpha \in I}$ are linear with respect to their second variables. This connection
∇ is completely determined by its Christoffel symbols:

$$\Gamma_\alpha : \psi_\alpha(U_\alpha) \longrightarrow L(\mathbb{E}, L(\mathbb{E}, \mathbb{E})) \equiv L_s^2(\mathbb{E} \times \mathbb{E}, \mathbb{E}) \ ; \ \alpha \in I$$

defined by $\Gamma_\alpha(y)[u] = \omega_\alpha(y, u)$ for each $(y, u) \in \psi_\alpha(U_\alpha) \times \mathbb{E}$.

The necessary condition for ∇ to be well defined on chart overlaps of M
is that the Christoffel symbols satisfy the following compatibility condition;

$$\Gamma_\alpha(\sigma_{\alpha\beta}(y))(d\sigma_{\alpha\beta}(y)(u), d\sigma_{\alpha\beta}(y)(v)) + (d^2 \sigma_{\alpha\beta}(y)(v))(u)$$
$$= d\sigma_{\alpha\beta}(y)(\Gamma_\beta(y)(u, v))$$

for all $(y, u, v) \in \psi_\alpha(U_\alpha) \times \mathbb{E} \times \mathbb{E}$. Here $\sigma_{\alpha\beta} = \psi_\alpha \circ \psi_\beta^{-1}$, and d, d^2 denote
the first and the second order differentials respectively.

Recalling our above definition of C_x, we define the equivalence relation $\approx_x$ as follows, for $f, g \in C_x$,

$$f \approx_x g \iff f'(0) = g'(0) \text{ and } f''(0) = g''(0).$$

Then $T_x^2 M = C_x / \approx_x$ and $T^2 M = \bigcup_{x \in M} T_x^2 M$. Here we see that $T_x^2 M$ is a topological vector space isomorphic to $\mathbb{E} \times \mathbb{E}$ under the isomorphism:

$$\phi_x : T_x^2 M \longrightarrow \mathbb{E} \times \mathbb{E}$$
$$[f, x]_2 \longmapsto ((\psi_\alpha \circ f)'(0), (\psi_\alpha \circ f)''(0)).$$

However, this identification cannot be extended to a vector bundle structure on $T^2 M$. This can nevertheless be achieved by the use of a linear connection ∇ of M by means of the following local trivializations:

$$\Phi_\alpha : \pi_2^{-1}(U_\alpha) \longrightarrow U_\alpha \times \mathbb{E} \times \mathbb{E}$$
$$[f, x]_2 \longmapsto (x, (\psi_\alpha \circ f)'(0), (\psi_\alpha \circ f)''(0) + \Gamma_\alpha(\psi_\alpha(x))((\psi_\alpha \circ f)'(0),$$
$$(\psi_\alpha \circ f)'(0))),$$

where $\pi_2 : T^2 M \longrightarrow M$ sending $[f, x]_2$ to x. In this way we see that $T^2 M$ becomes a vector bundle over M with fibres of type $\mathbb{E} \times \mathbb{E}$ and the structure group $GL(\mathbb{E} \times \mathbb{E})$.

Let $\Phi_{\alpha,x}$ be the restriction of Φ_α to the fibres $T_x^2 M$. Then the transition functions of $T^2 M$ will be:

$$T_{\alpha\beta} : U_\alpha \cap U_\beta \longrightarrow L(\mathbb{E} \times \mathbb{E}, \mathbb{E} \times \mathbb{E})$$
$$x \longmapsto \Phi_{\alpha,x} \circ \Phi_{\beta,x}^{-1}$$

More precisely, they have the form $T_{\alpha\beta} = (d(\sigma_{\alpha\beta} \circ \phi_\beta), d(\sigma_{\alpha\beta} \circ \phi_\beta))$, for more details see Dodson and Galanis.[2]

3. Classification for vector bundle structures of $T^2 M$

Here we turn to a class of Fréchet manifolds that are obtained as projective limits of Banach manifolds. Let $\{M^i, \varphi^{ji}\}_{i,j \in \mathbb{N}}$ be a projective system of Banach manifolds modelled on the Banach spaces $\{\mathbb{E}^i\}_{i \in \mathbb{N}}$ respectively; we require the model spaces also to form a projective system. Suppose that for $x = (x^i) \in M = \varprojlim M^i$ there exists a projective system of charts $\{(U_\alpha^i, \psi_\alpha^i)\}_{i \in \mathbb{N}}$ such that $x^i \in U_\alpha^i$ and the limit $\varprojlim U_\alpha^i$ is open in M. Then, the projective limit $M = \varprojlim M_i$ has a Fréchet manifold structure modelled on $\mathbb{F} = \varprojlim E_i$ with the atlas $\mathcal{A} = \{(\varprojlim U_\alpha^i, \varprojlim \psi_\alpha^i)\}_{\alpha \in I}$. Let $\{M^i, \phi^{ji}\}_{i,j \in \mathbb{N}}$ and $\{N^i, \phi'^{ji}\}_{i,j \in \mathbb{N}}$ be two projective systems of manifolds, with smooth maps $g^i : M^i \longrightarrow N^i$ such that $\varprojlim g^i = g$ exists. Suppose that for each

$i \in \mathbb{N}$; M^i and N^i are endowed with linear connections ∇_{M^i} and ∇_{N^i} which form the projective limits $\nabla_M = \varprojlim \nabla_{M^i}$ and $\nabla_N = \varprojlim \nabla_{N^i}$. The latter are then linear connections over M and N respectively.
Moreover, the next result holds:

Proposition 3.1. *Let ∇_{M^i} and ∇_{N^i} be g^i-conjugate for each $i \in \mathbb{N}$ then ∇_M and ∇_N are g-conjugate.*

Proof. We have to show that $\nabla_N o T(Tg) = Tgo\nabla_M$.

First we prove that $\varprojlim \nabla_{N^i} o T(Tg^i)$ exists. Let $i \leq j$, then:

$$T\phi'^{ji}(\nabla_{N^j} \circ T(Tg^j)) = (\nabla_{N^i} \circ T(T\phi'^{ji}))o(T(Tg^i))$$
$$= \nabla_{N^i} o[(T(Tg^i)) \circ T(T\phi^{ji})]$$
$$= (\nabla_{N^i} \circ T(Tg^i)) \circ T(T\phi^{ji}),$$

hence the limit $\varprojlim \nabla_{N^i} \circ T(Tg^i)$ exists. Furthermore for each $i \in \mathbb{N}$ we have:

$$T\phi'^{i} \circ (\nabla_N \circ T(Tg)) = (\nabla_{N^i} \circ T(T\phi'^{i})) \circ T(Tg) = \nabla_{N^i} \circ T(Tg^i)$$

where $\phi'^i : N \longrightarrow N^i$ are the canonical projections. As a result,

$$\nabla_N \circ T(Tg) = \varprojlim \nabla_{N^i} \circ T(Tg^i).$$

On the other hand, $\{Tg^i \circ \nabla_{M^i}\}_{i \in \mathbb{N}}$ is a projective system of maps. Indeed for $i \leq j$:

$$T\phi'^{ji} \circ (Tg^j \circ \nabla_{M^j}) = (Tg^i \circ T\phi^{ji}) \circ \nabla_{M^j}$$
$$= Tg^i \circ (\nabla_{M^i} \circ T(T\phi^{ji}))$$
$$= (Tg^i \circ \nabla_{M_i}) \circ T(T\phi^{ji}).$$

Hence $\varprojlim Tg^i \circ \nabla_{M^i}$ exists. Moreover:

$$T\phi'^{i} \circ (Tg \circ \nabla_M) = (Tg^i \circ T\phi^i) \circ \nabla_M = Tg^i \circ \nabla_{M^i}, \ i \in \mathbb{N},$$

where $\phi^i : M \longrightarrow M^i$ are the canonical projections of M. Hence

$$Tg \circ \nabla_M = \varprojlim(Tg^i \circ \nabla_{M^i}).$$

Based on the fact that each pair $(\nabla_{M^i}, \nabla_{N^i})$, $i \in \mathbb{N}$, consists of g^i-conjugate connections, we conclude that

$$\nabla_N \circ T(Tg) = \varprojlim(\nabla_{N^i} \circ T(Tg^i)) = \varprojlim(Tg^i \circ \nabla_{M^i}) = Tg \circ \nabla_M,$$

hence ∇_M and ∇_N are indeed g-conjugate. $\square$

Lemma 3.2. *If $\nabla_M = \varprojlim \nabla_{M^i}$, $\nabla_N = \varprojlim \nabla_{N^i}$ and ∇_{M^i} and ∇_{N^i} are g^i-conjugate, then $T^2g : T^2M \longrightarrow T^2N$ is linear on the fibres.*

Proof. Since ∇_{M^i} and ∇_{N^i} are g^i-conjugate[3] then $T^2_{x^i} g^i$ is linear for each $x^i \in M^i$. Since $T^2_x g = \varprojlim T^2_{x^i} g^i$ the result follows. □

Proposition 3.3. *Let $g^i : M^i \longrightarrow N^i$ be smooth maps and ∇_{M^i} and ∇_{N^i} be g^i-conjugate for each $i \in \mathbb{N}$. Then, $T^2 g : T^2 M \longrightarrow T^2 N$ is a vector bundle morphism.*

Sketch of proof. Each $T^2 g^i : T^2 M^i \longrightarrow T^2 N^i$ is a vector bundle morphism for each $i \in \mathbb{N}$;[3] and since $T^2 g = \varprojlim T^2 g^i$, we get the result.

In view of the above discussion, we deduce the following main result:

Theorem 3.4. *Let $g^i : M^i \longrightarrow M^i$ be a diffeomorphism and ∇^i and ∇'^i g^i-conjugate linear connections on M^i, for each $i \in \mathbb{N}$. If $\nabla = \varprojlim \nabla^i$ and $\nabla' = \varprojlim \nabla'^i$, then the vector bundle structures on $T^2 M$ induced by ∇ and ∇' are isomorphic.*

Let (M, ∇) denote the vector bundle structure of $T^2 M$ induced by ∇. For a diffeomorphism $g : M \longrightarrow M$ we define the equivalence relation $\sim_g$ as follows:

$$(M, \nabla) \sim_g (M, \nabla') \iff \nabla \text{ and } \nabla' \text{ are } g-conjugate.$$

Hence if (M, ∇) and (M, ∇') are in the same g-conjugate class $[(M, \nabla)]_g$, their induced vector bundle structures on $T^2 M$ are isomorphic.

Corollary 3.5. *All the elements of the class $[(M, \nabla)]_g$ have isomorphic induced vector bundle structures on $T^2 M$.*

4. Connections and ordinary differential equations

Let $\mathbb{E}$ be a Banach space and $L = (\mathbb{R} \times \mathbb{E}, \mathbb{R}, pr_1)$ be the trivial bundle over $\mathbb{R}$ with fibres of type $\mathbb{E}$. Vassiliou[15] showed that we can correspond an ordinary differential equation to a connection over the trivial bundle with the solution ξ being the horizontal global section of the obtained connection. Furthermore, it is shown that connections ∇ and ∇' over L are conjugate iff the corresponding differential equations $dx/dt = A(t)x$ and $dx/dt = B(t)y$ are equivalent.

Here we extend these concepts to Fréchet spaces. Let $\mathbb{F}$ be a Fréchet space with $\mathbb{F} = \varprojlim \{\mathbb{E}^i, \rho^{ji}\}_{i,j \in \mathbb{N}}$. Consider the trivial bundle $L = (\mathbb{R} \times \mathbb{F}, \mathbb{R}, pr_1)$ with respect to the usual atlas $\mathcal{A}$ for $\mathbb{R}$ formed by the global chart $(\mathbb{R}, id_\mathbb{R})$.

Assume that ∇^i is a linear connection over $L^i = (\mathbb{R} \times \mathbb{E}^i, \mathbb{R}, pr_1)$ and that the corresponding Christoffel symbols commute with the connecting morphisms ρ^{ji}. Then, $\nabla = \varprojlim \nabla^i$ a linear connection on $L = \varprojlim L^i$ characterized by a single Christoffel symbol:

$$\Gamma : \mathbb{R} \longrightarrow L^2(\mathbb{R} \times \mathbb{F}, \mathbb{F}).$$

Let $A(t) = \Gamma(t)(., 1)$ where 1 is the unit of $\mathbb{R}$. Then the following result holds true:

Theorem 4.1. *Linear connections of the above type are in one-to-one correspondence with the ordinary differential equations $dx/dt = A(t)x$ where the factor A is obtained as a projective limit. Moreover for each $t_0 \in \mathbb{R}$ there exists a unique horizontal global section*

$$\xi : \mathbb{R} \longrightarrow \mathbb{R} \times \mathbb{F}$$

as solution with $\xi_p(t_0) = f_0$, where $\xi_p : \mathbb{R} \longrightarrow \mathbb{F}$ is the principal part of ξ.

Proof. We know that $\nabla = \varprojlim \nabla^i$ such that each ∇^i is a linear connection over L^i. Also, each ∇^i corresponds bijectively to an ordinary differential equation $dx^i/dt = A^i(t)x^i$.[15] Furthermore, every solution of $dx^i/dt = A^i(t)x^i$ is the principal part of the horizontal global section of ∇^i, which we call ξ^i.

We notice firstly that $A^i(t^i) = \Gamma^i(t^i)(., 1)$ where Γ^i is the Christoffel symbol ∇^i over L^i assigned to the chart $(\mathbb{E}^i, id_{\mathbb{E}^i})$. Since $\nabla = \varprojlim \nabla^i$ we get $\Gamma(t)(., 1) = \varprojlim \Gamma^i(t)(., 1)$, $t \in \mathbb{R}$. Hence, $A(t) = \varprojlim A^i(t^i)$ is well defined and consequently $dx/dt = A(t)x$ is an ordinary differential equation on $\mathbb{E}^i$. For more details see Galanis.[6]

Let ξ_p^i be the solution of $dx/dt = A_i(t)x$, satisfying $\xi_p^i(t_0) = f_0^i$. We claim that $\{\xi_p^i\}_{i \in \mathbb{N}}$ is a projective system of maps and $\xi_p = \varprojlim \xi_p^i$ is the solution of $dx/dt = A(t)x$. One has first to check that:

$$\rho^{ji} \circ \xi_p^j = \xi_p^i$$

for $i \leq j$. To this end, we see that

$$(\rho^{ji} \circ \xi_p^j)'(t) = \rho^{ji} \circ (\xi_p^j)'(t) = \rho^{ji} \circ [A^j(t)](\xi_p^j(t))$$
$$= [\rho^{ji} \circ A^j(t)](\xi_p^j(t))$$
$$= [A^i(t) \circ \rho^{ji}](\xi_p^j(t))$$
$$= [A^i(t)](\xi_p^i(t))$$

Moreover, $\xi_p^i(t_0^i) = f_0^i$ and $(\rho^{ji} \circ \xi_p^j)(t_0) = \rho^{ji}(f_0^j) = f_0^i$. Based on the uniqueness of the solutions of differential equations on Banach spaces over given initial conditions, we conclude that $\rho^{ji} \circ \xi_p^j = \xi_p^i$. This implies that

$\xi_p = \varprojlim \xi_p^i$ exists. Furthermore, it is the solution of the above-mentioned differential equation:

$$\xi_p'(t) = \left(\xi_p^{i\,\prime}(t^i)\right)_{i\in\mathbb{N}} = \left(A^i(t^i)(\xi_p^{i\,\prime}(t^i))\right)_{i\in\mathbb{N}} = A(t)(\xi_p(t)).$$

Similar calculations ensure that $\xi_p = \varprojlim \xi_p^i$ is also the unique horizontal global section of ∇ as a projective limit of global sections. $\square$

Let $\nabla = \varprojlim \nabla^i$ and $\nabla' = \varprojlim \nabla'^i$ be two linear connections over L such that for each $i \in \mathbb{N}$, ∇^i and ∇'^i are g^i-related connections on L^i and $g = \varprojlim g^i$.

Theorem 4.2. *With the same assumptions, let $\nabla = \varprojlim \nabla^i$ and $\nabla' = \varprojlim \nabla'^i$ be two linear connections over L. Then ∇ and ∇' are $(g, id_\mathbb{R})$-related iff their corresponding differential equations, given by $dx/dt = A(t)x$ and $dy/dt = C(t)y$, are equivalent i.e. there exists a smooth transformation $Q : \mathbb{B} \longrightarrow \mathcal{H}^0(\mathbb{F})$ such that $x(t) = Q(t)y(t)$ or equivalently,*

$$C(t) = Q^{-1}(t) \circ (A(t) \circ Q(t) - \dot{Q}(t))$$

for each $t \in \mathbb{R}$.

Proof. By Vassiliou[15] ∇^i and ∇'^i are g^i-related connections over L^i iff $dx^i/dt = A^i(t)x^i$ and $dy^i/dt = C^i(t)y^i$ are equivalent i.e.

$$C^i(t) = \left(Q^i\right)^{-1}(t) \circ \left(A^i(t) \circ Q^i(t) - \dot{Q}^i(t)\right) : \; i \in \mathbb{N}, \tag{1}$$

where $Q = \epsilon \circ Q^*$, $Q^* = (Q^i)_{i\in\mathbb{N}}$ and ϵ is the natural morphism

$$\epsilon : \mathcal{H}^0(\mathbb{F}) \longrightarrow \mathcal{L}(\mathbb{F})$$
$$(l^i)_{i\in\mathbb{N}} \longmapsto \varprojlim l^i$$

Hence (1) implies that: ∇^i and ∇'^i are g^i-related iff $x(t) = Q(t)y(t)$. $\square$

Note that the existence of intrinsic obstacles in the structure of the space of continuous linear mappings $\mathcal{L}(\mathbb{F})$, which drops out of the category of Fréchet spaces, leads us to replace it with the Fréchet space $\mathcal{H}(\mathbb{F})$.

$$\mathcal{H}(\mathbb{F}) = \{(l^i)_{i\in\mathbb{N}} \in \prod_{i=1}^{\infty} \mathcal{L}(\mathbb{E}^i) : \varprojlim l^i \text{ exists}\}$$

More precisely, $\mathcal{H}(\mathbb{F})$ can be considered as the projective limit of the Banach spaces:

$$\mathcal{H}_i(\mathbb{F}) := \{(l^1, ..., l^i) \in \prod_{j=1}^{i} \mathcal{L}(\mathbb{E}^j) : \rho^{jk} \circ l^j = l^k \circ \rho^{jk}; \text{ for } k \leq j \leq i\}.$$

The work presented in this paper can be applied in order to obtain a potentially useful *Floquet − Liapunov* theorem in Fréchet spaces:

Corollary 4.3. *Let* $\nabla = \varprojlim \nabla^i$ *be a linear connection over* L *with periodic coefficient* A. *Then there exists a linear connection* ∇' *with constant Christoffel symbols, where* ∇ *and* ∇' *are* $(g, id_\mathbb{R}) - related$.

Proof. According to Galanis[8] the differential equation $\dot{x}(t) = A(t)x(t)$ with periodic coefficient A is equivalent with the differential equation $\dot{y}(t) = B(t)y(t)$ such that B is constant. Let ∇' be the linear connection over L assigned to B then, by Theorem 3.2, ∇ and ∇' are $(g, id_\mathbb{R}) - related$. □

5. The Earle and Eells foliation theorem in Fréchet spaces

The target here is to indicate a possible generalization to a wide class of Fréchet manifolds for the following result:

Theorem 5.1 (Earle and Eells[5]). *Let* (X, α), (Y, β) *be Finsler* C^1-*manifolds modeled on Banach spaces, and suppose that* (X, α) *is complete. Let* $f : X \to Y$ *be a surjective* C^1-*map which foliates* X. *If there is a, locally bounded over* Y, *Lipschitz splitting of the sequence*

$$0 \longrightarrow Ker f_* \longrightarrow TX \xrightarrow{f_*} f^{-1}(TY) \longrightarrow 0,$$

where f_* *stands for the differential of* f *at* x, *then* f *is a locally* C^0-*trivial fibration.*

The proof of Theorem 5.1 is strongly based on properties of differential equations in Banach spaces, used to construct coherent liftings of paths. As a result, any attempt to generalize it to the Fréchet framework encounters serious difficulties. This is because the local structure of the space models do not admit a general solvability theory for ordinary differential equations analogous to that of the Banach case. Indeed, in a Fréchet space an initial value problem may have no solution, a single one or multiple solutions.

A way out of these difficulties is proposed here for a wide class of Fréchet manifolds: those that can be obtained as projective limits of Banach corresponding factors (see Galanis and Dodson[2,7]). To be more precise, we consider the manifolds X, Y to be limits of a projective system of Banach Finsler manifolds: $X = \varprojlim\{(X^i, \alpha^i)\}_{i \in \mathbb{N}}$, $Y = \varprojlim\{(Y^i, \beta^i)\}_{i \in \mathbb{N}}$. Then, suppose that X and Y can be endowed with generalized Fréchet-Finsler structures in which the induced norms on the tangent spaces are replaced by sequences of semi-norms.

If this is the case, then a corresponding limit of mappings $f = \varprojlim f^i$: $X \to Y$ satisfying the properties of the previous theorem, can be realized as a projective limit of C^1-factors $f^i : X^i \to Y^i$, $i \in \mathbb{N}$, where each of factor satisfies also the assumptions of Theorem 5.1. It follows that a sequence of C^0-trivial fibrations will be obtained and it will projectively converge to f. Taking into account that the notion of triviality on the fibers is compatible with projective limits, one obtains this property also for f, avoiding the use of the pathological differential equations on Fréchet spaces.

Note that if the domain manifold X is assumed to be Banach modeled and Y is a Finsler manifold as above, then the result obtains for every surjective C^1-map f which foliates X.

References

1. D.E. Blair, *Handbook of differential geometry*, Vol. I, 153-185, North-Holland, Amsterdam, 2000.
2. C.T.J. Dodson and G.N. Galanis, *J. Geom. Phys.* **52** (2004), 127-136.
3. C.T.J. Dodson, G.N. Galanis and E. Vassiliou, *Math. Proc. Camb. Phil. Soc.* **141** (2006), 489-496.
4. C.T.J. Dodson and M.S.Radivoiovici, *An. st. Univ. "Al. I. Cuza"* **28** (1982), 63-71.
5. C. J. Earle and J. Eells Jr., *J. Diff. Geom.* **1** (1967), 33-41.
6. G.N. Galanis, *Rend. Sem. Mat. Padova* **112** (2004), 103-115.
7. G. Galanis, *Portugal. Math.* **55** (1998), 11-24.
8. G.N. Galanis and E. Vassiliou, *Ann. Scuola Norm. Sup. Pisa Cl. Sci. (4)* **27** (1998), 427-436.
9. K. Kinateder and P. McDonald, *Probab. Theory Related Fields* **124** (2002), 73-99
10. S. Lang, *Differential manifolds*, Addison-Wesley Publishing Co., Inc., Reading, Mass.-London-Don Mills, Ont., 1972.
11. J. A. Leslie, *J. Diff. Geom.* **42** (1968), 279-297.
12. P. McDonald, *Potential Anal.* **16** (2002), 115-138.
13. D. Minic and C-H. Tze, *Phys. Lett. B* **581** (2004), 111-118.
14. A. Sergeev, *Stochastic processes, physics and geometry: new interplays, II (Leipzig, 1999)*, 573-588, CMS Conf. Proc., **29**, Amer. Math. Soc., Providence, RI, 2000.
15. E. Vassiliou, *Period. Math. Hungar.* **13** (1982), 286-308.
16. J. Vilms, *J. Diff. Geom.* **41** (1967), 235-243.

TOTALLY BIHARMONIC SUBMANIFOLDS

Debora Impera* and Stefano Montaldo**

Dipartimento di Matematica e Informatica, Università di Cagliari
Via Ospedale 72, 09124 Cagliari, Italy
* E-mail: debora.impera@hotmail.it
** E-mail:montaldo@unica.it

We introduce the notion of totally biharmonic submanifolds and characterize totally biharmonic hypersurfaces in terms of their extrinsic geometry. Then, we classify the totally biharmonic surfaces in a three-dimensional space form.

Keywords: Biharmonic curves, geodesics

1. Introduction

Biharmonic curves $\gamma : I \subset \mathbb{R} \to (M, g)$ of a Riemannian manifold are the solutions of the fourth order differential equation

$$\nabla^3_{\gamma'} \gamma' - R(\gamma', \nabla_{\gamma'} \gamma') \gamma' = 0.$$

As we shall detail in the next section, they arise from a variational problem and are a natural generalization of geodesics.
In the last decade several articles (see, for example, Refs. 1,2,4–7,9) have appeared on the construction and classification of biharmonic curves, starting with Ref. 3, where the authors studied the case of biharmonic curves on a surface.

In this paper we consider the following problem: *under which conditions on the extrinsic geometry of a submanifold M in a Riemannian manifold (N, h) all the geodesics of M are biharmonic curves of N.*

We call *totally biharmonic* a submanifold M that satisfies the above property.

We first write down the system that characterizes the shape operator of a totally biharmonic submanifold. Next, we use this system to find all totally biharmonic surfaces in a space form $N^3(C)$ of constant sectional curvature C, i.e.:

(a) if $C \leq 0$, then M is a totally geodesic surface;
(b) if $C = 1$, and we take $N^3(1) = \mathbb{S}^3$, then M is either totally geodesic or one of the following surfaces:
   (i) part of the Clifford torus $\mathbb{S}^1(\frac{1}{\sqrt{2}}) \times \mathbb{S}^1(\frac{1}{\sqrt{2}})$;
   (ii) part of the sphere $\mathbb{S}^2(\frac{1}{\sqrt{2}})$.

In the last section we explore an example of a non-totally biharmonic surface in the Heisenberg group $\mathbb{H}_3$ admitting geodesics which are biharmonic in the ambient space.

2. Biharmonic maps

Harmonic maps $\varphi : (M,g) \to (N,h)$ between smooth Riemannian manifolds are critical points of the energy functional $E(\varphi) = \frac{1}{2} \int_M |d\varphi|^2 \, v_g$. The corresponding Euler-Lagrange equation is given by the vanishing of the tension field $\tau(\varphi) = trace \nabla d\varphi$. Biharmonic maps (as suggested by J. Eells and J.H. Sampson in Ref. 8) are the critical points of the *bienergy* functional $E_2(\varphi) = \frac{1}{2} \int_M |\tau(\varphi)|^2 \, v_g$. In Ref. 10, G.Y. Jiang derived the first variation formula of the bienergy showing that the Euler-Lagrange equation for E_2 is

$$\tau_2(\varphi) = -\Delta \tau(\varphi) - trace R^N(d\varphi, \tau(\varphi))d\varphi = 0,$$

where Δ is the rough Laplacian defined on sections of $\varphi^{-1}(TN)$ and $R^N(X,Y) = \nabla_X \nabla_Y - \nabla_Y \nabla_X - \nabla_{[X,Y]}$ is the curvature operator on (N,h). The equation $\tau_2(\varphi) = 0$ is called the *biharmonic equation*.

Let now $\gamma : I \subset \mathbb{R} \to (N,h)$ be a curve parametrized by arc-length. If we denote by $T = \gamma'$ its unit tangent vector field, then the biharmonic equation becomes

$$\nabla_T^3 T - R^N(T, \nabla_T T) T = 0. \tag{1}$$

For a survey on biharmonic maps and submanifolds we refer the rider to Ref. 11.

3. Totally biharmonic hypersurfaces

Let $\varphi : (M^n, g) \to (N^{n+k}, h)$ be the inclusion of a submanifold M in N. We shall denote by $\overline{\nabla}$ and ∇ the Levi-Civita connections on N and M respectively.

Definition 3.1. The inclusion $\varphi : (M^n, g) \to (N^{n+k}, h)$ is called *totally biharmonic* if all geodesics of M are biharmonic curves of N.

We shall now find the conditions on the extrinsic geometry of the submanifold M in N under which a geodesic of M is biharmonic in N.

For simplicity, in the sequel, we shall consider the case of codimension-one submanifolds, although most of the results can be generalized to higher codimension. We shall denote by η a unit section in the normal bundle of M in N, by $B(X,Y)$, X,Y vector fields tangent to M, the second fundamental form with values in the normal bundle and by S_η the shape operator with respect to the section η. The curvature operators of (M,g) and (N,h) will be denoted by R and $\overline{R}$ and the sectional curvatures by K and $\overline{K}$, respectively.

We have

Lemma 3.1. *Let $\varphi : (M^n, g) \to (N^{n+1}, h)$ be the canonical inclusion. Then a geodesic $\gamma : I \to M^n$ is biharmonic in N if and only if, along the curve γ, the shape operator S_η satisfies one of the following conditions:*

(a) $\langle S_\eta(T), T\rangle = 0$ and γ is a geodesic of N;
(b) S_η is a solution of the following system

$$\begin{cases} \langle S_\eta(T), T\rangle = \text{constant} \neq 0 \\ \|S_\eta(T)\|^2 = \overline{K}(T, \eta) \\ \langle \nabla_T S_\eta(T), T_i^\perp \rangle = -\langle \overline{R}(T, \eta)T, T_i^\perp\rangle, \quad i = 2, \ldots, n \end{cases} \quad (2)$$

where $T = \gamma'$, whilst $T_i^\perp$, $i = 2, \ldots, n$, are $n-1$ unit vector fields along γ, orthogonal to T.

Proof. Decomposing $B(T,T)$ with respect to η we have $B(T,T) = f\eta$, with $f = \langle B(T,T), \eta \rangle = \langle S_\eta(T), T\rangle$. Let now $\gamma : I \to M^n$ be a geodesic parametrized by arc-length, so that $\nabla_T T = 0$. Now, γ is a biharmonic curve of N if and only if

$$\overline{\nabla}_T^3 T - \overline{R}(T, \overline{\nabla}_T T)T = 0.$$

Taking into account that $\nabla_T T = 0$, the first addend becomes

$$\begin{aligned}\overline{\nabla}_T^3 T &= \overline{\nabla}_T^2(\nabla_T T + B(T,T)) \\ &= \overline{\nabla}_T^2 B(T,T) \\ &= \overline{\nabla}_T^2 f\eta \\ &= \overline{\nabla}_T(f'\eta + f\overline{\nabla}_T \eta) \\ &= \overline{\nabla}_T(f'\eta - fS_\eta(T)) \\ &= f''\eta + f'\overline{\nabla}_T \eta - f'S_\eta(T) - f(\nabla_T S_\eta(T) + B(S_\eta T, T)) \\ &= f''\eta - 2f'S_\eta(T) - f\nabla_T S_\eta(T) - fB(S_\eta(T), T).\end{aligned}$$

On the other hand, the the second addend is
$$\begin{aligned}\overline{R}(T,\overline{\nabla}_T T)T &= \overline{R}(T,\nabla_T T + B(T,T))T \\ &= \overline{R}(T, B(T,T))T \\ &= f\overline{R}(T,\eta)T.\end{aligned}$$

Thus γ is biharmonic in N if and only if
$$f''\eta - 2f'S_\eta(T) - f\nabla_T S_\eta(T) - fB(S_\eta(T),T) - f\overline{R}(T,\eta)T = 0. \qquad (3)$$

Next, taking the inner product of (3) with T, η and $n-1$ unit vector fields, along γ, $T_i^\perp$, $i = 2, \ldots, n$, orthogonal to T, we get

$$-3f'f = 0 \qquad (4)$$

$$f'' - f\langle B(S_\eta(T),T),\eta\rangle + f\overline{K}(T,\eta) = 0, \qquad (5)$$

$$-2f'\langle S_\eta(T), T_i^\perp\rangle - f\langle \nabla_T S_\eta(T) + \overline{R}(T,\eta)T, T_i^\perp\rangle = 0, \quad i = 2,\ldots,n. \ (6)$$

In Condition (4) we have two possibilities:

(a) $f = 0$,
(b) $f = \text{constant} \neq 0$.

In the first case
$$\langle \overline{\nabla}_T T, \eta\rangle = \langle S_\eta(T), T\rangle = f = 0, \qquad (7)$$
from which it follows that $\overline{\nabla}_T T = 0$ and thus γ is a geodesic of N^{n+1}. If we assume that $f = \langle S_\eta(T), T\rangle = \text{constant} \neq 0$, then (5) becomes

$$-f\langle B(T, S_\eta(T)), \eta\rangle + f\overline{K}(T,\eta) = -f\langle S_\eta(T), S_\eta(T)\rangle + f\overline{K}(T,\eta) = 0,$$

which gives the second equation of (2).
Finally, (6) reduces to

$$\langle \nabla_T S_\eta(T), T_i^\perp\rangle = -\langle \overline{R}(T,\eta)T, T_i^\perp\rangle, \quad i = 2,\ldots,n. \qquad \square$$

Since for any point $p \in M$ and for every vector $X \in T_p M$ there exists a geodesic starting at p with velocity vector X, we have the following

Theorem 3.1. *Let $\varphi : (M^n, g) \to (N^{n+1}, h)$ be the inclusion. Then φ is totally biharmonic if and only if, for any point $p \in M$, one of the following is satisfied:*

(a) $S_\eta \equiv 0$ and φ is totally geodesic;

(b) S_η is a solution of the following system:
$$\begin{cases} \langle S_\eta(X), X\rangle = \text{constant} \neq 0 \\ \|S_\eta(X)\|^2 = \overline{K}(X, \eta) \\ \langle \nabla_X S_\eta(X), Y\rangle = -\langle \overline{R}(X, \eta)X, Y\rangle, \end{cases} \tag{8}$$
for any orthonormal vectors $X, Y \in T_pM$.

4. Totally biharmonic surfaces of space forms

Let now M^2 be a surface in a three-dimensional space form $N^3(C)$ of constant sectional curvature C. Then, for arbitrary orthonormal vectors $X, Y \in T_pM$, we have

$$\langle \overline{R}(X, \eta)Y, X\rangle = C(\langle X, X\rangle\langle \eta, Y\rangle - \langle X, Y\rangle\langle \eta, X\rangle) = 0$$
$$\langle \overline{R}(X, \eta)\eta, X\rangle = C(\langle X, X\rangle\langle \eta, \eta\rangle - \langle X, \eta\rangle\langle \eta, X\rangle) = C.$$

Therefore, (8) becomes

$$\begin{cases} \langle S_\eta(X), X\rangle = \text{constant} \\ \|S_\eta(X)\|^2 = C \\ \langle \nabla_X S_\eta(X), Y\rangle = 0. \end{cases} \tag{9}$$

In particular, from the second condition in (9), we have immediately the following

Proposition 4.1. *Let N^3 be a three-dimensional space form of constant sectional curvature $C \leq 0$. Then the only totally biharmonic surfaces are the totally geodesic ones.*

We are then left to study the case when the sectional curvature of N^3 is positive. Without loss of generality we can assume that N^3 is the three-dimensional sphere $\mathbb{S}^3$ of sectional curvature 1. In this case we have the following characterization

Theorem 4.1. *Let M be a totally biharmonic surface of $\mathbb{S}^3$. Then M is either a totally geodesic or one of the following surfaces:*

(i) part of the Clifford torus $\mathbb{S}^1(\frac{1}{\sqrt{2}}) \times \mathbb{S}^1(\frac{1}{\sqrt{2}})$;
(ii) part of the sphere $\mathbb{S}^2(\frac{1}{\sqrt{2}})$.

Proof. Let M be a totally biharmonic surface of $\mathbb{S}^3$. Since $C = 1$, System (9) becomes

$$\begin{cases} \langle S_\eta(X), X\rangle = \text{constant} \\ \|S_\eta(X)\|^2 = 1 \\ \langle \nabla_X S_\eta(X), Y\rangle = 0. \end{cases} \tag{10}$$

Let e_1, e_2 be the principal curvatures of M^2 in $\mathbb{S}^3$, that is $S_\eta(e_i) = \lambda_i e_i$, $i = 1, 2$, $\lambda_i \in C^\infty(M)$. Then, applying the second equation of (10) to e_1 and e_2, we have

$$\lambda_i^2 = \|S_\eta(e_i)\|^2 = 1, \quad i = 1, 2.$$

If $\lambda_1 = \lambda_2 = 1$, S_η is the shape operator of the sphere $\mathbb{S}^2(\frac{1}{\sqrt{2}})$, while, if $\lambda_1 = 1$ and $\lambda_2 = -1$, S_η is the shape operator of the Clifford torus $\mathbb{S}^1(\frac{1}{\sqrt{2}}) \times \mathbb{S}^1(\frac{1}{\sqrt{2}})$. To end the proof we need to show that the sphere $\mathbb{S}^2(\frac{1}{\sqrt{2}})$ and the Clifford torus satisfy all conditions in (10). For the sphere $\mathbb{S}^2(\frac{1}{\sqrt{2}})$, where $S_\eta = \mathrm{Id}$, taking for any unit vector $T \in T_pM$ the geodesic $\gamma : I \to \mathbb{S}^2$ starting at p with velocity vector T, we have

$$\begin{cases} \langle S_\eta(T), T \rangle = \langle T, T \rangle = 1, \\ \|S_\eta(T)\|^2 = \|T\|^2 = 1, \\ \langle \nabla_T S_\eta(T), T^\perp \rangle = \langle \nabla_T T, T^\perp \rangle = 0. \end{cases}$$

Thus $\mathbb{S}^2\left(\frac{1}{\sqrt{2}}\right)$ is totally biharmonic. Let now consider the Clifford torus. Using the canonical inclusion of $\mathbb{S}^3$ in $\mathbb{R}^4$, a geodesic of the Clifford torus can be parametrized by

$$\gamma(s) = \frac{1}{\sqrt{2}}\left(\cos\left(\frac{\sqrt{2}}{\sqrt{1+a^2}}s\right), \sin\left(\frac{\sqrt{2}}{\sqrt{1+a^2}}s\right),\right.$$
$$\left.\cos\left(\frac{\sqrt{2}a}{\sqrt{1+a^2}}s + b\right), \sin\left(\frac{\sqrt{2}a}{\sqrt{1+a^2}}s + b\right)\right), \quad a, b \in \mathbb{R}.$$

With respect to the principal directions

$$\begin{cases} e_1 = \left(-\sin\left(\frac{\sqrt{2}}{\sqrt{1+a^2}}s\right), \cos\left(\frac{\sqrt{2}}{\sqrt{1+a^2}}s\right), 0, 0\right) \\ e_2 = \left(0, 0, -\sin\left(\frac{\sqrt{2}a}{\sqrt{1+a^2}}s + b\right), \cos\left(\frac{\sqrt{2}a}{\sqrt{1+a^2}}s + b\right)\right) \end{cases}$$

the shape operator of the Clifford torus is

$$S_\eta = \begin{pmatrix} 1 & 0 \\ 0 & -1 \end{pmatrix},$$

while the velocity vector of γ is

$$T = \frac{1}{\sqrt{1+a^2}}e_1 + \frac{a}{\sqrt{1+a^2}}e_2.$$

Then, we have

$$S_\eta(T) = \frac{1}{\sqrt{1+a^2}}e_1 - \frac{a}{\sqrt{1+a^2}}e_2,$$

and
$$T^\perp = \frac{-a}{\sqrt{1+a^2}}e_1 + \frac{1}{\sqrt{1+a^2}}e_2.$$

Let now show that System (10) is satisfied. For the first two equations we have at once

$$\langle S_\eta(T), T\rangle = \frac{1-a^2}{1+a^2} = \text{constant},$$

$$\|S_\eta(T)\|^2 = \frac{1+a^2}{1+a^2} = 1.$$

To verify the third one, we have

$$\langle \overline{\nabla}_T S_\eta(T), T^\perp\rangle = \langle \frac{1}{\sqrt{1+a^2}}\overline{\nabla}_T e_1 - \frac{a}{\sqrt{1+a^2}}\overline{\nabla}_T e_2, \frac{-a}{\sqrt{1+a^2}}e_1 + \frac{1}{\sqrt{1+a^2}}e_2\rangle$$
$$= \frac{1-a^2}{1+a^2}\langle \overline{\nabla}_T e_1, e_2\rangle.$$

Now, using the Weingarten equation of $\mathbb{S}^3$ in $\mathbb{R}^4$ we see that

$$\langle \overline{\nabla}_T e_1, e_2\rangle = \langle \nabla_T^{\mathbb{R}^4} e_1, e_2\rangle = 0,$$

where the last equality comes from the fact that $\nabla_T^{\mathbb{R}^4} e_1$ is tangent to the first factor of the Clifford torus while e_2 is tangent to the second one. □

5. Biharmonic curves in $\mathbb{H}_3$

The three-dimensional Heisenberg group $\mathbb{H}_3$ is the two-step nilpotent Lie group whose elements can be represented in $Gl_3(\mathbb{R})$ by the matrices

$$\begin{bmatrix} 1 & x_1 & x_3 + \frac{1}{2}x_1x_2 \\ 0 & 1 & x_2 \\ 0 & 0 & 1 \end{bmatrix}$$

with $x_i \in \mathbb{R}$, $i = 1, 2, 3$. Endowed with the left-invariant metric

$$g = dx_1^2 + dx_2^2 + \left(dx_3 + \frac{1}{2}x_2 dx_1 - \frac{1}{2}x_1 dx_2\right)^2, \tag{11}$$

$(\mathbb{H}_3, g)$ has a rich geometric structure. In fact its group of isometries is of dimension 4, which is the maximal possible dimension for a non constant curvature metric on a three-manifold. Also, from the algebraic point of view, $(\mathbb{H}_3, g)$ is a 2-step nilpotent Lie group, i.e. "almost Abelian". An orthonormal basis of left-invariant vector fields is given, with respect to the coordinates vector fields, by

$$\begin{cases} E_1 = \frac{\partial}{\partial x_1} - \frac{x_2}{2}\frac{\partial}{\partial x_3} \\ E_2 = \frac{\partial}{\partial x_2} + \frac{x_1}{2}\frac{\partial}{\partial x_3} \\ E_3 = \frac{\partial}{\partial x_3} \end{cases} \quad (12)$$

and with respect to these left invariant vector fields the non-zero component of the curvature tensor field are

$$R_{1221} = -\frac{3}{4}, \quad R_{1331} = R_{2332} = \frac{1}{4}. \quad (13)$$

Let now $X : U \subset \mathbb{R}^2 \to \mathbb{H}_3$, $X(u,v) = (\cos(u), \sin(u), v)$ be the immersion of a right cylinder in $\mathbb{H}_3$. This is invariant under the action of the one-parameter subgroup of isometries of $\mathbb{H}_3$ generated by rotation about the x_3-axes and the induced metric is $ds^2 = \frac{3}{4}du^2 - \frac{1}{2}dudv + dv^2$. A geodesic of the cylinder, with respect to the induced metric, can be parametrized by

$$\gamma(t) = (\cos(at+b), \sin(at+b), ct+d), \quad a,b,c,d \in \mathbb{R}. \quad (14)$$

With respect to the left invariant vector fields E_i, the velocity vector of γ is

$$T = -a\sin(at+b)E_1 + a\cos(at+b)E_2 + \left(c - \frac{a}{2}\right)E_3,$$

the normal to the cylinder along γ is

$$\eta = \cos(at+b)E_1 + \sin(at+b)E_2,$$

and

$$T^\perp = -\left(c - \frac{a}{2}\right)\sin(at+b)E_1 + \left(c - \frac{a}{2}\right)\cos(at+b)E_2 - \left(c - \frac{a}{2}\right)E_3.$$

We now want to find under which conditions on the parameters $a, b, c, d \in \mathbb{R}$, a geodesic of the cylinder is biharmonic in $\mathbb{H}_3$. We have

Proposition 5.1. *Let $X : U \to \mathbb{H}_3$, $X(u,v) = (\cos(u), \sin(u), v)$, the immersion of a right cylinder in $\mathbb{H}_3$. Then all the meridians and the geodesics parametrized by (14) with $2c = 5a$ are biharmonic curves of $\mathbb{H}_3$.*

Proof. We shall compute for which values of the parameters $a, b, c, d \in \mathbb{R}$ a geodesic of the cylinder is a solution of (2). A straightforward computation gives

$$S_\eta(T) = \left(\frac{5}{4}a - \frac{c}{2}\right)\sin(at+b)E_1 - \left(\frac{5}{4}a - \frac{c}{2}\right)\cos(at+b)E_2 + \frac{a}{2}E_3.$$

Then, the first of (2) reduces to
$$\langle S_\eta(T), T \rangle = -a\left(\frac{5}{4}a - \frac{c}{2}\right) + \frac{a}{2}\left(c - \frac{a}{2}\right) = \text{constant}$$
and is always satisfied. Before computing the second of (2), note that
$$K(T, \eta) = -\frac{3}{4}a^2 + \frac{1}{4}\left(c - \frac{a}{2}\right)^2.$$
Thus the second condition of (2) becomes
$$\frac{40}{16}a^2 - ac = 0$$
which is satisfied if either $a = 0$, where the geodesic is a meridian, or $c = \frac{5}{2}a$. Let now show that the third condition of (2) is always satisfied. From
$$\nabla_T S_\eta(T) = \left[\left(\frac{5}{4}a - \frac{c}{2}\right)^2 + \frac{a^2}{4}\right]\left(\cos(at+b)E_1 + \sin(at+b)E_2\right),$$
we have immediately that
$$\langle \nabla_T S_\eta(T), T^\perp \rangle = -\left[\left(\frac{5}{4}a - \frac{c}{2}\right)^2 + \frac{a^2}{4}\right]\left(c - \frac{a}{2}\right)\cos(at+b)\sin(at+b)$$
$$+ \left[\left(\frac{5}{4}a - \frac{c}{2}\right)^2 + \frac{a^2}{4}\right]\left(c - \frac{a}{2}\right)\cos(at+b)\sin(at+b)$$
$$= 0.$$

Moreover, a straightforward computation, taking into account (13), gives
$$\langle R(T, \eta)T, T^\perp \rangle = 0. \qquad \square$$

Acknowledgements

The second author wishes to thank the organizers of the "VIII International Colloquium on Differential Geometry, Santiago de Compostela, 7-11 July 2008" for their exquisite hospitality and the opportunity of presenting a lecture.

Research of the second author was supported by the MIUR Project: *Riemannian metrics and differentiable manifolds*–P.R.I.N. 2005.

References

1. R. Caddeo, S. Montaldo, C. Oniciuc. *Internat. J. Math.* **12** (2001), 867–876.
2. R. Caddeo, S. Montaldo, C. Oniciuc. *Israel J. Math.* **130** (2002), 109–123.
3. R. Caddeo, S. Montaldo, P. Piu. *Rend. Mat. Appl.* **21** (2001), 143–157.
4. R. Caddeo, C. Oniciuc, P. Piu. *Rend. Sem. Mat. Univ. Politec. Torino* **62** (2004), 265–277.

5. R. Caddeo, S. Montaldo, C. Oniciuc, P. Piu. *Mediterr. J. Math.* **3** (2006), 449–465.
6. J.T. Cho, J. Inoguchi and J. Lee. *Ann. Mat. Pura Appl.* **186** (2007), 685–701.
7. I. Dimitric. *Bull. Inst. Math. Acad. Sinica* **20** (1992), 53–65.
8. J. Eells, J.H. Sampson. *Amer. J. Math.* **86** (1964), 109–160.
9. J. Inoguchi. *Int. J. Math. Math. Sci.* **21** (2003) 1365–1368.
10. G.Y. Jiang. *Chinese Ann. Math. Ser. A* **7** (1986), 389–402.
11. S. Montaldo, C. Oniciuc. *Rev. Union Mat. Argent.* **47** (2006), 1–22.

THE BIHARMONICITY OF UNIT VECTOR FIELDS ON THE POINCARÉ HALF-SPACE H^n

M. K. MARKELLOS

Department of Mathematics, University of Patras
Rion, 26500, Greece
E-mail: mark@upatras.gr

We find examples of unit vector fields on the n-dimensional Poincaré half-space H^n which define non-harmonic biharmonic maps into the unit tangent sphere bundle equipped with the Sasaki metric.

Keywords: Harmonic maps, biharmonic maps, homogeneous spaces, homogeneous structures

1. Introduction

The energy functional of a map $\phi : (M, g) \mapsto (N, h)$ between Riemannian manifolds has been widely investigated by several researchers[5,16] and is given by

$$E_1(\phi) = \int_M \|d\phi\|^2 v_g$$

where $d\phi$ denotes the differential of the map ϕ. Critical points for the energy functional are called *harmonic maps* and have been characterized by the vanishing of the *tension field* $\tau_1(\phi) = tr\nabla d\phi$.

Let (M, g) be a Riemannian manifold and denote by $(T_1 M, g_S)$ its unit tangent sphere bundle equipped with the Sasaki metric g_S. Every smooth unit vector field V on M determines a mapping from M into $T_1 M$, embedding M into its unit tangent bundle. If M is compact and orientable, we can define the *energy* of V as the energy of the corresponding map ([18]). A unit vector field which is a critical point of the energy functional restricted to the set of unit vector fields on (M, g) is called a *harmonic vector field* and the corresponding critical point condition has been determined in ([17]) and ([18]). It should be pointed out that a harmonic vector field determines a harmonic map if an additional condition involving the curvature is satisfied

(⁶). The notion of harmonic vector fields can be extended to unit vector fields on possibly non-compact or non-orientable manifolds.

A natural generalization of harmonic maps can be given by considering the functional obtained integrating the square of the norm of the tension field. More precisely, J.Eells and J.H.Sampson (⁵) define the *bienergy* of ϕ as the functional

$$E_2(\phi) = \frac{1}{2}\int_M \|\tau_1(\phi)\|^2 v_g,$$

and a map is *biharmonic* if it is a critical point of E_2. In (³), Caddeo et al. classified biharmonic curves and surfaces of the unit 3-sphere S^3. In fact, they found that they are circles, helices which are geodesics in the Clifford minimal torus and small hyperspheres. The same authors in (⁴) constructed examples of proper biharmonic submanifolds of $S^n, n > 3$.

In (¹²), the author studies the existence and classification of left-invariant unit vector fields which define nonharmonic biharmonic maps on Lie groups equipped with a left-invariant Riemannian metric extending the results of (⁸). More generally, O. Gil-Medrano et al. (⁷) constructed new examples of G-invariant unit vector fields which are either harmonic or which determine harmonic maps on homogeneous Riemannian manifolds $(M = G/G_0, g)$ by means of *homogeneous structures* and *infinitesimal models*. We use these notions in order to construct examples of unit vector fields which determine nonharmonic biharmonic maps into the unit tangent sphere bundle of the Poincaré half-space H^n.

More specifically, the paper is organized in the following way. Section 2 contains the basic notions about biharmonic maps and the Sasaki metric g_S of the unit tangent sphere bundle.

In Section 3 we recall some basic facts about the homogeneous structures. Especially, the classification of homogeneous structures of type $\mathcal{G}_1$ provides us with examples of unit vector fields which define nonharmonic (*proper*) biharmonic maps into the unit tangent sphere bundle of hyperbolic spaces (see Theorem 3.1).

2. Preliminaries

2.1. *Biharmonic maps*

Let $(M^m, g), (N^n, h)$ be Riemannian manifolds and let $\phi : (M^m, g) \mapsto (N^n, h)$ be a smooth map between them. We denote by ∇^ϕ the connection of the vector bundle $\phi^{-1}TN$ induced from the Levi-Civita connection $\bar{\nabla}$ of (N, h) and ∇ the Levi-Civita connection of (M, g). The tension field $\tau_1(\phi)$

of ϕ is a section of the vector bundle $\varphi^{-1}TN$ defined by

$$\tau_1(\phi) = tr(\nabla d\phi) = \sum_{i=1}^{m}\{\nabla^{\phi}_{e_i}d\phi(e_i) - d\phi(\nabla_{e_i}e_i)\},$$

where $\{e_i\}$ is a local orthonormal frame field of M^m. For a compact subset D of M, the energy of ϕ is defined by

$$E_1(\phi) = \frac{1}{2}\int_D \|d\phi\|^2 dv_g.$$

The smooth map ϕ is said to be a *harmonic map* if it is a critical point of the energy over every compact subset of M and the corresponding Euler-Lagrange equation is

$$\tau_1(\phi) = 0$$

which is called *harmonic equation* ([16]).

Definition 2.1. A smooth map $\varphi : (M^m, g) \mapsto (N^n, h)$ is said to be *biharmonic* if it is a critical point of the bienergy functional:

$$E_2(\phi) = \frac{1}{2}\int_D \|\tau_1(\phi)\|^2 dv_g, \qquad (1)$$

over every compact domain D of M.

G. Jiang obtained ([10,11]) the Euler-Lagrange equation associated to the bienergy. More precisely, a smooth map $\phi : (M^m, g) \mapsto (N^n, h)$ is biharmonic if and only if it satisfies the following biharmonic equation

$$\tau_2(\phi) = -\bar{\Delta}_\phi \tau_1(\phi) - tr R^N(d\phi, \tau_1(\phi))d\phi = 0,$$

where $\bar{\Delta}_\phi = -tr(\nabla^\phi \nabla^\phi - \nabla^\phi_\nabla)$ is the rough Laplacian acting on sections of $\phi^{-1}TN$ and $R^N(X,Y) = [\nabla_X, \nabla_Y] - \nabla_{[X,Y]}$ is the curvature operator on N. The section $\tau_2(\phi)$ is called the *bitension field* of ϕ. It is clear that a harmonic map is automatically a biharmonic map, in fact a minimum of the bienergy. Non-harmonic biharmonic maps are called *proper* biharmonic maps. For more details about biharmonic maps, we refer to ([13]).

2.2. The tangent bundle and the unit sphere bundle

We recall some basic facts and formulas about the geometry of the tangent bundle and the unit tangent sphere bundle. For a more elaborate exposition, we refer to the survey ([9]).

Let (M, g) be an m-dimensional Riemannian manifold with Levi-Civita connection ∇. The tangent bundle TM consists of pairs (x, u) where x is a point in M and u a tangent vector to M at x. The mapping $\pi : TM \mapsto M : (x, u) \mapsto x$ is the natural projection from TM onto M.

It is well-known that the tangent space to TM at (x, u) splits into the direct sum of the vertical subspace $\mathcal{V}_{(x,u)} = Ker(d\pi|_{(x,u)})$ and the horizontal subspace $\mathcal{H}_{(x,u)}$ with respect to ∇:

$$T_{(x,u)}TM = \mathcal{H}_{(x,u)} \oplus \mathcal{V}_{(x,u)}.$$

For any vector $w \in T_xM$, there exists a unique vector $w^h \in \mathcal{H}_{(x,u)}$ at the point $(x, u) \in TM$, which is called the *horizontal lift* of w to (x, u), such that $d\pi(w^h) = w$ and a unique vector $w^v \in \mathcal{V}_{(x,u)}$, which is called the *vertical lift* of w to (x, u), such that $w^v(df) = w(f)$ for all functions f on M. In a similar way, one can lift vector fields on M to horizontal or vertical vector fields on TM.

The tangent bundle TM of a Riemannian manifold (M, g) can be endowed in a natural way with a Riemannian metric g_S, the *Sasaki metric*, depending only on the Riemannian structure g of the base manifold M. It is uniquely determined by

$$g_S(X^h, Y^h) = g_S(X^v, Y^v) = g(X, Y) \circ \pi, \quad g_S(X^h, Y^v) = 0 \qquad (2)$$

for all vector fields X and Y on M. More intuitively, the metric g_S is constructed in such a way that the vertical and horizontal subbundles are orthogonal and the bundle map $\pi : (TM, g_S) \mapsto (M, g)$ is a Riemannian submersion.

Next, we consider the unit tangent sphere bundle T_1M which is an embedded hypersurface of TM defined by the equation $g_x(u, u) = 1$. The vector field $N = u^i \frac{\partial}{\partial u^i}$ is a unit normal of T_1M as well as the position vector for a point $(x, u) \in T_1M$. For $X \in T_xM$, we define the *tangential lift* of X to $(x, u) \in T_1M$ by (^{2})

$$X^t_{(x,u)} = X^v_{(x,u)} - g(X, u)N_{(x,u)}.$$

Clearly, the tangent space to T_1M at (x, u) is spanned by vectors of the form X^h and X^t where $X \in T_xM$. The *tangential lift* of a vector field X on M to T_1M is the vector field X^t on T_1M whose value at the point $(x, u) \in T_1M$ is the tangential lift of X_x to (x, u).

A unit vector field V on M can be regarded as the immersion $V : (M, g) \mapsto (T_1M, g_S) : x \mapsto (x, V_x) \in T_1M$ into its unit tangent sphere bundle T_1M equipped with the Sasaki metric g_S. The pull-back metric V^*g_S is given by

$$(V^*g_S)(X, Y) = g(X, Y) + g(\nabla_X V, \nabla_Y V).$$

As a consequence, the unit vector field V determines an isometric immersion into its unit tangent sphere bundle if and only if it is parallel.

The tension field $\tau_1(V)$ is given by ([6])

$$\tau_1(V) = (\sum_{i=1}^{m} R(\nabla_{e_i}V,V)e_i)^h + (tr\nabla^2 V - g(tr\nabla^2 V, V)V)^v,$$

or, equivalently,

$$\tau_1(V) = (\sum_{i=1}^{m} R(\nabla_{e_i}V,V)e_i)^h + (-\bar{\Delta}V)^t. \tag{3}$$

where $\{e_i\}$ is a local orthonormal frame field of (M,g). The term $-tr\nabla^2 V$ is equal to the *rough Laplacian* $\bar{\Delta}V$, that is

$$\bar{\Delta}V = -tr\nabla^2 V = \sum_{i=1}^{m}\{\nabla_{\nabla_{e_i}e_i}V - \nabla_{e_i}\nabla_{e_i}V\}.$$

For the sake of convenience, we set $\tilde{A} = \sum_{i=1}^{m} R(\nabla_{e_i}V,V)e_i$, where V is a unit vector field.

3. Homogeneous structures

In this section we consider the notion of homogeneous Riemannian structure on a Riemannian manifold (M,g). More specifically, we study homogeneous structures of type $\mathcal{G}_1$.

A *homogeneous structure* on a Riemannian manifold (M,g) is a tensor field T of type $(1,2)$ satisfying $\tilde{\nabla}g = \tilde{\nabla}R = \tilde{\nabla}T = 0$ where $\tilde{\nabla}$ is the metric connection determined by $\tilde{\nabla} = \nabla - T$. Moreover, we have $\tilde{\nabla}\tilde{T} = \tilde{\nabla}\tilde{R} = 0$, where $\tilde{T}$ and $\tilde{R}$ denote the torsion and the curvature tensor of $\tilde{\nabla}$, respectively. Any homogeneous Riemannian manifold (M,g) admits a homogeneous structure ([15, Theorem 1.12]). On the other hand, Ambrose and Singer ([1]) proved that a connected, complete and simply connected Riemannian manifold (M,g) is homogeneous if and only if it admits a homogeneous structure T. In particular, given the tensor field T, they constructed the Lie group G (which is called the *transvection group*) which acts transitively and effectively on M ([15, Theorem 1.18]). First, we set $\tilde{\nabla} = \nabla - T$. The Lie algebra $\mathfrak{g}$ of G splits into the direct sum

$$\mathfrak{g} = m \oplus k$$

and called the *transvection algebra*, where m coincides with the tangent space T_oM and k is the algebra generated by all curvature transformations

$\tilde{R}(X,Y), X, Y \in m$. In this case, we have the representation of M as the reductive space $(M, g) = G/K$ with the canonical connection $\tilde{\nabla}$ ([14]) and K being isomorphic to the restricted holonomy group of $(M, \tilde{\nabla})$ at the origin.

Moreover, in ([15]) the authors classified the homogeneous structures into eight classes. In this paper, we are interested only in the first class $\mathcal{G}_1$. Let (M, g) be a connected Riemannian manifold of dimension n which admits a non-trivial homogeneous structure T ($T \neq 0$) of type $\mathcal{G}_1$, i.e. there exists a tensor field T of type $(1,2)$ on M given by

$$g(T(X,Y), Z) = g(T_X Y, Z) = g(X,Y)\varphi(Z) - g(X,Z)\varphi(Y),$$

where X, Y, Z are vector fields on M and φ is a non-zero 1-form on M. Equivalently, if we consider the vector field ξ dual to the 1-form φ, T is given by

$$T_X Y = g(X,Y)\xi - g(\xi, Y)X,$$

for every vector fields X, Y on M. In this case, the manifold M is of constant negative curvature and furthermore, if M is complete and simply connected then it is isometric to the hyperbolic space and conversely. In the sequel, we follow the terminology given in ([7]). We consider the n-dimensional Poincaré half-plane $H^n = \{(x_1, x_2, \ldots, x_n) \in \mathcal{R}^n | x_1 > 0\}$ with the metric

$$g = (cx_1)^{-2} \sum_{i=1}^n (dx_i)^2.$$

It has constant curvature equal to $-c^2$. According to ([15]), the vector field $\xi = c^2 x_1 \frac{\partial}{\partial x_1}$ determines a non-trivial homogeneous structure of type $\mathcal{G}_1$ on (H^n, g). More precisely, the Poincaré half plane (H^2, g) has, up to isomorphisms, only two homogeneous structures, namely $T = 0$ (corresponding to the symmetric case) and the other corresponding to ξ. In this case, the canonical connection $\tilde{\nabla}$ is flat and hence the holonomy algebra k of $\tilde{\nabla}$ is trivial. The computation of the transvection algebra $\mathfrak{g}$, yields the representation of H^n as the subgroup G of $GL(n, \mathcal{R})$ of the form

$$\alpha = \begin{pmatrix} e^{x_1} & 0 & \cdots & 0 & x_2 \\ 0 & e^{x_1} & \cdots & 0 & x_3 \\ \vdots & \vdots & \ddots & \vdots & \vdots \\ 0 & 0 & \cdots & e^{x_1} & x_n \\ 0 & 0 & \cdots & 0 & 1 \end{pmatrix}.$$

Furthermore, it is a solvable Lie group which is a semi-direct product of the multiplicative group $\mathcal{R}_0^+ = \{x \in \mathcal{R} | x > 0\}$ and the additive group $\mathcal{R}^{n-1}$. A basis of left-invariant vector fields is given by

$$X_1 = c\frac{\partial}{\partial x_1}, X_i = ce^{x_1}\frac{\partial}{\partial x_i}, (2 \leq i \leq n)$$

and the Lie bracket satisfies $[X_1, X_i] = cX_i$, the remainder brackets being zero. On G, we consider the Riemannian metric $<,>$. Then, these vector fields constitute an orthonormal basis at each point with respect to this inner product. With this metric, the group G is locally isometric to (H^n, g). The Levi-Civita connection with respect to this metric is determined by

$$\nabla_{X_i} X_i = cX_1, \nabla_{X_i} X_1 = -cX_i, (2 \leq i \leq n) \quad (4)$$

and the remaining covariant derivatives vanish. In the sequel, we completely determine the unit left-invariant vector fields on (H^n, g) which define proper biharmonic maps into the unit tangent sphere bundle equipped with the Sasaki metric. Furthermore, let $\mathcal{S}$ be the unit sphere of $\mathfrak{g}$ with respect to $<,>$. For $V \in \mathcal{S}, \nabla V, \tilde{A}, \bar{\Delta}V$, the tension field $\tau_1(V)$ and the bitension field $\tau_2(V)$ are invariant by left translation. Therefore, these can be viewed as tensors on $\mathfrak{g}$ which are determined by the Lie algebra structure and its inner product $<,>$.

Now, let $V \in \mathcal{S}$ and put $V = V_1 X_1 + \sum_{i=2}^n V_i X_i$. Then, using (4), we easily get

$$\nabla_{X_1} V = 0, \nabla_{X_i} V = c(V_i X_1 - V_1 X_i), (2 \leq i \leq n). \quad (5)$$

By straightforward calculation, using (4) and (5), we deduce

$$R(\nabla_{X_i} V, V) X_i = (-c^2)(<V, X_i> \nabla_{X_i} V - <\nabla_{X_i} V, X_i> V)$$
$$= (-c^2)(V_i \nabla_{X_i} V + cV_1 V) = (-c^3)[(V_i^2 + V_1^2) X_1$$
$$+ V_1 \sum_{\substack{j=2 \\ j \neq i}}^n V_j X_j], \quad (2 \leq i \leq n)$$

and, as a consequence, we get

$$\tilde{A} = R(\nabla_{X_1} V, V) X_1 + \sum_{i=2}^n R(\nabla_{X_i} V, V) X_i$$
$$= (-c^3)[(\sum_{i=2}^n V_i^2 + (n-1)V_1^2) X_1 + (n-2)V_1 \sum_{i=2}^n V_i X_i]. \quad (6)$$

Once again, using (4) and (5), we compute

$$\bar{\Delta}V = -\sum_{i=2}^n \nabla_{X_i} \nabla_{X_i} V = c^2[(n-1)V_1 X_1 + \sum_{i=2}^n V_i X_i]. \quad (7)$$

Combining (2), (3), (6) and (7), we obtain

$$\|\tau_1(V)\|_g^2 = \|\tau_1(V)\|_e^2 = \|\tilde{A}\|_e^2 + \|\bar{\Delta}V - <\bar{\Delta}V, V> V\|_e^2$$

$$= c^6[(\sum_{i=2}^n V_i^2 + (n-1)V_1^2)^2 + (n-2)^2 V_1^2 \sum_{i=2}^n V_i^2] + c^4[(n-1)^2 V_1^2$$

$$+ \sum_{i=2}^n V_i^2] - c^4(\sum_{i=2}^n V_i^2 + (n-1)V_1^2)^2,$$

$$= f(V_1, V_2, \ldots, V_n),$$

since the Riemannian metric $<,>$ is left invariant. The Lie group G is non-compact and, so, we choose an open neighbourhood $\mathcal{U}$ of the neutral element e with compact closure. Working on $\mathcal{U}$ and using (1), we get

$$E_2(V) = \frac{1}{2} \int_{\mathcal{U}} \|\tau_1(V)\|_g^2 dv_g$$
$$= \frac{1}{2} \int_{\mathcal{U}} \|\tau_1(V)\|_e^2 dv_g = \frac{1}{2} f(V_1, V_2, \ldots, V_n) Vol(\mathcal{U}). \qquad (8)$$

From the above analysis we have the following Proposition:

Proposition 3.1. *A left-invariant unit vector field V on (H^n, g) defines a biharmonic map into its unit tangent sphere bundle equipped with the Sasaki metric g_S if and only if V is a critical point of the function f on $\mathcal{S}$.*

We define the Lagrange function $L(V_1, V_2, \ldots, V_n, \lambda) = f(V_1, V_2, \ldots, V_n) + \lambda(\sum_{i=2}^n V_i^2 + V_1^2 - 1), (V_1, V_2, \ldots, V_n) \in \mathcal{R}^n$, where λ is the Lagrange multiplier. The critical points of the function f are given by the following system:

$$V_i \Big\{ c^6[2(\sum_{i=2}^n V_i^2 + (n-1)V_1^2) + (n-2)^2 V_1^2] + c^4$$

$$- 2c^4(\sum_{i=2}^n V_i^2 + (n-1)V_1^2) + \lambda \Big\} = 0, \quad i = 2, \ldots, n$$

$$V_1 \Big\{ c^6[2(n-1)(\sum_{i=2}^n V_i^2 + (n-1)V_1^2) + (n-2)^2 \sum_{i=2}^n V_i^2] + (n-1)^2 c^4$$

$$- 2(n-1)c^4(\sum_{i=2}^n V_i^2 + (n-1)V_1^2) + \lambda \Big\} = 0,$$

$$\sum_{i=2}^n V_i^2 + V_1^2 = 1$$

The solutions of the above system lead to the following Theorem:

Theorem 3.1. *Let $(H^n, g) \simeq G$ be the n-dimensional Poincaré half-space. For $n = 2$ every left-invariant unit vector field defines a proper biharmonic map into the unit tangent sphere bundle $T_1 H^2$ equipped with the Sasaki metric g_S. For $n > 2$, the set of left-invariant unit vector fields V which define proper biharmonic maps into the unit tangent sphere bundle is given by: $\pm \frac{1}{c}\xi, \mathcal{S} \cap \xi^\perp$ and the $(n-2)$-dimensional hyperspheres: $C_1 = \{\sum_{i=2}^{n} V_i^2 = \frac{n-2-c^2 n}{2(n-2)}, V_1 = \sqrt{\frac{n-2+c^2 n}{2(n-2)}}\}$, $C_2 = \{\sum_{i=2}^{n} V_i^2 = \frac{n-2-c^2 n}{2(n-2)}, V_1 = -\sqrt{\frac{n-2+c^2 n}{2(n-2)}}\}$ with the assumption that $c^2 < 1 - \frac{2}{n}$, where $\xi^\perp$ denotes the distribution defined by the vector fields orthogonal to ξ.*

Remark 3.1. According to Proposition 5.2 in (^{7}), the unit vector field ξ of $H^n, n > 2$ and those which are orthogonal to ξ are harmonic vector fields and the corresponding maps into the unit tangent sphere bundle are not harmonic. On the contrary, they define proper biharmonic maps into $T_1 H^n$.

Remark 3.2. The non harmonic unit vector fields whose coordinates satisfy the equations of the hyperspheres C_1, C_2 mentioned in Theorem 3.1 define proper biharmonic maps and they do not determine harmonic maps.

Remark 3.3. In (^{12}), the author completely determines the left-invariant unit vector fields which define proper biharmonic maps into the unit tangent sphere bundle of the generalized Heisenberg groups $H(1, r), r \geq 2$. Furthermore, the generalized Heisenberg groups are unimodular Lie groups. On the other hand, since $trad_{X_1} = c(n-1) \neq 0$, the transvection group G is non-unimodular. As a consequence, Theorem 3.1 provides us with examples of unit vector fields which define proper biharmonic maps into the unit tangent sphere bundle of non-unimodular Lie groups of dimension greater than three (if we choose $n > 3$).

Acknowledgements

The author was partially supported by the Greek State Scholarships Foundation (I.K.Y.) and by the C. Caratheóodory grant no.C.161 2007 - 10, University of Patras. Also, the author expresses his thanks to Professor Vassilis Papantoniou for his helpful comments in the preparation of this work.

References

1. W. Ambrose and I. M. Singer, *Duke Math. J.* **25**, 647 (1958).

2. E. Boeckx and L. Vanhecke, *Houston J. Math.* **23**, 427 (1997).
3. R. Caddeo and S. Montaldo and C. Oniciuc, *Intern. J. Math* **12**, 867 (2001).
4. R. Caddeo and S. Montaldo and C. Oniciuc, *Israel J.Math.* **130**, 109 (2002).
5. J. Eells and J. H. Sampson, *Amer. J. Math.* **86**, 109 (1964).
6. O. Gil-Medrano, *Differential Geom. Appl.* **15**, 137 (2001).
7. O. Gil-Medrano and J. C. González-Dávila and L. Vanhecke, *Houston J.Math.* **27**, 377 (2001).
8. J. C. González-Dávila and L. Vanhecke, *Boll. Unione Mat. Ital. Sez. B Artic. Ric. Mat. (8)* **5**, 377 (2002).
9. S. Gudmundsson and E. Kappos, *Expo. Math.* **20**, 1 (2002).
10. G. Y. Jiang, *Chinese Ann. Math. Ser. A* **7**, 389 (1986).
11. G. Y. Jiang, *Chinese Ann. Math. Ser. A* **7**, 130 (1986).
12. M. Markellos, *submitted*.
13. S. Montaldo and C. Oniciuc, *Rev. Un. Mat. Argentina* **47**, 1 (2006).
14. K. Nomizu, *Amer. J. Math* **76**, 33 (1954).
15. F. Tricerri and L. Vanhecke, in *Homogeneous structures on Riemannian manifolds* (London Math. Soc., Cambridge University Press, 1989).
16. H. Urakawa, in *Calculus of Variations and Harmonic Maps* (American Mathematical Society, Providence, 1993).
17. G. Wiegmink, *Math. Ann.* **303**, 325 (1995).
18. C. M. Wood, *Geom. Dedicata* **64**, 319 (1997).

PERSPECTIVES ON BIHARMONIC MAPS AND SUBMANIFOLDS

A. Balmuș

Faculty of Mathematics, "Al.I. Cuza" University
Iași, Bd. Carol I Nr. 11, 700506, Romania
E-mail: adina.balmus@uaic.ro
www.math.uaic.ro

This paper is intended to present the main results of the author's PhD Thesis[2] concerning the geometry of biharmonic maps:

(i) we present several new methods, inspired by the Baird-Kamissoko method[1], for constructing proper biharmonic maps starting with harmonic maps and using warped product manifolds;

(ii) we obtain classification and non-existence results for proper biharmonic submanifolds in space forms;

(iii) we study the biharmonicity of the Gauss map for submanifolds in the Euclidean space, in the intent of generalizing the celebrated Ruh Vilms Theorem[24].

Keywords: Biharmonic maps and submanifolds, warped products, Gauss map

1. Introduction

Denote by $C^\infty(M, N)$ the space of smooth maps $\phi : (M, g) \to (N, h)$ between two Riemannian manifolds. A map $\phi \in C^\infty(M, N)$ is called *harmonic* if it is a critical point of the *energy* functional[16]

$$E : C^\infty(M, N) \to \mathbb{R}, \quad E(\phi) = \frac{1}{2} \int_M |d\phi|^2 \, v_g,$$

and is characterized by the vanishing of the *tension field* $\tau(\phi) = \operatorname{trace} \nabla d\phi$. We recall that, if $\phi : (M, g) \to (N, h)$ is a Riemannian immersion, then it is harmonic if and only if it is a *minimal* immersion[16], i.e. a critical point of the volume functional.

One can generalize harmonic maps by considering the functional obtained by integrating the squared norm of the tension field. More precisely, *biharmonic maps* are the critical points of the *bienergy* functional

$$E_2 : C^\infty(M, N) \to \mathbb{R}, \quad E_2(\phi) = \frac{1}{2} \int_M |\tau(\phi)|^2 \, v_g .$$

The Euler-Lagrange equation for E_2 (see Ref. 19), called the *biharmonic equation*, is given by the vanishing of the *bitension field*

$$\tau_2(\phi) = -J(\tau(\phi)) = -\Delta\tau(\phi) - \mathrm{trace} R^N(d\phi, \tau(\phi))d\phi = 0,$$

where J is (formally) the Jacobi operator of ϕ, Δ is the rough Laplacian defined on sections of $\phi^{-1}(TN)$ and $R^N(X,Y) = \nabla_X \nabla_Y - \nabla_Y \nabla_X - \nabla_{[X,Y]}$ is the curvature operator on (N, h).

Biharmonic maps have been extensively studied in the last decade and there are two main research directions. On one hand, in differential geometry, a special attention has been payed to the construction of examples and classification results (see, for example, Refs. 1,9,13,15,18,21). On the other hand, from the analytic point of view, biharmonic maps are solutions of a fourth order strongly elliptic semilinear PDE and the study of their regularity is nowadays a well-developed field (see, for example, Refs. 10,20,22,25).

We are interested in the geometric aspect and, since any harmonic map is trivially biharmonic, we concentrate on biharmonic non-harmonic maps and biharmonic non-minimal submanifolds, which are called *proper biharmonic*.

In his studies on finite type submanifolds[11], B-Y. Chen defined biharmonic submanifolds $M \subset \mathbb{R}^n$ of the Euclidean space as those with harmonic mean curvature vector field, that is $\Delta H = 0$, where Δ is the rough Laplacian. By considering the definition of biharmonic map for Riemannian immersions into the Euclidean space one recovers the notion of biharmonic submanifold in the sense of B-Y. Chen. Thus biharmonic Riemannian immersions can also be thought of as a generalization of Chen's biharmonic submanifolds. All the results obtained by B-Y. Chen and his collaborators on proper biharmonic submanifolds in Euclidean spaces are non-existence results. Nevertheless, by adapting their techniques it is possible to produce, as we shall see in the sequel, several classification results for proper biharmonic submanifolds in spheres.

For an up-to-date bibliography on biharmonic maps see Ref. 26.

2. Bihamonic maps and warped product manifolds

We shall present in this section a series of new methods for constructing proper biharmonic maps by using harmonic ones.

A natural way for constructing proper biharmonic maps is the following: given a harmonic map $\varphi : (M,g) \to (N,h)$ one can conformally change the metric g, or h, in order to render φ proper biharmonic. The behavior of the bitension field under the conformal change of the domain metric was studied by P. Baird and D. Kamissoko[1]. They established an interesting connection between conformally equivalent biharmonic metrics and isoparametric functions on Einstein manifolds.

We extend the idea of studying the effect of conformal changes of metric to the biharmonic equation by considering the situation of warped products. The first problem consists in the study of the biharmonicity of the inclusion of a Riemannian manifold N into the warped product $M \times_{f^2} N$ and we obtain

Theorem 2.1 ([3]). *The inclusion map $\mathbf{i}_{x_0} : N \to M \times_{f^2} N$ is a proper biharmonic map if and only if x_0 is not a critical point for f^2, but is a critical point for $|\mathrm{grad} f^2|^2$.*

With this setting we obtain new examples of proper biharmonic maps and recover some of the examples first obtained in Refs. 7 and 9.

Affine functions play an important role in our study

Theorem 2.2 ([3]). *Let (M,g) be a Riemannian manifold with a positive nontrivial affine function f^2 and let (N,h) be an arbitrary Riemannian manifold. Then any inclusion $\mathbf{i}_x : N \to M \times_{f^2} N$, $x \in M$, is a proper biharmonic map.*

We then consider the product of two harmonic maps $\phi = \mathbf{1}_M \times \psi : M \times N \to M \times N$. By warping the metric on the domain or codomain we lose the harmonicity; nevertheless, under certain conditions on the warping function the product map remains biharmonic. We obtain

Theorem 2.3 ([3]). *Let $\psi : N \to N$ be a harmonic map and let $f \in C^\infty(M)$ be a positive function. Then $\overline{\phi} = \overline{\mathbf{1}_M \times \psi} : M \times_{f^2} N \to M \times N$ is a proper biharmonic map if and only if f is a non-constant solution of*

$$\mathrm{trace}_g \nabla^2 \mathrm{grad}\ln f + \mathrm{Ricci}^M(\mathrm{grad}\ln f) + \frac{n}{2}\mathrm{grad}(|\mathrm{grad}\ln f|^2) = 0.$$

In the case that the product map is the identity map $\overline{\mathbf{1}} : M \times_{f^2} N \to M \times N$ we shall call the warping function, which is a solution of the above problem, a *biharmonic warping function* and if M is an Einstein manifold we show that the isoparametric functions provide examples of biharmonic warping ones

Proposition 2.1 ([3]). *Let M be an Einstein space. If $\rho \in C^\infty(M)$ is an isoparametric function, then it admits, away from its critical points, a local reparameterization f which is a biharmonic warping function.*

For the projection $\bar{\pi}: M \times_{f^2} N \to M$ we obtain

Proposition 2.2 ([3]). *The projection $\bar{\pi}: M \times_{f^2} N \to M$ is proper biharmonic if and only if f is a biharmonic warping function.*

We also give the complete classification of the biharmonic warping functions when $M = \mathbb{R}$. Similar results are obtained when the codomain metric is warped and ψ is harmonic with constant energy density.

Finally, we use the warped product setting to study axially symmetric biharmonic maps from $\mathbb{R}^m \setminus \{0\}$ to an n-dimensional space form. We get a general characterization result

Theorem 2.4 ([3]). *Let $\varphi: \mathbb{S}^{m-1} \to \mathbb{S}^{n-1}$ be an eigenmap of eigenvalue $2k \geq 0$. Then $\phi = \rho \times \varphi: \mathbb{R}^m \setminus \{0\} \to \mathbb{R} \times_{f^2} \mathbb{S}^{n-1}$ is biharmonic if and only if ρ is a solution of*

$$F'' + \frac{m-1}{t}F' - \frac{2k}{t^2}\left(f'^2(\rho) + f(\rho)f''(\rho)\right)F = 0,$$

where

$$F = \rho'' + \frac{m-1}{t}\rho' - \frac{2k}{t^2}f(\rho)f'(\rho).$$

We discuss several examples and, when the target manifold is $\mathbb{R}^n \setminus \{0\}$, we give the complete classification of biharmonic axially symmetric maps. From the former classification we also deduce that the generalized Kelvin transformation $\phi: \mathbb{R}^m \setminus \{0\} \to \mathbb{R}^m \setminus \{0\}$, $\phi(y) = y/|y|^\ell$ is a proper biharmonic map if and only if $m = \ell + 2$ or $\ell = -2$.

3. Biharmonic submanifolds in space forms

The attempt to classify the biharmonic submanifolds in space forms was initiated in Refs. 14 and 23. It was proved that an m-dimensional submanifold M in an n-dimensional space form $\mathbb{E}^n(c)$ is biharmonic if and only if

$$\begin{cases} \Delta^\perp H + \text{trace}B(\cdot, A_H \cdot) - mcH = 0, \\ 2\text{trace}A_{\nabla^\perp_{(\cdot)} H}(\cdot) + \frac{m}{2}\text{grad}(|H|^2) = 0, \end{cases}$$

where A denotes the Weingarten operator, B the second fundamental form, H the mean curvature vector field, $\nabla^\perp$ and $\Delta^\perp$ the connection and the Laplacian in the normal bundle of M in $\mathbb{E}^n(c)$.

All the known results on proper biharmonic submanifolds in non-positively curved spaces are non-existence results. There are no proper biharmonic hypersurfaces in $\mathbb{R}^3$ and $\mathbb{R}^4$ (see Refs. 14,17). Also, it was proved that the biharmonicity implies the minimality for the following classes of submanifolds in the Euclidean space: curves, submanifolds of finite type, submanifolds of constant mean curvature, pseudo-umbilical submanifolds of dimension $m \neq 4$, hypersurfaces with at most two distinct principal curvatures, conformally flat hypersurfaces (see Refs. 14,15). Similar results were obtained when the ambient space is the hyperbolic space (see, for example, Ref. 7).

The non-existence theorems for the case of non-positive sectional curvature codomains, as well as the

Generalized Chen Conjecture. *Biharmonic submanifolds of a non-positive sectional curvature manifold are minimal,*
encouraged the study of proper biharmonic submanifolds in spheres and other curved spaces (see, for example, Ref. 21).

The first achievement towards the classification problem is represented by the complete classification of proper biharmonic submanifolds of the 3-dimensional unit Euclidean sphere[7]. Then, inspired by the 3-dimensional case, two methods for constructing proper biharmonic submanifolds in $\mathbb{S}^n$ were given in Ref. 8. Although important results and examples were obtained, the classification of proper biharmonic submanifolds in spheres is still an open problem.

This section is devoted to the classification results for proper biharmonic submanifolds in space forms contained in Refs. 4, 5.

A partial classification result for biharmonic submanifolds in spheres was given in Ref. 23. It was proved that a proper biharmonic submanifold M of constant mean curvature $|H|$ in $\mathbb{S}^n$ satisfies $|H| \in (0,1]$. Moreover, $|H| = 1$ if and only if M is minimal in $\mathbb{S}^{n-1}(\frac{1}{\sqrt{2}})$. Taking this further, we study the type of compact proper biharmonic submanifolds of constant mean curvature in $\mathbb{S}^n$ and prove that, depending on the value of the mean curvature, they are of 1-type or of 2-type as submanifolds of $\mathbb{R}^{n+1}$

Theorem 3.1 ([4]). *Let M^m be a compact constant mean curvature, $|H|^2 = k$, submanifold in $\mathbb{S}^n$. Then M is proper biharmonic if and only if either*

(i) $|H|^2 = 1$ and M is a 1-type submanifold of $\mathbb{R}^{n+1}$ with eigenvalue $\lambda = 2m$, or
(ii) $|H|^2 = k \in (0,1)$ and M is a 2-type submanifold of $\mathbb{R}^{n+1}$ with the eigenvalues $\lambda_{1,2} = m(1 \pm \sqrt{k})$.

We then study proper biharmonic hypersurfaces with at most two distinct principal curvatures in space forms. We first prove the following

Theorem 3.2 ([4]). *Let M be a hypersurface with at most two distinct principal curvatures in $\mathbb{E}^{m+1}(c)$. If M is proper biharmonic in $\mathbb{E}^{m+1}(c)$, then it has constant mean curvature.*

By using this result, we obtain the full classification of such hypersurfaces in spheres

Theorem 3.3 ([4]). *Let M^m be a proper biharmonic hypersurface with at most two distinct principal curvatures in $\mathbb{S}^{m+1}$. Then M is an open part of $\mathbb{S}^m(\frac{1}{\sqrt{2}})$ or of $\mathbb{S}^{m_1}(\frac{1}{\sqrt{2}}) \times \mathbb{S}^{m_2}(\frac{1}{\sqrt{2}})$, $m_1 + m_2 = m$, $m_1 \neq m_2$.*

A similar classification is obtained for conformally flat biharmonic hypersurfaces in spheres. In contrast, for the hyperbolic space $\mathbb{H}^{m+1}$ we prove a non-existence result. The problem of the biharmonic hypersurfaces with at most 3 distinct principal curvatures is also investigated. We first prove

Theorem 3.4 ([5]). *There exist no compact proper biharmonic hypersurfaces of constant mean curvature and three distinct curvatures in the unit Euclidean sphere.*

The key of the proof was to show that such hypersurfaces are isoparametric and to use the expressions of the principal curvatures. Further, Theorem 3.4 allows us to fully classify the proper biharmonic compact hypersurfaces in $\mathbb{S}^4$

Theorem 3.5 ([5]). *The only proper biharmonic compact hypersurfaces of $\mathbb{S}^4$ are the hypersphere $\mathbb{S}^3(\frac{1}{\sqrt{2}})$ and the torus $\mathbb{S}^1(\frac{1}{\sqrt{2}}) \times \mathbb{S}^2(\frac{1}{\sqrt{2}})$.*

Moreover, we recover the non-existence result for proper biharmonic hypersurfaces in $\mathbb{R}^4$ and prove that it also holds for hypersurfaces of $\mathbb{H}^4$.

We prove that the pseudo-umbilical biharmonic submanifolds in spheres have constant mean curvature and we give an estimate for their scalar curvature. Using these results, we classify the proper biharmonic pseudo-umbilical submanifolds of codimension 2

Theorem 3.6 ([4]). *Let M^m be a pseudo-umbilical submanifold of $\mathbb{S}^{m+2}$, $m \neq 4$. Then M is proper biharmonic if and only if it is minimal in $\mathbb{S}^{m+1}(\frac{1}{\sqrt{2}})$.*

We also investigate surfaces with parallel mean curvature vector field in $\mathbb{S}^n$

Theorem 3.7 ([4]). *Let M^2 be a proper biharmonic surface with parallel mean curvature vector field in $\mathbb{S}^n$. Then M is minimal in $\mathbb{S}^{n-1}(\frac{1}{\sqrt{2}})$.*

In view of the above results we proposed the following

Conjecture 3.1 ([4]). *The only proper biharmonic hypersurfaces in $\mathbb{S}^{m+1}$ are the open parts of hyperspheres $\mathbb{S}^m(\frac{1}{\sqrt{2}})$ or of generalized Clifford tori $\mathbb{S}^{m_1}(\frac{1}{\sqrt{2}}) \times \mathbb{S}^{m_2}(\frac{1}{\sqrt{2}})$, $m_1 + m_2 = m$, $m_1 \neq m_2$.*

Conjecture 3.2 ([4]). *Any biharmonic submanifold in $\mathbb{S}^n$ has constant mean curvature.*

4. On the biharmonicity of the Gauss map

The study of submanifolds with harmonic Gauss map in Euclidean spaces has been a classical problem in harmonic maps theory, ever since the characterization result obtained by E. Ruh and J. Vilms in Ref. 24 and the remarkable link with constant mean curvature hypersurfaces. In Ref. 6 we propose a generalization of this problem: the characterization of submanifolds with proper biharmonic Gauss map in Euclidean spaces.

By using the expression of the curvature operator for the Grassmannian $G(m,n)$ of oriented m-planes in $\mathbb{R}^{m+n}$, we first characterize the biharmonicity of the Gauss map $\gamma : M \to G(m,n)$ associated to a submanifold M in the Euclidean space $\mathbb{R}^{m+n}$.

Theorem 4.1 ([6]). *The Gauss map associated to a m-dimensional orientable submanifold M of $\mathbb{R}^{m+n}$ is proper biharmonic if and only if $\nabla^\perp H \neq 0$ and*

$$\nabla_X^\perp \Delta^\perp H - m\nabla_{A_H(X)}^\perp H + \operatorname{trace} B\big(2A_{\nabla_{(\cdot)}^\perp H}(X) - A_{\nabla_X^\perp H}(\cdot), \cdot\big)$$
$$-2\operatorname{trace} R^\perp(\cdot, X)\nabla_\cdot^\perp H - \operatorname{trace}(\nabla_\cdot^\perp R^\perp)(\cdot, X)H = 0,$$

for all $X \in C(TM)$, where A denotes the Weingarten operator and H the mean curvature vector field of M in $\mathbb{R}^{m+n}$.

Although rather technical, this condition simplifies in the case of hypersurfaces, since the Grassmannian is the m-dimensional sphere and the tension field of the Gauss map relies completely on the gradient of the mean curvature function of the hypersurface

Theorem 4.2 ([6]). *The Gauss map of a nowhere zero mean curvature hypersurface M^m in $\mathbb{R}^{m+1}$ is proper biharmonic if and only if $\operatorname{grad} f \neq 0$*

and

$$\Delta \mathrm{grad} f + A^2(\mathrm{grad} f) - |A|^2 \mathrm{grad} f = 0,$$

where Δ denotes the rough Laplacian on $C(TM)$ and f and A denote the mean curvature function and, respectively, the shape operator of M in $\mathbb{R}^{m+1}$.

In order to obtain examples of hypersurfaces with proper biharmonic Gauss map we study the hypercones generated by hypersurfaces in spheres.

Theorem 4.3 ([6]). *Let $\overline{M}$ be a nowhere zero mean curvature hypersurface of $\mathbb{S}^{m+1}$ with constant norm of the shape operator. If $\overline{M}$ is compact, then the Gauss map associated to the hypercone $(0, \infty) \times_{t^2} \overline{M}$ is proper biharmonic if and only if $\overline{M}$ has constant mean curvature in $\mathbb{S}^{m+1}$, $m > 2$, and $|\overline{A}|^2 = 3(m-2)$, where $\overline{A}$ is the shape operator of $\overline{M}$ in $\mathbb{S}^{m+1}$.*

By considering constant mean curvature hypersurfaces in spheres we obtain

Theorem 4.4 ([6]). *Let $\overline{M}$ be a constant non-zero mean curvature hypersurface of $\mathbb{S}^{m+1}$. The Gauss map associated to the hypercone $(0, \infty) \times_{t^2} \overline{M}$ is proper biharmonic if and only if $m > 2$ and $|\overline{A}|^2 = 3(m-2)$.*

The previous result is used in order to provide examples of hypersurfaces with proper biharmonic associated Gauss map in any $(m+2)$-dimensional Euclidean space, with $m > 2$. In this respect, we study the biharmonicity of the Gauss map associated to hypercones over isoparametric hypersurfaces in spheres, and we only present here the following

Proposition 4.1 ([6]). *Consider the hypercone $(0, \infty) \times_{t^2} \mathbb{S}^m(a)$, $a \in (0, 1)$, in $\mathbb{R}^{m+2}$. Its associated Gauss map is proper biharmonic if and only if $m > 2$ and $a = \sqrt{\frac{m}{4m-6}}$.*

Finally, non-existence results for hypercones with biharmonic Gauss map in $\mathbb{R}^3$ and $\mathbb{R}^4$ are obtained

Theorem 4.5 ([6]). *There exist no cones in $\mathbb{R}^3$ with proper biharmonic Gauss map.*

Theorem 4.6 ([6]). *There exist no hypercones in $\mathbb{R}^4$, over compact nowhere zero mean curvature surfaces $\overline{M}^2 \subset \mathbb{S}^3$, with proper biharmonic Gauss map.*

Acknowledgements

The author is grateful to the Istituto Nazionale di Alta Matematica "Francesco Severi" for the financial support during the PhD Thesis preparation and to the Department of Mathematics, University of Cagliari, for hospitality.

The author wishes to thank the organizers of the VIII International Colloquium on Differential Geometry, Santiago de Compostela, 7-11 July 2008, for the Conference Grant.

The author was supported by Grant CEEX ET 5871/2006, ROMANIA.

References

1. P. Baird, D. Kamissoko, *Ann. Global Anal. Geom.* **23** (2003), 65–75.
2. A. Balmuş, *Biharmonic maps and submanifolds*. PhD Thesis, University of Cagliari, Italy, 2008.
3. A. Balmuş, S. Montaldo, C. Oniciuc, *J. Geom. Phys.* **57** (2007), 449–466.
4. A. Balmuş, S. Montaldo, C. Oniciuc, *Israel J. Math.* **168** (2008), 201–220.
5. A. Balmuş, S. Montaldo, C. Oniciuc, *Math. Nachr.*, to appear (arXiv:0709.2023).
6. A. Balmuş, S. Montaldo, C. Oniciuc, (arXiv:0809.1298)
7. R. Caddeo, S. Montaldo, C. Oniciuc, *Internat. J. Math.* **12** (2001), 867–876.
8. R. Caddeo, S. Montaldo, C. Oniciuc, *Israel J. Math.* **130** (2002), 109–123.
9. R. Caddeo, S. Montaldo, P. Piu, *Rend. Mat. Appl. (7)* **21** (2001), 143–157.
10. S.-Y. A. Chang, L. Wang, P.C. Yang, *Comm. Pure Appl. Math.* **52** (1999), 1113–1137.
11. B-Y. Chen, *Total mean curvature and submanifolds of finite type*. Series in Pure Mathematics **1**. World Scientific Publishing Co., Singapore, 1984.
12. B-Y. Chen, *Soochow J. Math.* **17** (1991), 169–188.
13. B-Y. Chen, *Soochow J. Math.* **22** (1996), 117–337.
14. B-Y. Chen, S. Ishikawa, *Kyushu J. Math.* **52** (1998), 167–185.
15. I. Dimitric, *Bull. Inst. Math. Acad. Sinica* **20** (1992), 53–65.
16. J. Eells, J.H. Sampson, *Amer. J. Math.* **86** (1964), 109–160.
17. Th. Hasanis, Th. Vlachos, *Math. Nachr.* **172** (1995), 145–169.
18. J-I. Inoguchi, *Colloq. Math.* **100** (2004), 163–179.
19. G.Y. Jiang, *Chinese Ann. Math. Ser. A* **7** (1986), 130-144.
20. T. Lamm, *Calc. Var.* **22** (2005), 421–445.
21. S. Montaldo, C. Oniciuc, *Rev. Un. Mat. Argentina* **47** (2006), 1–22.
22. R. Moser, *Proc. London Math. Soc.* **96** (2008), 199–226.
23. C. Oniciuc, *An. Stiint. Univ. Al.I. Cuza Iasi Mat (N.S.)* **48** (2002), 237–248.
24. E. Ruh, J. Vilms, *Trans. Amer. Math. Soc.* **149** (1970), 569–573.
25. C. Wang, *Calc. Var.* **21** (2004), 221–242.
26. *The Bibliography of Biharmonic Maps*, http://people.unica.it/biharmonic.

CONTACT PAIR STRUCTURES AND ASSOCIATED METRICS

G. Bande

Dipartimento di Matematica e Informatica, Università degli Studi di Cagliari
Via Ospedale 72, 09124 Cagliari, Italy
E-mail: gbande@unica.it

A. Hadjar

Laboratoire de Mathématiques, Informatique et Applications, Univ. de Haute Alsace
4 Rue des Frères Lumière, 68093 Mulhouse Cedex, France
E-mail: Amine.Hadjar@uha.fr

We introduce the notion of contact pair structure and the corresponding associated metrics, in the same spirit of the geometry of almost contact structures. We prove that, with respect to these metrics, the integral curves of the Reeb vector fields are geodesics and that the leaves of the Reeb action are totally geodesic. Moreover, we show that, in the case of a metric contact pair with decomposable endomorphism, the characteristic foliations are orthogonal and their leaves carry induced contact metric structures.

Keywords: Contact pairs, foliation, associated metrics, contact metric structure

1. Introduction

A *contact pair* on a smooth manifold M is a pair of one-forms α_1 and α_2 of constant and complementary classes, for which α_1 restricted to the leaves of the characteristic foliation of α_2 is a contact form and vice versa. This definition was introduced in Refs. 1,3 and is the analogous to that of *contact-symplectic pairs* and *symplectic pairs* (see Refs. 2,7,10).

In this paper we introduce the notion of *contact pair structure*, that is a contact pair on a manifold M endowed with a tensor field ϕ, of type $(1,1)$, such that $\phi^2 = -Id + \alpha_1 \otimes Z_1 + \alpha_2 \otimes Z_2$ and $\phi(Z_1) = \phi(Z_2) = 0$, where Z_1 and Z_2 are the Reeb vector fields of the pair.

We define *compatible* and *associated* metrics and we prove several properties of these metrics. For example, we show that the orbits of the locally free $\mathbb{R}^2$-action generated by the two commuting *Reeb vector fields* are totally geodesic with respect to these metrics. Another important feature is that, given a metric which is associated with respect to a contact pair structure $(\alpha_1, \alpha_2, \phi)$, if we assume that the endomorphism ϕ preserves the tangent spaces of the leaves of the two characteristic foliations, then the contact forms induced on the leaves are *contact metric structures*.

In what follows we denote by $\Gamma(B)$ the space of sections of a vector bundle B. For a given foliation $\mathcal{F}$ on a manifold M, we denote by $T\mathcal{F}$ the subbundle of TM whose fibers are given by the distribution tangent to the leaves. All the differential objects considered are supposed smooth.

2. Preliminaries on contact pairs

We gather in this section the notions concerning contact pairs that will be needed in the sequel. We refer the reader to Refs. 2,3,6,7 for further informations and several examples of such structures.

Definition 2.1. A pair (α_1, α_2) of 1-forms on a manifold is said to be a contact pair of type (h, k) if:

i) $\alpha_1 \wedge (d\alpha_1)^h \wedge \alpha_2 \wedge (d\alpha_2)^k$ is a volume form,
ii) $(d\alpha_1)^{h+1} = 0$,
iii) $(d\alpha_2)^{k+1} = 0$.

Since the form α_1 (resp. α_2) has constant class $2h+1$ (resp. $2k+1$), the distribution $\mathrm{Ker}\,\alpha_1 \cap \mathrm{Ker}\,d\alpha_1$ (resp. $\mathrm{Ker}\,\alpha_2 \cap \mathrm{Ker}\,d\alpha_2$) is completely integrable and then it determines the so-called characteristic foliation $\mathcal{F}_1$ (resp. $\mathcal{F}_2$) whose leaves are endowed with a contact form induced by α_2 (resp. α_1).

To a contact pair (α_1, α_2) of type (h, k) are associated two commuting vector fields Z_1 and Z_2, called *Reeb vector fields* of the pair, which are uniquely determined by the following equations:

$$\alpha_1(Z_1) = \alpha_2(Z_2) = 1, \quad \alpha_1(Z_2) = \alpha_2(Z_1) = 0,$$
$$i_{Z_1} d\alpha_1 = i_{Z_1} d\alpha_2 = i_{Z_2} d\alpha_1 = i_{Z_2} d\alpha_2 = 0,$$

where i_X is the contraction with the vector field X. In particular, since the Reeb vector fields commute, they determine a locally free $\mathbb{R}^2$-action, called the *Reeb action*.

The kernel distribution of $d\alpha_1$ (resp. $d\alpha_2$) is also integrable and then it defines a foliation whose leaves inherit a contact pair of type $(0, k)$ (resp. $(h, 0)$).

A contact pair of type (h, k) has a local model (see Refs. 2,3), which means that for every point of the manifold there exists a coordinate chart on which the pair can be written as follows:

$$\alpha_1 = dx_{2h+1} + \sum_{i=1}^{h} x_{2i-1} dx_{2i}, \quad \alpha_2 = dy_{2k+1} + \sum_{i=1}^{k} y_{2i-1} dy_{2i},$$

where $(x_1, \cdots, x_{2h+1}, y_1, \cdots, y_{2k+1})$ are the standard coordinates on $\mathbb{R}^{2h+2k+2}$.

The tangent bundle of a manifold M endowed with a contact pair (α_1, α_2) can be split in different ways. For $i = 1, 2$, let $T\mathcal{F}_i$ be the subbundle determined by the characteristic foliation of α_i, $T\mathcal{G}_i$ the subbundle of TM whose fibers are given by $\operatorname{Ker} d\alpha_i \cap \operatorname{Ker} \alpha_1 \cap \operatorname{Ker} \alpha_2$ and $\mathbb{R}Z_1, \mathbb{R}Z_2$ the line bundles determined by the Reeb vector fields. Then we have the following splittings:

$$TM = T\mathcal{F}_1 \oplus T\mathcal{F}_2 = T\mathcal{G}_1 \oplus T\mathcal{G}_2 \oplus \mathbb{R}Z_1 \oplus \mathbb{R}Z_2.$$

Moreover we have $T\mathcal{F}_1 = T\mathcal{G}_1 \oplus \mathbb{R}Z_2$ and $T\mathcal{F}_2 = T\mathcal{G}_2 \oplus \mathbb{R}Z_1$.

3. Contact pair structures and almost contact structures

In this section we firstly recall the basic definitions of almost contact structures and their associated metrics. Next we introduce a new structure, namely the *contact pair structure*. More details on almost contact structures can be found in Ref. 8.

3.1. *Almost contact structures*

An almost contact structure on a manifold M is a triple (α, Z, ϕ) of a one-form α, a vector field Z and a field of endomorphisms ϕ of the tangent bundle of M, such that $\phi^2 = -Id + \alpha \otimes Z$, $\phi(Z) = 0$ and $\alpha(Z) = 1$. In particular, it follows that $\alpha \circ \phi = 0$ and that the rank of ϕ is $\dim M - 1$.

In the study of almost contact structures, there are two types of metrics which are particularly interesting and we recall their definition:

Definition 3.1. For a given almost contact structure (α, Z, ϕ) on a manifold M, a Riemannian metric g is called:

i) compatible if $g(\phi X, \phi Y) = g(X, Y) - \alpha(X)\alpha(Y)$ for every $X, Y \in \Gamma(TM)$;
ii) associated if $g(X, \phi Y) = d\alpha(X, Y)$ and $g(X, Z) = \alpha(X)$ for every $X, Y \in \Gamma(TM)$.

In particular if g is an associated metric with respect to (α, Z, ϕ), then α must be a contact form. For a given almost contact structure there always exists a compatible metric and for a given contact form there always exists an associated metric.

Definition 3.2. A contact metric structure (α, Z, ϕ, g) is an almost contact structure (α, Z, ϕ), where α is a contact form and Z its Reeb vector field, together with an associated metric g.

3.2. Contact pair structures

Now we want to generalize the notion of contact metric structure to the contact pairs. To do that, we begin with the following definition:

Definition 3.3. A *contact pair structure* on a manifold M is a triple $(\alpha_1, \alpha_2, \phi)$, were (α_1, α_2) is a contact pair and ϕ a tensor field of type $(1, 1)$ such that:

$$\phi^2 = -Id + \alpha_1 \otimes Z_1 + \alpha_2 \otimes Z_2, \quad (1)$$

$$\phi(Z_1) = \phi(Z_2) = 0. \quad (2)$$

where Z_1 and Z_2 are the Reeb vector fields of (α_1, α_2).

As in the case of an almost contact structure, it is easy to check that $\alpha_i \circ \phi = 0$, $i = 1, 2$ and the rank of ϕ is equal to $\dim M - 2$.

Remark 3.1. Actually, the condition $\phi(Z) = 0$ in the definition of almost contact structure is not needed, since it follows from $\phi^2 = -I + \alpha \otimes Z$ (see Ref. 8, Theorem 4.1 p. 33). In the case of a contact pair, it is possible to construct an endomorphism satisfying (1) but not (2). This can be done, for example, by taking ϕ to be an almost complex structure on $T\mathcal{G}_1 \oplus T\mathcal{G}_2$ and extending it by setting $\phi(Z_1) = -\phi(Z_2) = Z_1 + Z_2$.

On a manifold M endowed with a contact pair, there always exists an endomorphism ϕ verifying (1) and (2), since on the subbundle $T\mathcal{G}_1 \oplus T\mathcal{G}_2$ of TM, the 2-form $d\alpha_1 + d\alpha_2$ is symplectic. Then one can choose an almost complex structure on $T\mathcal{G}_1 \oplus T\mathcal{G}_2$ and extend it to be zero on the Reeb vector fields, to produce an endomorphism ϕ of TM with the desired properties.

The following definition is justified by the fact that we are interested on the contact structures induced on the leaves of the characteristic foliations of a contact pair.

Definition 3.4. The endomorphism ϕ is said to be *decomposable* if $\phi(T\mathcal{F}_i) \subset T\mathcal{F}_i$, for $i = 1, 2$.

The condition for ϕ to be decomposable is equivalent to $\phi(T\mathcal{G}_i) = T\mathcal{G}_i$, because ϕ vanishes on Z_1, Z_2 and has rank $2k$ (resp. $2h$) on $T\mathcal{F}_1$ (resp. $T\mathcal{F}_2$).

By Definition 2.1, α_1 induces a contact form on the leaves of the characteristic foliation of α_2 and vice versa. Then it is easy to prove the following:

Proposition 3.1. *If ϕ is decomposable, then (α_1, Z_1, ϕ) (resp. (α_2, Z_2, ϕ)) induces an almost contact structure on the leaves of $\mathcal{F}_2$ (resp. $\mathcal{F}_1$).*

Here is a simple example of contact pair on a manifold which is just a product of two contact manifolds and ϕ is not decomposable:

Example 3.1. Let us consider standard coordinates $(x_1, y_1, x_2, y_2, z_1, z_2)$ on $\mathbb{R}^6$ and the contact pair (α_1, α_2) given by:

$$\alpha_1 = dz_1 - x_1 dy_1 \quad \alpha_2 = dz_2 - x_2 dy_2,$$

with Reeb vector fields $Z_1 = \frac{\partial}{\partial z_1}$, $Z_2 = \frac{\partial}{\partial z_2}$. Let us define ϕ as follows:

$$\phi\left(a_1 \frac{\partial}{\partial x_1} + b_1 \frac{\partial}{\partial y_1} + a_2 \frac{\partial}{\partial x_2} + b_2 \frac{\partial}{\partial y_2} + c_1 \frac{\partial}{\partial z_1} + c_2 \frac{\partial}{\partial z_2}\right)$$
$$= \left(-a_2 \frac{\partial}{\partial x_1} - b_2 \frac{\partial}{\partial y_1} + a_1 \frac{\partial}{\partial x_2} + b_1 \frac{\partial}{\partial y_2} - x_1 b_2 \frac{\partial}{\partial z_1} + x_2 b_1 \frac{\partial}{\partial z_2}\right),$$

where $a_1, a_2, b_1, b_2, c_1, c_2$ are smooth functions on $\mathbb{R}^6$.

Then $(\alpha_1, \alpha_2, \phi)$ is a contact pair structure and ϕ is not decomposable. This because, for example, $\frac{\partial}{\partial x_1} \in \text{Ker}(\alpha_2 \wedge d\alpha_2)$, $\frac{\partial}{\partial x_2} \in \text{Ker}(\alpha_1 \wedge d\alpha_1)$ but $\phi(\frac{\partial}{\partial x_1}) = \frac{\partial}{\partial x_2}$.

Remark 3.2. A more general structure, similar to the almost contact structures, is obtained by considering a 5-tuple $(\alpha_1, \alpha_2, Z_1, Z_2, \phi)$, where $\phi^2 = -Id + \alpha_1 \otimes Z_1 + \alpha_2 \otimes Z_2$, $\phi(Z_1) = \phi(Z_2) = 0$ and the α_i's are just non-vanishing 1-forms verifying $\alpha_i(Z_j) = \delta_{ij}$.

4. Compatible and associated metrics

For a given contact pair structure on a manifold M, it is natural to consider the following metrics:

Definition 4.1. Let $(\alpha_1, \alpha_2, \phi)$ be a contact pair structure on a manifold M, with Reeb vector fields Z_1 and Z_2. A Riemannian metric g on M is called:

i) *compatible* if $g(\phi X, \phi Y) = g(X, Y) - \alpha_1(X)\alpha_1(Y) - \alpha_2(X)\alpha_2(Y)$ for all $X, Y \in \Gamma(TM)$,

ii) associated if $g(X, \phi Y) = (d\alpha_1 + d\alpha_2)(X, Y)$ and $g(X, Z_i) = \alpha_i(X)$, for $i = 1, 2$ and for all $X, Y \in \Gamma(TM)$.

It is clear that an associated metric is also compatible, but the converse is not true. In the Example 3.1, the metric $g = \sum_{i=1}^{2}(dx_i^2 + dy_i^2 + \alpha_i^2)$ is compatible but not associated.

Definition 4.2. A *metric contact pair* (MCP) on a manifold M is a 4-tuple $(\alpha_1, \alpha_2, \phi, g)$ where $(\alpha_1, \alpha_2, \phi)$ is a contact pair structure and g an associated metric with respect to it.

Like for compatible metrics on almost contact manifolds, we have:

Proposition 4.1. *For every contact pair structure on a manifold M there exists a compatible metric.*

Proof. Let $(\alpha_1, \alpha_2, \phi)$ be a contact pair structure on M. Pick any Riemannian metric h on M and define k as

$$k(X, Y) = h(\phi^2 X, \phi^2 Y) + \alpha_1(X)\alpha_1(Y) + \alpha_2(X)\alpha_2(Y).$$

Now put

$$g(X, Y) = \frac{1}{2}(k(X, Y) + k(\phi X, \phi Y) + \alpha_1(X)\alpha_1(Y) + \alpha_2(X)\alpha_2(Y)).$$

It is straightforward to show that g is a compatible metric with respect to $(\alpha_1, \alpha_2, \phi)$. □

Compatible and associated metrics have several interesting properties given by the following

Theorem 4.1. *Let M be a manifold endowed with a contact pair structure $(\alpha_1, \alpha_2, \phi)$, with Reeb vector fields Z_1, Z_2. Let g be a compatible metric with respect to it, with Levi-Civita connection ∇. Then we have:*

(a) $g(Z_i, X) = \alpha_i(X)$, *for $i = 1, 2$ and for every $X \in \Gamma(TM)$;*
(b) $g(Z_i, Z_j) = \delta_i^j$, $i, j = 1, 2$;
(c) $\nabla_{Z_i} Z_j = 0$, $i, j = 1, 2$ *(in particular the integral curves of the Reeb vector fields are geodesics);*
(d) *the Reeb action is totally geodesic (i.e the orbits are totally geodesic 2-dimensional submanifolds).*

Moreover, if g is an associated metric and L_{Z_i} is the Lie derivative along Z_i, then $L_{Z_i}\phi = 0$ if and only if Z_i is a Killing vector field.

Proof. The first two properties are easy consequences of the definitions of compatible metric and of the Reeb vector fields of a contact pair. For the third property, let us remember that α_i is invariant by Z_j, $i,j = 1,2$, and then:

$$\begin{aligned} 0 &= (L_{Z_j}\alpha_i)(X) = Z_j(\alpha_i(X)) - \alpha_i(L_{Z_j}X) \\ &= Z_j(g(X,Z_i)) - g(Z_i, \nabla_{Z_j}X - \nabla_X Z_j) \\ &= g(\nabla_{Z_j}X, Z_i) + g(X, \nabla_{Z_j}Z_i) - g(Z_i, \nabla_{Z_j}X - \nabla_X Z_j) \\ &= g(X, \nabla_{Z_j}Z_i) + g(Z_i, \nabla_X Z_j) \,. \end{aligned}$$

Summing up with $(L_{Z_i}\alpha_j)(X)$ and recalling that $[Z_i, Z_j] = 0$, we get $g(X, \nabla_{Z_j}Z_i) = 0$ for all X.

To prove the fourth property, let us consider the second fundamental form B of an orbit $\tilde M$ of the Reeb action. Since the Reeb vector fields are tangent to the orbits of the Reeb action, we can choose $\{Z_1, Z_2\}$ (restricted to $\tilde M$) as a basis of the tangent space at every point of $\tilde M$. For a vector field X on M along $\tilde M$, let us denote by X^T its tangential component. Then, by the third property, we have:

$$B(Z_i, Z_j) = \nabla_{Z_i}Z_j - (\nabla_{Z_i}Z_j)^T = 0\,.$$

Finally, if g is associated, we want to prove that $[L_{Z_i}g](X,Y) = 0$ for all $X, Y \in \Gamma(TM)$ if and only if $(L_{Z_i}\phi)(Y) = 0$ for every $Y \in \Gamma(TM)$. This is clearly true for $Y = Z_1$ or Z_2, since in this case $[L_{Z_i}\phi](Z_j)$ vanishes identically and $L_{Z_i}\alpha_j = 0$ applied to a vector field X implies $[L_{Z_i}g](X,Z_j) = 0$. It remains to prove the assertion for Y in $\operatorname{Ker}\alpha_1 \cap \operatorname{Ker}\alpha_2$. In this case, for all vector fields X, Y, we have:

$$\begin{aligned} 0 &= L_{Z_i}(d\alpha_1 + d\alpha_2)(X,Y) \\ &= Z_i g(X, \phi Y) - (d\alpha_1 + d\alpha_2)(L_{Z_i}X, Y) - (d\alpha_1 + d\alpha_2)(X, L_{Z_i}Y) \\ &= [L_{Z_i}g](X, \phi Y) + g(X, [L_{Z_i}\phi](Y)) \end{aligned}$$

and this completes the proof since ϕ restricted to $\operatorname{Ker}\alpha_1 \cap \operatorname{Ker}\alpha_2$ is an isomorphism. □

Remark 4.1. Codimension two geodesible foliations of closed 4-manifolds have been studied by Cairns and Ghys in Ref. 9. In particular, they have proven that, in this situation, there exists a metric on the manifold for which the leaves of the foliation are minimal and have same constant curvature $K = 0, 1, -1$. Our case belongs to what they have called *parabolic case*, that is $K = 0$.

As for the metric contact structures (see Ref. 8 for example), by polarization one can show the existence of associated metrics, and in fact we have:

Proposition 4.2. *For every contact pair (α_1, α_2) on a manifold M, there exists an endomorphism ϕ of TM and a metric g such that $(\alpha_1, \alpha_2, \phi, g)$ is a metric contact pair. Moreover ϕ can be chosen to be decomposable.*

Proof. Take any Riemannian metric k. Since $d\alpha_1 + d\alpha_2$ is symplectic on the subbundle $T\mathcal{G}_1 \oplus T\mathcal{G}_2$, then it can be polarized by using k to obtain a metric $\tilde{g}$ and an almost complex structure $\tilde{\phi}$ compatible with $\tilde{g}$. Extending $\tilde{\phi}$ to be zero on the Reeb vector fields, we obtain ϕ. Defining g to be equal to $\tilde{g}$ on $T\mathcal{G}_1 \oplus T\mathcal{G}_2$ and putting $g(X, Z_i) = \alpha_i(X)$, gives the desired metric.

To obtain a decomposable ϕ, polarize $d\alpha_i$ on $T\mathcal{G}_i$ to obtain two metrics $\tilde{g}_i$ and two endomorphisms $\tilde{\phi}_i$ on $T\mathcal{G}_i$ ($i = 1, 2$) and then take the direct sums. Finally, extend them as before to obtain the desired tensors. □

We end this section with two results concerning the structures induced on the leaves of the characteristic foliations:

Theorem 4.2. *Let M be a manifold endowed with a MCP $(\alpha_1, \alpha_2, \phi, g)$ and suppose that ϕ is decomposable. Then (α_i, ϕ, g) induces a contact metric structure on the leaves of the characteristic foliation of α_j for $i \neq j$, $i, j = 1, 2$.*

Proof. We will prove the assertion only for (α_1, ϕ, g), the other case being completely similar. If F is a leaf of the foliation $\mathcal{F}_2$, then (α_1, ϕ) induces an almost contact structure on it and α_1 is a contact form. Since g is an associated metric with respect to $(\alpha_1, \alpha_2, \phi)$, by Definition 4.1, when restricted to vectors X, Y which are tangent to F we have:

$$g(X, \phi Y) = d\alpha_1(X, Y) \ , \ g(X, Z_1) = \alpha_1(X),$$

showing that its restriction to F is an associated metric with respect to the contact form induced by α_1. □

In a similar way, recalling that the characteristic foliation of $d\alpha_i$ for $i = 1, 2$ is given by $\operatorname{Ker} d\alpha_i$, we can prove:

Theorem 4.3. *Let M be a manifold endowed with a MCP $(\alpha_1, \alpha_2, \phi, g)$ and suppose that ϕ is decomposable. Then $(\alpha_1, \alpha_2, \phi, g)$ induces a metric contact pairs of type $(h, 0)$ on the leaves of the characteristic foliation of $d\alpha_2$ and one of type $(0, k)$ on the leaves of the characteristic foliation of $d\alpha_1$.*

4.1. Orthogonal foliations

The following theorem explains the link between decomposability of ϕ and orthogonality of the characteristic foliations when the metric is an associated one:

Theorem 4.4. *For a MCP $(\alpha_1, \alpha_2, \phi, g)$, the tensor ϕ is decomposable if and only if the foliations $\mathcal{F}_1, \mathcal{F}_2$ are orthogonal.*

Proof. Suppose that $\phi(T\mathcal{F}_i) \subset T\mathcal{F}_i$ for $i = 1, 2$ and let $X \in \Gamma(T\mathcal{F}_1), Y \in \Gamma(T\mathcal{F}_2)$. Because g is associated, by the choice of X, Y, we have:

$$g(X, \phi Y) = d\alpha_1(X, Y) + d\alpha_2(X, Y) = 0 \ , \ g(X, Z_1) = \alpha_1(X) = 0,$$

which proves that X is orthogonal to $T\mathcal{F}_2$ and then the two foliations are orthogonal.

Conversely, suppose that the two foliations are orthogonal. Then, for $X \in \Gamma(T\mathcal{F}_1)$ and every $Y \in \Gamma(T\mathcal{F}_2)$, we have

$$g(Y, \phi X) = (d\alpha_1 + d\alpha_2)(Y, X) = 0$$

which implies that $\phi X \in \Gamma(T\mathcal{F}_2^\perp) = \Gamma(T\mathcal{F}_1)$, that is ϕ is decomposable. $\square$

In the Example 3.1 the foliations are orthogonal with respect to the metric $g = \sum_{i=1}^{2}(dx_i^2 + dy_i^2 + \alpha_i^2)$ which is compatible but not associated because ϕ is not decomposable.

Here is an example of MCP with decomposable ϕ on a nilpotent Lie group and its corresponding nilmanifolds:

Example 4.1. Consider the simply connected nilpotent Lie group G with structure equations:

$$d\omega_1 = d\omega_6 = 0 \ , \ d\omega_2 = \omega_5 \wedge \omega_6,$$
$$d\omega_3 = \omega_1 \wedge \omega_4 \ , \ d\omega_4 = \omega_1 \wedge \omega_5 \ , \ d\omega_5 = \omega_1 \wedge \omega_6,$$

where the ω_i's form a basis for the cotangent space of G at the identity.

The pair (ω_2, ω_3) is a contact pair of type $(1, 1)$ with Reeb vector fields (X_2, X_3), the X_i's being dual to the ω_i's. Now define ϕ to be zero on the Reeb vector fields and

$$\phi(X_5) = -X_6 \ , \ \phi(X_6) = X_5 \ , \ \phi(X_1) = -X_4 \ , \ \phi(X_4) = X_1.$$

Then ϕ is easy verified to be decomposable and the metric $g = \sum_{i=1}^{6} \omega_i^2$ is an associated metric with respect to $(\omega_2, \omega_3, \phi)$.

Since the structure constants of the group are rational, there exist lattices Γ such that G/Γ is compact. Since the MCP on G is left invariant, it descends to all quotients G/Γ and we obtain nilmanifolds carrying the same type of structure.

Final comments

Further metric properties of the contact pairs structures are studied in Ref. 5, where we prove, for example, that the characteristic foliations are minimal with respect to an associated metric.

In Ref. 4 we study the analog for a contact pair structure of the notion of normality for almost contact structures. We give there several constructions involving Boothby-Wang fibrations and flat bundles, which can be used to produce more examples of metric contact pairs.

Acknowledgements

The first author was supported by the MIUR Project: *Riemannian metrics and differentiable manifolds*–P.R.I.N. 2005.

References

1. G. Bande, *Formes de contact généralisé, couples de contact et couples contacto-symplectiques*, Thèse de Doctorat, Université de Haute Alsace, Mulhouse 2000.
2. G. Bande, *Trans. Amer. Math. Soc.* **355** (2003), 1699–1711.
3. G. Bande and A. Hadjar, *Tohoku Math. J.* **57** (2005), 247–260.
4. G. Bande and A. Hadjar, arXiv:math.DG/0805.0193
5. G. Bande and A. Hadjar, On the Riemannian geometry of contact pairs, in progress.
6. G. Bande, P. Ghiggini and D. Kotschick, *Int. Math. Res. Not.* **68** (2004), 3673–3688.
7. G. Bande and D. Kotschick, *Trans. Amer. Math. Soc.* **358** (2006), 1643–1655.
8. D. E. Blair, *Riemannian geometry of contact and symplectic manifolds*, Progress in Mathematics, vol. 203, Birkhäuser, 2002.
9. G. Cairns and É. Ghys, *J. Differential Geom.* **23** (1986), 241–254
10. D. Kotschick and S. Morita, *Topology* **44** (2005), no. 1, 131–149.

PARAQUATERNIONIC MANIFOLDS AND MIXED 3-STRUCTURES

S. Ianuş

Department of Mathematics, University of Bucharest
C.P. 10-119, Post. Of. 10, Bucharest 72200, Romania
E-mail: ianus@gta.math.unibuc.ro

G. E. Vîlcu

Department of Mathematics and Computer Science
Petroleum-Gas University of Ploieşti
Bulevardul Bucureşti, Nr. 39, Ploieşti, Romania
E-mail: gvilcu@mail.upg-ploiesti.ro

In this paper we review basic properties of manifolds endowed with paraquaternionic structures and mixed 3-structures. Also, we investigate the existence of mixed 3-structures on normal semi-invariant submanifolds of paraquaternionic Kähler manifolds.

Keywords: Paraquaternionic Kähler manifold, mixed 3-structure, Einstein space

1. Introduction

The paraquaternionic structures, firstly named quaternionic structures of second kind, have been introduced by P. Libermann (Ref. 1) in 1952. The theory of paraquaternionic manifolds parallels the theory of quaternionic manifolds, but uses the algebra of paraquaternionic numbers, in which two generators have square 1 and one generator has square -1. Accordingly, such manifolds are equipped with a subbundle of rank 3 in the bundle of the endomorphisms, locally spanned by two almost product structures and one almost complex structure. From the metric point of view, the almost paraquaternionic Hermitian manifolds have neutral signature.

The counterpart in odd dimension of paraquaternionic geometry was introduced by the present authors and R. Mazzocco (Ref. 2) in 2006. It is called mixed 3-structure, which appears in a natural way on lightlike hypersurfaces in paraquaternionic manifolds. We review properties of manifolds endowed with mixed 3-structures and give some examples. Particularly, we obtain some conditions for the existence of a metric mixed 3-structure on a normal semi-invariant submanifold of a paraquaternionic Kähler manifold.

2. Paraquaternionic structures on manifolds

Following Refs. 3–6 we first recall some basic facts concerning paraquaternionic Kähler manifolds.

An almost product structure on a smooth manifold M is a tensor field P of type (1,1) on M, $P \neq \pm Id$, such that:

$$P^2 = Id. \qquad (1)$$

An almost complex structure on a smooth manifold M is a tensor field J of type (1,1) on M such that:

$$J^2 = -Id. \qquad (2)$$

An almost para-hypercomplex structure on a smooth manifold M is a triple $H = (J_\alpha)_{\alpha=\overline{1,3}}$, where J_1 is an almost complex structure on M and J_2, J_3 are almost product structures on M, satisfying:

$$J_2 J_1 = -J_1 J_2 = J_3. \qquad (3)$$

In this case (M, H) is said to be an almost para-hypercomplex manifold.

A semi-Riemannian metric g on (M, H) is said to be para-hyperhermitian if it satisfies:

$$g(J_\alpha X, J_\alpha Y) = \epsilon_\alpha g(X, Y), \qquad (4)$$

for all vector fields X,Y on M and $\alpha \in \{1, 2, 3\}$, where $\epsilon_1 = 1, \epsilon_2 = \epsilon_3 = -1$. Moreover, (M, g, H) is said to be an almost para-hyperhermitian manifold.

An almost para-hypercomplex manifold (M, H) is said to be a para-hypercomplex manifold if each J_α, $\alpha = 1, 2, 3$, is integrable, that is, if the corresponding Nijenhuis tensors:

$$N_\alpha(X,Y) = [J_\alpha X, J_\alpha Y] - J_\alpha[X, J_\alpha Y] - J_\alpha[J_\alpha X, Y] - \epsilon_\alpha[X, Y] \qquad (5)$$

$\alpha = 1, 2, 3$, vanish for all vector fields X,Y on M. In this case H is said to be a para-hypercomplex structure on M.

An almost hermitian paraquaternionic manifold is a triple (M, σ, g), where M is a smooth manifold, σ is a rank 3-subbundle of $End(TM)$ which is locally spanned by an almost para-hypercomplex structure $H = (J_\alpha)_{\alpha=\overline{1,3}}$ and g is a para-hyperhermitian metric with respect to H. It is easy to see that any almost hermitian paraquaternionic manifold is of dimension $4n$ and g is necessarily of neutral signature $(2n, 2n)$.

If the bundle σ is parallel with respect to the Levi-Civita connection ∇ of g, then (M, σ, g) is said to be a paraquaternionic Kähler manifold. Equivalently, locally defined 1-forms $\omega_1, \omega_2, \omega_3$ exist on M such that we have:

$$\nabla_X J_\alpha = \epsilon_\alpha \{\omega_{\alpha+1}(X) J_{\alpha+2} - \omega_{\alpha+2}(X) J_{\alpha+1}\} \qquad (6)$$

for any vector field X on M, where the indices are taken from $\{1, 2, 3\}$ modulo 3. We can remark that an important example in this theory is the paraquaternionic projective space as described by Blazić (see Ref. 3). Moreover, if $\omega_1 = \omega_2 = \omega_3 = 0$, then (M, σ, g) is said to be a locally para-hyper-Kähler manifold.

Definition 2.1. (Ref. 4) Let (M, σ, g) be a paraquaternionic Kähler manifold.

(a) A subspace π of $T_p M$, $p \in M$ is called non-degenerate if the restriction of g to π is no degenerate. In particular a 2-plane π in $T_p M$ is non-degenerate if and only if it has a basis $\{X, Y\}$ satisfying:

$$\Delta(\pi) = g(X, X) g(Y, Y) - g(X, Y)^2 \neq 0.$$

(b) If $X \in T_p M$, $p \in M$, the 4-plane $PQ(X)$ spanned by $\{X, J_1 X, J_2 X, J_3 X\}$ is called a paraquaternionic 4-plane.

(c) A 2-plane in $T_p M, p \in M$, spanned by $\{X, Y\}$ is called half-paraquaternionic if $PQ(X) = PQ(Y)$.

Note that if X is a non-lightlike vector of $T_p M$, then $PQ(X)$ is a non-degenerate subspace.

Definition 2.2. (Ref. 4) Let (M, σ, g) be a paraquaternionic Kähler manifold and let π be a non-degenerate 2-plane in $T_p M, p \in M$, spanned by $\{X, Y\}$. The sectional curvature $K(\pi)$ is defined by:

$$K(\pi) = \frac{g(R(X, Y) Y, X)}{\Delta(\pi)},$$

where the Riemannian curvature tensor R is taken with the sign convention $R(X, Y) = [\nabla_X, \nabla_Y] - \nabla_{[X, Y]}$, for all vector fields X, Y on M.

In particular, the sectional curvature for a half-paraquaternionic plane is called paraquaternionic sectional curvature. A paraquaternionic Kähler manifold of constant paraquaternionic sectional curvature is called a paraquaternionic space form.

We have the following fundamental properties concerning paraquaternionic Kähler manifolds (see Refs. 3–6):

Theorem 2.1. *Any paraquaternionic Kähler manifold (M, σ, g) is an Einstein manifold, provided that $dim M > 4$.*

Theorem 2.2. *A paraquaternionic Kähler manifold (M, σ, g) with $dim M > 4$ and zero scalar curvature is a locally para-hyper-Kähler manifold.*

Theorem 2.3. *The curvature tensor R of a paraquaternionic space form (M, σ, g) is given by:*

$$R(X,Y)Z = \frac{c}{4}\{g(Z,Y)X - g(X,Z)Y + \sum_{\alpha=1}^{3} \epsilon_\alpha[g(Z, J_\alpha Y)J_\alpha X - $$
$$-g(Z, J_\alpha X)J_\alpha Y + 2g(X, J_\alpha Y)J_\alpha Z]\} \quad (7)$$

for all vector fields X, Y, Z on M and any local basis $\{J_1, J_2, J_3\}$ of σ.

3. Manifolds endowed with mixed 3-structures

The counterpart in odd dimension of a paraquaternionic structure, called a mixed 3-structure, was introduced in Ref. 2.

Let M be a differentiable manifold equipped with a triple (ϕ, ξ, η), where ϕ is a a field of endomorphisms of the tangent spaces, ξ is a vector field and η is a 1-form on M such that:

$$\phi^2 = -\epsilon Id + \eta \otimes \xi, \quad \eta(\xi) = \epsilon. \quad (8)$$

If $\epsilon = 1$ then (ϕ, ξ, η) is said to be an almost contact structure on M (see Ref. 7), and if $\epsilon = -1$ then (ϕ, ξ, η) is said to be an almost paracontact structure on M (see Ref. 8).

Definition 3.1. (Ref. 2) Let M be a differentiable manifold which admits an almost contact structure (ϕ_1, ξ_1, η_1) and two almost paracontact structures (ϕ_2, ξ_2, η_2) and (ϕ_3, ξ_3, η_3), satisfying the following conditions:

$$\eta_\alpha(\xi_\beta) = 0, \forall \alpha \neq \beta, \quad (9)$$

$$\phi_\alpha(\xi_\beta) = -\phi_\beta(\xi_\alpha) = \epsilon_\gamma \xi_\gamma, \quad (10)$$

$$\eta_\alpha \circ \phi_\beta = -\eta_\beta \circ \phi_\alpha = \epsilon_\gamma \eta_\gamma, \quad (11)$$

$$\phi_\alpha \phi_\beta - \eta_\beta \otimes \xi_\alpha = -\phi_\beta \phi_\alpha + \eta_\alpha \otimes \xi_\beta = \epsilon_\gamma \phi_\gamma, \quad (12)$$

where (α, β, γ) is an even permutation of (1,2,3) and $\epsilon_1 = 1, \epsilon_2 = \epsilon_3 = -1$.

Then the manifold M is said to have a mixed 3-structure $(\phi_\alpha, \xi_\alpha, \eta_\alpha)_{\alpha=\overline{1,3}}$.

Definition 3.2. (Ref. 9) If a manifold M with a mixed 3-structure $(\phi_\alpha, \xi_\alpha, \eta_\alpha)_{\alpha=\overline{1,3}}$ admits a semi-Riemannian metric g such that:

$$g(\phi_\alpha X, \phi_\alpha Y) = \epsilon_\alpha g(X,Y) - \eta_\alpha(X)\eta_\alpha(Y), \quad g(X, \xi_\alpha) = \eta_\alpha(X) \qquad (13)$$

for all vector fields X, Y on M and $\alpha = 1, 2, 3$, then we say that M has a metric mixed 3-structure and g is called a compatible metric. Moreover, if $(\phi_1, \xi_1, \eta_1, g)$ is a Sasakian structure, i.e. (see Ref. 7):

$$(\nabla_X \phi_1)Y = g(X,Y)\xi_1 - \eta_1(Y)X \qquad (14)$$

and $(\phi_2, \xi_2, \eta_2, g)$, $(\phi_3, \xi_3, \eta_3, g)$ are LP-Sasakian structures, i.e. (see Ref. 8):

$$(\nabla_X \phi_2)Y = g(\phi_2 X, \phi_2 Y)\xi_2 + \eta_2(Y)\phi_2^2 X, \qquad (15)$$

$$(\nabla_X \phi_3)Y = g(\phi_3 X, \phi_3 Y)\xi_3 + \eta_3(Y)\phi_3^2 X, \qquad (16)$$

then $((\phi_\alpha, \xi_\alpha, \eta_\alpha)_{\alpha=\overline{1,3}}, g)$ is said to be a mixed Sasakian 3-structure on M.

It is easy to see that any manifold M with a mixed 3-structure admits a compatible semi-Riemannian metric g. Moreover, the signature of g is $(2n+1, 2n+2)$ and the dimension of the manifold M is $4n+3$.

If M^{4n+3} is a manifold endowed with a mixed 3-Sasakian structure $((\phi_\alpha, \xi_\alpha, \eta_\alpha)_{\alpha=\overline{1,3}}, g)$, then we can define a para-hyper-Kähler structure $\{J_\alpha\}_{\alpha=\overline{1,3}}$ on the cone $(C(M), \overline{g}) = (M \times \mathbb{R}_+, dr^2 + r^2 g)$, by:

$$\begin{cases} J_\alpha X = \phi_\alpha X - \eta_\alpha(X)\Phi \\ J_\alpha \Phi = \xi_\alpha \end{cases} \qquad (17)$$

for any vector field X on M and $\alpha = 1, 2, 3$, where $\Phi = r\partial_r$ is the Euler field on $C(M)$.

Conversely, if a cone $(C(M), \overline{g}) = (M \times \mathbb{R}_+, dr^2 + r^2 g)$ admits a para-hyper-Kähler structure $\{J_\alpha\}_{\alpha=\overline{1,3}}$, then we can identify M with $M \times \{1\}$ and we have a mixed 3-Sasakian structure $((\phi_\alpha, \xi_\alpha, \eta_\alpha)_{\alpha=\overline{1,3}}, g)$ on M given by:

$$\xi_\alpha = J_\alpha(\partial_r), \quad \phi_\alpha X = -\epsilon_\alpha \nabla_X \xi_\alpha, \quad \eta_\alpha(X) = g(\xi_\alpha, X), \qquad (18)$$

for any vector field X on M and $\alpha = 1, 2, 3$.

Finally, since a para-hyper-Kähler manifold is Ricci-flat, we obtain the main property of a manifold endowed with a mixed 3-Sasakian structure (see Ref. 9).

Theorem 3.1. *Any* $(4n+3)$-*dimensional manifold endowed with a mixed 3-Sasakian structure is an Einstein space with Einstein constant* $\lambda = 4n+2$.

We give now some examples of manifolds endowed with such kind of structures.

1. It is easy to see that if we define $(\phi_\alpha, \xi_\alpha, \eta_\alpha)_{\alpha=\overline{1,3}}$ in $\mathbb{R}^3$ by their matrices:

$$\phi_1 = \begin{pmatrix} 0 & 0 & 1 \\ 0 & 0 & 0 \\ -1 & 0 & 0 \end{pmatrix}, \phi_2 = \begin{pmatrix} 0 & 0 & 0 \\ 0 & 0 & 1 \\ 0 & 1 & 0 \end{pmatrix}, \phi_3 = \begin{pmatrix} 0 & -1 & 0 \\ -1 & 0 & 0 \\ 0 & 0 & 0 \end{pmatrix},$$

$$\xi_1 = \begin{pmatrix} 0 \\ 1 \\ 0 \end{pmatrix}, \xi_2 = \begin{pmatrix} 1 \\ 0 \\ 0 \end{pmatrix}, \xi_3 = \begin{pmatrix} 0 \\ 0 \\ 1 \end{pmatrix},$$

$$\eta_1 = \begin{pmatrix} 0 & 1 & 0 \end{pmatrix}, \eta_2 = \begin{pmatrix} -1 & 0 & 0 \end{pmatrix}, \eta_3 = \begin{pmatrix} 0 & 0 & -1 \end{pmatrix},$$

then $(\phi_\alpha, \xi_\alpha, \eta_\alpha)_{\alpha=\overline{1,3}}$ is a mixed 3-structure on $\mathbb{R}^3$.

We define now $(\phi'_\alpha, \xi'_\alpha, \eta'_\alpha)_{\alpha=\overline{1,3}}$ in $\mathbb{R}^{4n+3}$ by:

$$\phi'_\alpha = \begin{pmatrix} \phi_\alpha & 0 \\ 0 & J_\alpha \end{pmatrix}, \xi'_\alpha = \begin{pmatrix} \xi_\alpha \\ 0 \end{pmatrix}, \eta'_\alpha = \begin{pmatrix} \eta_\alpha & 0 \end{pmatrix},$$

for $\alpha = 1, 2, 3$, where J_1 is the almost complex structure on $\mathbb{R}^{4n}$ given by:

$$J_1((x_i)_{i=\overline{1,4n}}) = (-x_2, x_1, -x_4, x_3, ..., -x_{4n-2}, x_{4n-3}, -x_{4n}, x_{4n-1}), \quad (19)$$

and J_2, J_3 are almost product structures on $\mathbb{R}^{4n}$ defined by:

$$J_2((x_i)_{i=\overline{1,4n}}) = (-x_{4n-1}, x_{4n}, -x_{4n-3}, x_{4n-2}, ..., -x_3, x_4, -x_1, x_2), \quad (20)$$

$$J_3((x_i)_{i=\overline{1,4n}}) = (x_{4n}, x_{4n-1}, x_{4n-2}, x_{4n-3}, ..., x_4, x_3, x_2, x_1). \quad (21)$$

Since $J_2 J_1 = -J_1 J_2 = J_3$, it is easily checked that $(\phi'_\alpha, \xi'_\alpha, \eta'_\alpha)_{\alpha=\overline{1,3}}$ is a mixed 3-structure on $\mathbb{R}^{4n+3}$.

2. The unit sphere S^{4n+3}_{2n+1} is the canonical example of manifold with a mixed Sasakian 3-structure. This structure is obtained by taking S^{4n+3}_{2n+1} as hypersurface of $(\mathbb{R}^{4n+4}_{2n+2}, \overline{g})$. It is easy to see that on the tangent spaces $T_p S^{4n+3}_{2n+1}$, $p \in S^{4n+3}_{2n+1}$, the induced metric g is of signature $(2n+1, 2n+2)$.

If $(J_\alpha)_{\alpha=\overline{1,3}}$ is the canonical paraquaternionic structure on the $\mathbb{R}^{4n+4}_{2n+2}$ and N is the unit normal vector field to the sphere, we can define three vector fields on S^{4n+3}_{2n+1} by:

$$\xi_\alpha = -J_\alpha N, \ \alpha = 1, 2, 3.$$

If X is a tangent vector to the sphere then $J_\alpha X$ uniquely decomposes onto the part tangent to the sphere and the part parallel to N. Denote this decomposition by:

$$J_\alpha X = \phi_\alpha X + \eta_\alpha(X) N.$$

This defines the 1-forms η_α and the tensor fields ϕ_α on S^{4n+3}_{2n+1}, where $\alpha = 1, 2, 3$. Now we can easily see that that $(\phi_\alpha, \xi_\alpha, \eta_\alpha)_{\alpha=\overline{1,3}}$ is a mixed Sasakian 3-structure on S^{4n+3}_{2n+1}.

3. Since we can recognize the unit sphere S^{4n+3}_{2n+1} as the projective space $P^{4n+3}_{2n+1}(\mathbb{R})$, by identifying antipodal points, we have also that $P^{4n+3}_{2n+1}(\mathbb{R})$ admits a mixed Sasakian 3-structure.

4. Let $(\overline{M}, \overline{g})$ be a $(m+2)$-dimensional semi-Riemannian manifold with index $q \in \{1, 2, \ldots, m+1\}$ and let (M, g) be a hypersurface of $\overline{M}$, with $g = \overline{g}_{|M}$. We say that M is a lightlike hypersurface of $\overline{M}$ if g is of constant rank m (see Ref. 10). Unlike the classical theory of non-degenerate hypersurfaces, in case of lightlike hypersurfaces, the induced metric tensor field g is degenerate.

We consider the vector bundle $TM^\perp$ whose fibres are defined by:

$$T_p M^\perp = \{Y_p \in T_p\overline{M} | \overline{g}_p(X_p, Y_p) = 0, \forall X_p \in T_p M\}, \forall p \in M.$$

If $S(TM)$ is the complementary distribution of $TM^\perp$ in TM, which is called the screen distribution, then there exists a unique vector bundle $ltr(TM)$ of rank 1 over M so that for any non-zero section ξ of $TM^\perp$ on a coordinate neighborhood $U \subset M$, there exists a unique section N of $ltr(TM)$ on U satisfying:

$$\overline{g}(N, \xi) = 1, \ \overline{g}(N, N) = \overline{g}(W, W) = 0, \ \forall W \in \Gamma(S(TM)_{|U}).$$

In a lightlike hypersurface M of an almost hermitian paraquaternionic manifold $(\overline{M}, \sigma, \overline{g})$ such that ξ and N are globally defined on M, there is a mixed 3-structure (see Ref. 2).

4. Normal semi-invariant submanifolds and mixed 3-structures

The study of submanifolds of a paraquaternionic Kähler manifold is a very interesting subject and several types of such submanifolds we can find in the recent literature: Kähler and para-Kähler submanifolds (Refs. 11,12), normal semi-invariant submanifolds (Ref. 13–15), F-invariant submanifolds (Ref. 16), paraquaternionic CR-submanifolds (Ref. 17), lightlike submanifolds (Refs. 2,18). Next we investigate the existence of mixed 3-structures on a normal semi-invariant submanifold of a paraquaternionic Kähler manifold.

Let (M,g) be a non-degenerate submanifold of a paraquaternionic Kähler manifold $(\overline{M}, \sigma, \overline{g})$, with $g = \overline{g}_{|M}$. Then (M,g) is called a normal semi-invariant submanifold of $(\overline{M}, \sigma, \overline{g})$ if there exists a non-degenerate vector subbundle Q of the normal bundle $TM^\perp$ such that:

(i) $J_\alpha(Q_p) = Q_p, \forall p \in M, \ \forall \alpha \in \{1,2,3\}$;
(ii) $J_\alpha(Q_p^\perp) \subset T_pM, \forall p \in M, \ \forall \alpha \in \{1,2,3\}$, where $Q^\perp$ is the complementary orthogonal bundle to Q in $TM^\perp$ (see Ref. 13).

If (M,g) is a normal semi-invariant submanifold of a paraquaternionic Kähler manifold $(\overline{M}, \sigma, \overline{g})$, then we set $D_{\alpha p} = J_\alpha(Q_p^\perp)$. We consider $D_{1p} \oplus D_{2p} \oplus D_{3p} = D_p^\perp$ and 3s-dimensional distribution $D^\perp : p \mapsto D_p^\perp$ globally defined on M, where $s = \dim D_p^\perp$. We denote by D the complementary orthogonal distribution to $D^\perp$ in TM. We remark that D is called the paraquaternionic distribution because it is invariant with respect to the action of J_α, $\alpha \in \{1,2,3\}$ (see Ref. 13). Let (M,g) be a normal semi-invariant submanifold of a paraquaternionic Kähler manifold $(\overline{M}, \sigma, \overline{g})$.

We recall the following result concerning the integrability of the distribution $D^\perp$ (see Ref. 13).

Theorem 4.1. *Let (M,g) be a normal semi-invariant submanifold of a paraquaternionic Kähler manifold $(\overline{M}, \sigma, \overline{g})$. Then the following assertions are equivalent:*

(i) The distribution $D^\perp$ is integrable.
(ii) The second fundamental form h of M satisfies

$$h(X,Y) \in \Gamma(Q), \ \forall \ X \in \Gamma(D), \ Y \in \Gamma(D^\perp).$$

Next we show the existence of a mixed 3-structure on a normal semi-invariant submanifold of a paraquaternionic Kähler manifold under some conditions. We suppose that (M,g) is a normal semi-invariant submanifold of a paraquaternionic Kähler manifold $(\overline{M}, \sigma, \overline{g})$ such that $\dim Q^\perp = 1$ and $\overline{g}_{|Q^\perp}$ is positive definite.

For any vector field X on M and $\alpha \in \{1,2,3\}$ we have the decomposition:

$$J_\alpha X = \phi_\alpha X + F_\alpha X,$$

where $\phi_\alpha X$ and $F_\alpha X$ are the tangent part and the normal part of $J_\alpha X$, respectively. We can remark that, in fact, $F_\alpha X \in \Gamma(Q^\perp)$, for any vector field X on M.

Since $dim Q^\perp = 1$ and $\overline{g}_{|Q^\perp}$ is positive definite, we have $Q^\perp = <N>$, where N is a unit space-like vector field and we can see that we have the decomposition:

$$J_\alpha X = \phi_\alpha X + \eta_\alpha(X)N, \tag{22}$$

where:

$$\eta_\alpha(X) = \overline{g}(J_\alpha X, N). \tag{23}$$

We define now the vector field ξ_α by:

$$\xi_\alpha = -J_\alpha N,$$

for any $\alpha \in \{1,2,3\}$.

From (22) we have for all $\alpha \in \{1,2,3\}$:

$$\phi_\alpha^2 X = J_\alpha(\phi_\alpha X) - \eta_\alpha(\phi_\alpha X)N$$

and thus we conclude:

$$\phi_\alpha^2 = -\epsilon_\alpha X + \eta_\alpha(X)\xi_\alpha. \tag{24}$$

On the other hand, from (23) we deduce:

$$\eta_\alpha(\xi_\alpha) = \epsilon_\alpha, \ \forall \alpha \in \{1,2,3\} \tag{25}$$

and

$$\eta_\alpha(\xi_\beta) = 0, \ \forall \alpha \neq \beta \tag{26}$$

From (24) and (25) we conclude that (ϕ_1, ξ_1, η_1) is an almost contact structure on M, while (ϕ_2, ξ_2, η_2) and (ϕ_3, ξ_3, η_3) are Lorentzian almost paracontact structures on M. Moreover, by straightforward computations we obtain the following result (see Ref. 15).

Theorem 4.2. *Let (M,g) be a normal semi-invariant submanifold of a paraquaternionic Kähler manifold $(\overline{M}, \sigma, \overline{g})$ such that $dim Q^\perp = 1$ and $\overline{g}_{|Q^\perp}$ is positive definite. Then M admits a metric mixed 3-structure.*

Acknowledgements

We would like to thank the referees for carefully reading the paper and making valuable comments.

References

1. P. Libermann, *C.R. Acad. Sc. Paris*, **234**, 1030 (1952).
2. S. Ianuş, R. Mazzocco and G. E. Vîlcu, *Mediterr. J. Math.*, **3(3-4)**, 581 (2006).
3. N. Blažić, *Publ. Inst. Math.* **60(74)**, 101 (1996).
4. E. García-Río, Y. Matsushita and R. Vázquez-Lorenzo, *Rocky Mount. J. Math.*, **31(1)**, 237 (2001).
5. S. Vukmirović, math. DG/03044424,
6. S. Ivanov and S. Zamkovoy, *Differential Geom. Appl.*, **23**, 205 (2005).
7. D. E. Blair, *Contact manifolds in Riemannian Geometry*, Lectures Notes in Math. 509 (Springer-Verlag, 1976).
8. K. Matsumoto, *Bull. Yamagata Univ. Nat. Sci.*, **12(2)**, 151 (1989).
9. S. Ianuş and G. E. Vîlcu, *Int. J. Geom. Methods Mod. Phys.*, **5(6)**, 893 (2008).
10. A. Bejancu and K. L. Duggal, *Lightlike submanifolds of semi-Riemannian manifolds and its application* (Kluwer, Dortrecht 1996).
11. D. Alekseevsky and V. Cortés, *Osaka J. Math.*, **45(1)**, 215 (2008).
12. M. Vaccaro, *Kaehler and para-Kaehler submanifolds of a para-quaternionic Kaehler manifold*, ph. d. thesis, Università degli Studi di Roma II "Tor Vergata" (2007).
13. A. Bejancu, *Kuwait J. Sci. Eng.*, **33(2)**, 33 (2006).
14. A. Al-Aqeel and A. Bejancu, *Toyama Math. J.*, **30**, 63 (2007).
15. G. E. Vîlcu, Normal semi-invariant submanifolds of paraquaternionic space forms and mixed 3-structures, in *Proc. Int. Conference "Differential Geometry and Dynamical Systems" (DGDS-2007)*, (Bucharest, Romania, 2007), BSG Proceedings 15 (Geometry Balkan Press, 2008).
16. G. E. Vîlcu, *Int. Math. Forum*, **2(15)**, 735 (2007).
17. S. Ianuş, S. Marchiafava and G. E. Vîlcu, *Paraquaternionic CR-submanifolds of paraquaternionic Kähler manifolds*, preprint (2008).
18. S. Ianuş and G. E. Vîlcu, *J. Gen. Lie Theory Appl.*, **2(3)**, 175 (2008).

ON TOPOLOGICAL OBSTRUCTION OF COMPACT POSITIVELY RICCI CURVED MANIFOLDS

Wen-Haw Chen

Department of Mathematics, Tunghai University
Taichung, 40704, Taiwan
E-mail: whchen@thu.edu.tw

In this short note, we consider topological obstruction of compact Riemannian manifolds with positive Ricci curvature. We survey the author and Wu's work in 2006 about an extension of the classical Myers' theorem. This work is a partially generalization of Fukaya and Yamaguchi's work in 1992.

Keywords: Ricci curvature, Myers theorem, Fundamental groups

1. Introduction

The investigation of the relationships between curvature and topology of a Riemannian manifold is a fundamental undertaking in Riemannian geometry. The classical Myers' theorem shows that

Myers' Theorem: *Let (M^n, g) be a compact Riemannian n-manifold. If $Ric_M \geq (n-1)r^{-2}$ then the diameter of M, $diam(M^n, g)$, satisfies $diam(M^n, g) \leq \pi r$. In particular, the fundamental group $\pi_1(M)$ of M is finite provided $Ric_M > 0$.*

It is interesting to investigate the problem that "*What can be said about the fundamental group of a compact positively Ricci curved manifold depending only on the dimension of the manifold, except that it is finite?*" In [2], the author and Wu give an extension of Myers' theorem by generalizing Fukaya and Yamaguchi's work in [3]. Moreover, our investigation gives a partial answer to Gromov's conjecture about almost nonnegative Ricci curved manifolds.

2. An Extension of Myers' Theorem

We consider compact Riemannian n-manifolds with positive Ricci curvature. When $n = 2$, it is known that only the projective plan $\mathbb{RP}^2$ and the

2-sphere admit metrics with positive curvature. Hamilton showed in [5] that a compact 3-manifold with positive Ricci curvature also admits a metric with a constant sectional curvature of +1, and is then covered by the 3-sphere. However, one has only Myers' theorem for the general cases. The following result is proved in [2].

Theorem 2.1. *Given $n \geq 2$, there exist constants p_n and C_n depending only on n such that if a compact Riemannian n-manifold M^n has the Ricci curvature $Ric_{M^n} > 0$, then*
(a) the first betti number $b_1(M^n, \mathbb{Z}_p)$ with p-cyclic group coefficient $\mathbb{Z}_p$ satisfies $b_1(M^n, \mathbb{Z}_p) \leq n - 1$ for all prime $p \geq p_n$, and
(b) the ratio of diameters satisfies

$$\frac{diam(\tilde{M}^n)}{diam(M^n)} < C_n,$$

where $\tilde{M}^n$ denotes the universal covering of M^n.

Remark 2.1. Considering the flat n-torus $\mathbb{T}^n$, $b_1(\mathbb{T}^n, \mathbb{Z}_p) = n$ is obtained for all prime p, and the canonical Euclidean n-space $\mathbb{R}^n$ is its universal covering space. Fukaya and Yamaguchi showed in [3] that if a compact Riemannian n-manifold M with sectional curvature K_M and diameter $diam(M)$ satisfies $K_M diam(M)^2 > -\epsilon_n$ for some constant ϵ_n depending only on n, then $b_1(M^n, \mathbb{Z}_p) \leq n$ for all $p \geq p_n$, and the maximal case $b_1(M^n, \mathbb{Z}_p) = n$ arise only when M^n is diffeomorphic to a torus. They also found that $diam(\tilde{M})/diam(M)$ is uniformly bounded by a constant depending only on n provided the fundamental group $\pi_1(M)$ is finite. Theorem 2.1 extends their results to manifolds with positive Ricci curvature. Notably positively Ricci curved n-manifolds with $n \leq 3$ are covered by spheres as discussed above. Hence Theorem 2.1 holds for these manifolds.

Gromov made a conjecture in [4] that a positive number ϵ_n exists which depends only on n such that if a compact Riemannian n-manifold M with almost nonnegative Ricci curvature $Ric_M diam(M)^2 > -\epsilon_n$, then the fundamental group $\pi_1(M)$ of M is *almost nilpotent*. It means that $\pi_1(M)$ contains a nilpotent subgroup of finite index. Fukaya and Yamaguchi have shown that Gromov's conjecture holds when the condition $K_M diam(M)^2 > -\epsilon_n$ in [3]. Cheeger and Colding proved in [1] that there exists an $\epsilon_n > 0$ such that if $Ric_M diam(M)^2 > -\epsilon_n$ and the first betti number $b_1(M, \mathbb{R}^n) = n$, then the manifold M is homeomorphic to the flat n-torus $\mathbb{T}^n$ if $n \neq 3$ and homotopically equivalent to $\mathbb{T}^n$ if $n = 3$. Under the additional assumptions on the conjugate radius of M, that is, consider compact Riemannian n-manifolds with $diam(M) < 1$ and conjugate radius $\geq c_0 > 0$. Then Paeng shows in [6] that there exists an $\epsilon_{n,c_0} > 0$ such that $\pi_1(M)$ is almost nilpotent. In this work, we give a partial answer to the conjecture and prove the fundamental group is almost solvable.

Theorem 2.2. *There exist ϵ_n, $w_n > 0$ depending only on n such that if a compact Riemannian n-manifold satisfies $Ric_M diam(M)^2 > -\epsilon_n$ then the fundamental group $\pi_1(M)$ of M contains a solvable subgroup of finite index less than w_n.*

The main tool in the proof of Theorem 2.1 and Theorem 2.2 are the equivariant pointed Hausdorff convergence developed by Fukaya and Yamaguchi in [3]. Use this notion, Fukaya and Yamaguchi prove the following solvability theorem. Recall that the *length of polycyclicity* of a solvable group G is the smallest integer m for which G admits a filtration

$$\{e\} = G_m \subset G_{m-1} \subset \ldots \subset G_1 \subset G_0 = G$$

such that each G_i/G_{i-1} is cyclic.

Theorem 2.3 (Solvability Theorem (Fukaya & Yamaguchi, 1992)).
For given positive integers n and k, $n \geq k$ and a positive number μ_0, there exist positive numbers $\epsilon = \epsilon_{n,k}(\mu_0)$, $w = w_{n,k}$ and a function $\tau(\epsilon) = \tau_{n,k,\mu_0}(\epsilon)$ with $\lim_{\epsilon \to 0} \tau(\epsilon) = 0$ such that if (M^n, p) and (N^k, q) are pointed-Riemannian manifolds of dimension n and k respectively such that
(2.3.1) the sectional curvatures of M and N satisfy $K_M > -1$, $|K_N| \leq 1$ and the injectivity radius of N with $inj(N) > \mu_0$;
(2.3.2) $d_H(M, N) < \epsilon$,
where d_H denotes the classical Hausdorff distance, then there exists a map $f : M \to N$ satisfying
(2.3.3) f is a continuous $\tau(\epsilon)$-Hausdorff approximation which is also a fiberation and a $\tau(\epsilon)$-Riemannian submersion (i.e. $e^{-\tau(\epsilon)} < \frac{|df(\xi)|}{|\xi|} < e^{\tau(\epsilon)}$ for each ξ is orthogonal to the fiber.)
(2.3.4) there exists a normal subgroup H of the fundamental group Γ of the fiber of f such that
(i) H is solvable with length of polycyclicity $\leq n - k$,
(ii) the index $[\Gamma : H] \leq w$.

By use of Theorem 2.3, Fukaya and Yamaguchi prove the following Margulis's lemma.

Theorem 2.4 (Margulis's Lemma; Fukaya & Yamaguchi, 1992).
There exists a positive number δ_n depending only on n and satisfying the following: Let (M^n, p) be a complete pointed Riemannian n-manifold with sectional curvature $K_{M^n} \geq -1$. Then there exists a point $p' \in B_p(1/2)$ such that the image of the inclusion homomorphism

$$\Gamma' = Im[\pi_1(B_{p'}(\delta_n)) \to \pi_1(B_p(1))]$$

is almost nilpotent and admits a subgroup $\Lambda' \subset \Gamma'$ with
(1) $[\Gamma' : \Lambda'] < w_n$, where w_n depends only on n;
(2) Λ' is solvable with length of polycyclicity $\leq n$.

The idea in the proof of Theorem 2.1 and 2.2 is to generalize Fukaya and Yamaguchi's work, especially the Solvability Theorem and the Margulis's Lemma to manifolds with lower Ricci curvature bound. There are two difficulties to this end.

- There is no fiberation with the property of almost Riemannian submersion as in (2.3.3) for manifolds with lower Ricci curvature bound.
- The Margulis lemma (Theorem 2.4) cannot be extended to manifolds under a lower Ricci curvature bound only.

The following result is a generalization of Fukaya and Yamaguchi's solvability theorem and is crucial to the proof of Theorem 2.1 and Theorem 2.2.

Theorem 2.5 (Chen & Wu, 2006). *For given positive integers n and k, $n \geq k$ and a positive number μ_0, there exist positive numbers $\epsilon = \epsilon_{n,k}(\mu_0)$, $w = w_{n,k}$ and a function $\tau(\epsilon) = \tau_{n,k,\mu_0}(\epsilon)$ with $\lim_{\epsilon \to 0} \tau(\epsilon) = 0$ such that if (M^n, p) and (N^k, q) are pointed-Riemannian manifolds of dimension n and k respectively such that*
(2.5.1) $Ric_M \geq -(n-1)$, $Ric_N \geq -(n-1)$ and $inj(N) > \mu_0 > 0$,
(2.5.2) $d_{GH}((M,p),(N,q)) < \epsilon$,
where d_{GH} denotes the pointed Gromov-Hausdorff distance, then there exists a map $f : M \to N$ with $f(p) = q$ satisfying the following:
(2.5.3) f is a continuous $\tau(\epsilon)$-Hausdorff approximation such that $f_* : \pi_1(M,p) \to \pi_1(N,q)$ is surjective;
(2.5.4) Let $V = B_q(\frac{\mu_0}{2})$ be the ball around q with radius $\frac{\mu_0}{2}$. Set $U = f^{-1}(V)$. Then there is a normal subgroup H of the fundamental group $\Gamma = \pi_1(U)$ of U such that

(i) H is a solvable subgroup of Γ with length of polycyclicity $\leq n - k$,
(ii) $[\Gamma : H] \leq w_{n,k}$.

The following Weak Margulis's Lemma under a lower Ricci curvature bound is proved in [2]. It is a weaker version of the Margulis Lemma for only "one point".

Theorem 2.6 (Weak Margulis's Lemma; Chen & Wu, 2006).
There exists a positive number δ_n depending only on n and satisfying the following: Let (M^n, p) be a complete pointed Riemannian n-manifold with $Ric_{M^n} \geq -(n-1)$. Then there exists a point $p' \in B_p(1/2)$ such that the image of the inclusion homomorphism

$$\Gamma' = Im[\pi_1(B_{p'}(\delta_n)) \to \pi_1(B_p(1))]$$

admits a subgroup $\Lambda' \subset \Gamma'$ with
(1) $[\Gamma' : \Lambda'] < w_n$, where w_n depends only on n;
(2) Λ' is solvable with length of polycyclicity $\leq n$.

Remark 2.2. Consider a compact Riemannian n-manifold M^n with $Ric_{M^n} \geq 0$. Scaling the metric of M so that $diam(M) \leq \frac{\delta_n}{2}$ still leaves $Ric_M \geq 0$, thus showing the fundamental group $\pi_1(M)$ of M admits a subgroup H such that
(1) $[H : \pi_1(M)] < w_n$.
(2) H is solvable with length of polycyclicity $\leq n$.

We refer the readers to [2] for the details of the proofs.

Acknowledgements

This work is partially supported by Taiwan NSC grants.

References

1. J. Cheeger and T. Colding, *J. Diff. Geom.* **Vol. 46** (1997), 406–480.
2. W. -H. Chen and J. -Y. Wu, *Bull. Belg. Math. Soc.* **13** (2006), 441–453.
3. K. Fukaya and T. Yamaguchi, *Ann. Math.* **136** (1992), 253–333.
4. M. Gromov, *Synthetic geometry in Riemannian manifolds*, In Proc. of International Congress of Mathematicians, Helsinki, 415–419, 1978.
5. R. Hamilton, *J. Diff. Geom.* **17** (1982), 255–306.
6. S. -H. Paeng, *Proc. Amer. Math. Soc.* **131** (2003), 2577–2583.

GRAY CURVATURE CONDITIONS AND THE TANAKA-WEBSTER CONNECTION

Raluca Mocanu

Faculty of Mathematics, University of Bucharest
Str. Academiei n.14, s 1, Bucharest, Romania
E-mail: xipita@yahoo.com

The purpose of this note is to establish Gray-type curvature conditions for a contact metric manifold endowed with the generalized Tanaka-Webster connection.

Keywords: Contact metric manifolds, Gray identities, Tanaka-Webster connection, strongly pseudoconvex CR-manifolds

1. Different types of Gray curvature conditions

The Gray curvature conditions, (the so-called Gray identities), were introduced by Alfred Gray[1] for almost Hermitian manifolds (M, g, J):

$$R(X,Y,Z,W) = R(X,Y,JZ,JW)$$
$$R(X,Y,Z,W) = R(JX,Y,Z,JW) + R(X,JY,Z,JW) + R(X,Y,JZ,JW)$$
$$R(X,Y,Z,W) = R(JX,JY,JZ,JW)$$

A natural question which arises is to find analogous curvature conditions in the case of odd dimensional manifolds, namely of almost contact and contact metric manifolds.

Some Gray-type curvature conditions for almost contact metric manifolds were later introduced by A. Bonome, L. Hervella and I. Rozas[2]

K1φ : $R(X,Y,Z,W) = R(X,Y,\varphi Z, \varphi W)$
K2φ : $R(X,Y,Z,W) = R(\varphi X, Y, Z, \varphi W)$
$\quad\quad\quad + R(X, \varphi Y, Z, \varphi W) + R(X, Y, \varphi Z, \varphi W)$
K3φ : $R(X,Y,Z,W) = R(\varphi X, \varphi Y, \varphi Z, \varphi W)$

and other Gray-type conditions by M.I. Munteanu and the author[3]

G1: $R(X,Y,Z,W) = R(X,Y,\varphi Z,\varphi W) - g(X,\varphi Z)g(Y,\varphi W)$
$\qquad + g(Y,\varphi Z)g(X,\varphi W) - g(Y,Z)g(X,W) + g(X,Z)g(Y,W)$

G2: $R(X,Y,Z,W) = R(\varphi X, Y, Z, \varphi W) + R(X, \varphi Y, Z, \varphi W)$
$\qquad + R(X,Y,\varphi Z, \varphi W) + g(X,Z)\eta(W)\eta(Y) - g(Z,Y)\eta(X)\eta(W)$

G3: $R(X,Y,Z,W) = R(\varphi X, \varphi Y, \varphi Z, \varphi W) + g(X,Z)\eta(W)\eta(Y)$
$\qquad - g(Z,Y)\eta(X)\eta(W) + g(Y,W)\eta(X)\eta(Z) - g(X,W)\eta(Y)\eta(Z)$

Remark 1. ([3]) It is easy to see that every contact metric manifold satisfying one of **G1, G2, G3** is Sasakian.

2. Gray curvature conditions for the Tanaka-Webster connection

Then it becomes necessary to establish other curvature conditions for contact metric manifolds, if necessary using special tools, like the Tanaka Webster connection instead of the Levi-Civita connection.

The generalized Tanaka-Webster connection of a contact manifold (M,φ,ξ,η) is given by the expression:[4]

$$\tilde{\nabla}_X Y = \nabla_X Y + \eta(X)\varphi Y + \eta(Y)(\varphi X + \varphi h X) + d\eta(X + hX, Y)\xi$$

where ∇ is the Levi-Civita connection of the manifold (M,φ,ξ,η) and the tensor field h is defined as $h = \frac{1}{2}L_\xi \varphi$ (see for example[5]).

The Tanaka-Webster $\tilde{\nabla}$ connection is the unique connection on (M,g,φ,ξ,η) which satisfies the following properties:[6]

$$\tilde{\nabla} g = 0 \quad \tilde{\nabla}\eta = 0 \quad \tilde{\nabla}\xi = 0$$
$$(\tilde{\nabla}_X \varphi)Y = (\nabla_X \varphi)Y - g(X + hX, Y)\xi + \eta(Y)(X + hX)$$
$$\tilde{T}(Z,Z') = 2d\eta(Z,Z')\xi \quad \forall Z, Z' \in Ker\,\eta$$

Remark 2. (cf.[7]) It is easy to see that for strongly pseudoconvex CR-manifolds ([8,9]), the generalized Tanaka-Webster connection coincides with the classical Tanaka connection introduced by N. Tanaka.[10]

Let $R(X,Y,Z,W)$ denote the curvature tensor of ∇ and $\tilde{R}(X,Y,Z,W)$ the curvature tensor of $\tilde{\nabla}$. We study the relation between the curvatures of the Tanaka-Webster and Levi-Civita connections.

$$\begin{aligned}
\tilde{R}(X,Y)Z + \varphi\tilde{R}(X,Y)\varphi Z &= R(X,Y)Z + \varphi R(X,Y)\varphi Z \\
&+ 2\eta(Y)\eta(Z)(X+hX) - 2\eta(X)\eta(Z)(Y+hY) + 2\eta(Y)(\nabla_X\varphi)Z \\
&- 2\eta(X)(\nabla_Y\varphi)Z + 2\eta(\nabla_Y\varphi Z)\eta(X)\xi - 2\eta(\nabla_X\varphi Z)\eta(Y)\xi \\
&- \eta(\nabla_Y\varphi Z)(X+hX) + \eta(\nabla_X\varphi Z)(Y+hY) + \eta(Z)(\nabla_X\varphi)Y \\
&- \eta(Z)(\nabla_Y\varphi)X + \eta(Z)(\nabla_X\varphi h)Y - \eta(Z)(\nabla_Y\varphi h)X \\
&+ \eta(X)g(Y+hY,Z)\xi - \eta(Y)g(X+hX,Z)\xi \\
&- g(\varphi X + \varphi hX, Z)(\varphi Y + \varphi hY) + g(\varphi Y + \varphi hY, Z)(\varphi X + \varphi hX) \\
&+ g(Y+hY, (\nabla_X\varphi)Z)\xi - g(X+hX, (\nabla_Y\varphi)Z)\xi \\
&+ g((\nabla_X h)Y - (\nabla_Y h)X, \varphi Z)\xi
\end{aligned}$$

In the particular case of Sasakian manifolds, as we have $(\nabla_X\varphi)Y = g(X,Y)\xi - \eta(Y)X$, we obtain

$$\begin{aligned}
R(X,Y,Z,W) &- R(X,Y,\varphi Z,\varphi W) - g(X,\varphi Z)g(\varphi Y,W) \\
&+ g(Y,\varphi Z)g(\varphi X,W) + g(Y,Z)g(X,W) - g(X,Z)g(Y,W) \\
&= \tilde{R}(X,Y,Z,W) - \tilde{R}(X,Y,\varphi Z,\varphi W)
\end{aligned}$$

As **G1** holds on Sasakian manifolds, as particular cases of $C(1)$-manifolds,[11] we obtain the annulation of the left member. Because of the fact that $\tilde{\nabla}\varphi = 0$ (see[12,13]) we obtain that

Theorem 2.1. *For strongly pseudoconvex CR manifolds, the following expression holds*

$$\begin{aligned}
0 &= R(X,Y)Z + \varphi R(X,Y)\varphi Z - g(Y+hY,Z)(X+hX) \\
&+ g(X+hX,Z)(Y+hY) + \eta(Z)(\nabla_X\varphi h)Y - \eta(Z)(\nabla_Y\varphi h)X \\
&- g(\nabla_X\xi, Z)\nabla_Y\xi + g(\nabla_Y\xi, Z)\nabla_X\xi + g((\nabla_X h)Y - (\nabla_Y h)X, \varphi Z)\xi.
\end{aligned}$$

In order to study the second condition, we obtain, by similar computations, that for Sasakian manifolds

$$\begin{aligned}
\tilde{R}(X,Y)Z &+ \varphi\tilde{R}(\varphi X, Y)Z + \varphi\tilde{R}(X,\varphi Y)Z + \varphi\tilde{R}(X,Y)\varphi Z \\
&= R(X,Y)Z + \varphi R(\varphi X,Y)Z + \varphi R(X,\varphi Y)Z + \varphi R(X,Y)\varphi Z \\
&+ \eta(Y)g(X,Z)\xi - \eta(X)g(Y,Z)\xi
\end{aligned}$$

finally obtaining **G2** in the right member. We obtain the identity **G3** by similar computations; the expression becomes for Sasakian manifolds

$$\begin{aligned}
R(X,Y,Z,W) &- R(\varphi X,\varphi Y,\varphi Z,\varphi W) = \tilde{R}(X,Y,Z,W) \\
&- \tilde{R}(\varphi X,\varphi Y,\varphi Z,\varphi W) + g(X,Z)\eta(W)\eta(Y) - g(Z,Y)\eta(X)\eta(W) \\
&+ g(Y,W)\eta(X)\eta(Z) - g(X,W)\eta(Y)\eta(Z)
\end{aligned}$$

Then obviously we can adapt the following conditions for every contact metric manifold $(M, g, \varphi, \xi, \eta)$:

(**K1**) $\tilde{R}(X,Y,Z,W) = \tilde{R}(X,Y,\varphi Z, \varphi W)$
(**K2**) $\tilde{R}(X,Y,Z,W) = \tilde{R}(\varphi X,Y,Z,\varphi W)$
$\qquad + \tilde{R}(X,\varphi Y, Z, \varphi W) + \tilde{R}(X,Y,\varphi Z, \varphi W)$
(**K3**) $\tilde{R}(X,Y,Z,W) = \tilde{R}(\varphi X, \varphi Y, \varphi Z, \varphi W)$

The natural question is if there are non-Sasakian manifolds on which these conditions hold.

Example 2.1. Example of a contact metric manifold satisfying the **Ki** curvature conditions without satisfying the **Gi** curvature conditions, with $i \in 1, 2, 3$.

Consider on R^3 the base composed by the following vector fields $X = 2\partial_y$, $Y = \cos y \partial_x - \sin y \partial_z$ and $\xi = \sin y \partial_x + \cos y \partial_z$. Let $\eta = \sin y dx + \cos y dz$. As $\eta \wedge d\eta = -dx \wedge dy \wedge dz$, (R^3, η) is a contact manifold, we endow it with an almost contact structure (φ, ξ, η) by defining φ by $\varphi X = -Y$, $\varphi Y = X$ and $\varphi \xi = 0$ and an associated metric $g = dx^2 + dz^2 + \frac{1}{4}dy^2$. $(R^3, g, \varphi, \xi, \eta)$ now becomes a contact metric manifold of dimension 3. Hence, $(R^3, g, \varphi, \xi, \eta)$ becomes a strongly pseudoconvex CR manifold.[5]

As $hX = X$ and $hY = -Y$ this manifold is not K-contact. The Levi-Civita connection is given by $\nabla_X Y = -2\xi$ and $\nabla_X \xi = 2Y$ and zero for the other terms. The Tanaka Webster connection is given by $\tilde{\nabla}_\xi X = -Y$ and $\tilde{\nabla}_\xi Y = X$ and zero for the other terms. As the manifold R^3 endowed with the upper structure is not Sasakian, it does not satisfy the Gi conditions (as we can directly see by computing the curvature $R(\xi, X)X$). But all of the conditions K_i introduced for the Tanaka Webster connection hold for this manifold.

References

1. A. Gray, *Tohôku Math. J.* **28**(1976), 601-612.
2. A. Bonome, L.M.Hervella and I.Rozas *Acta Math. Hung.* **56** (1990), 29-37.
3. R. Mocanu and M.-I. Munteanu, arXiv:0706.2570v1.
4. S. Tanno *Trans. Amer. Math. Soc.* **314** (1989), 349 379.
5. D. E. Blair, *Riemannian Geometry of Contact and Symplectic Manifolds*, Birkhäuser, 2002
6. B. Cappelletti-Montano, arXiv:0706.0888v1.
7. D. Perrone, *Monatsh. Math.* **114** (1992), 245-259.
8. C. Gherghe, S. Ianus and A. M. Pastore, *J. Geom.* **71**(2001), 42-53.
9. S. Dragomir and G. Tomassini, *Differential Geometry and Analysis on CR-manifolds*, Springer Verlag, 2006.

10. N.Tanaka, *A differential geometric study on strongly pseudoconvex manifolds*, Lectures in Math., Kyoto University, **9**, 1975.
11. F. Tricerri and L.Vanhecke, *Trans. Amer. Math. Soc.* **267** (1981), 365-398.
12. M.-I. Munteanu, *Invarianti geometrici asociati CR-structurilor pseudoconvexe*, Ph.D. thesis,Universitatea Al.I.Cuza, Iasi, 2003
13. K. Sakamoto and Y. Takemura, *Kodai Math. J.* **4** (1981), 251-265.

RIEMANNIAN STRUCTURES ON HIGHER ORDER FRAME BUNDLES FROM CLASSICAL LINEAR CONNECTIONS

J. Kurek

Institute of Mathematics, Maria Curie-Sklodowska University
pl. Marii Curie Skłodowskiej 1
20–031 Lublin, Poland
E-mail: kurek@hektor.umcs.lublin.pl

W. M. Mikulski

Institute of Mathematics Jagiellonian University
Lojasiewicza 6,
30–348 Kraków, Poland
E-mail: Wlodzimierz.Mikulski@im.uj.edu.pl

We describe all $\mathcal{M}f_m$-natural operators $A : Q \rightsquigarrow \mathcal{R}iem P^r$ transforming classical linear connections ∇ on m-dimensional manifolds M into Riemannian structures $A(\nabla)$ on the r-th order frame bundle $P^r M = inv J_0^r(\mathbb{R}^m, M)$ over M.

Keywords: Classical linear connection, Riemannian structure, higher order frame bundle, natural operator

1. Introduction

Let $\mathcal{M}f_m$ denote the category of m-dimensional manifolds and their embeddings (i.e. diffeomorphisms onto open subsets) and $\mathcal{F}\mathcal{M}$ denote the category of fibred manifolds and their fibred maps. Manifolds and maps are assumed to be of class C^∞.

For any m-manifold M we have the r-th order frame bundle $P^r M = inv J_0^r(\mathbb{R}^m, M)$ of M. This is a principal bundle with the corresponding Lie group $G_m^r = J_0^r(\mathbb{R}^m, \mathbb{R}^m)_0$ acting on the right on $P^r M$ via compositions of jets. Every $\mathcal{M}f_m$-map $\psi : M_1 \to M_2$ induces a principal bundle embedding $P^r \psi : P^r M_1 \to P^r M_2$ by $P^r \psi(j_0^r \varphi) = j_0^r(\psi \circ \varphi)$, where $\varphi : \mathbb{R}^m \to M_1$ is a $\mathcal{M}f_m$-map. The correspondence $P^r : \mathcal{M}f_m \to \mathcal{F}\mathcal{M}$ is a bundle functor in the sense of [2].

For any n-manifold N we have the Riemannian bundle $\mathcal{R}iem(N) = \bigcup_{y\in N} Met(T_yN)$ over N. The correspondence $\mathcal{R}iem : \mathcal{M}f_n \to \mathcal{F}\mathcal{M}$ is a bundle functor in the sense of [2].

For any m-manifold M we have the classical linear connection bundle $QM := (id_{T^*M} \otimes \pi^1)^{-1}(id_{TM}) \subset T^*M \otimes J^1TM$ of M, where $\pi^1 : J^1TM \to TM$ is the projection of the first jet prolongation of TM. The correspondence $Q : \mathcal{M}f_m \to \mathcal{F}\mathcal{M}$ is a bundle functor in the sense of [2].

In the present short note we study the problem how a classical linear connection ∇ on an m-dimensional manifold M can induce (canonically) a Riemannian structure $A(\nabla)$ on P^rM. This problem is reflected in the concept of $\mathcal{M}f_m$-natural operators $A : Q \rightsquigarrow \mathcal{R}iem P^r$. In the note we describe explicitly all $\mathcal{M}f_m$-natural operators A in question.

2. Natural operators

Definition 2.1. ([2]) An $\mathcal{M}f_m$-natural operator $A : Q \rightsquigarrow \mathcal{R}iem P^r$ is a family of $\mathcal{M}f_m$-invariant regular operators (functions)

$$A = A_M : \underline{Q}(M) \to \underline{\mathcal{R}iem}(P^rM)$$

for any $\mathcal{M}f_m$-object M, where $\underline{Q}(M)$ is the set of all classical linear connections on M (sections of $Q(M) \to M$) and $\underline{\mathcal{R}iem}(N)$ is the set of all Riemannian structures on N (sections of $\mathcal{R}iem(N) \to N$) for any manifold N. The invariance means that if $\nabla_1 \in \underline{Q}(M_1)$ and $\nabla_2 \in \underline{Q}(M_2)$ are related by an $\mathcal{M}f_m$-map $\psi : M_1 \to M_2$ (i.e. $\overline{Q(\psi)} \circ \nabla_1 = \nabla_2 \circ \psi$) then $A(\nabla_1)$ and $A(\nabla_2)$ are $P^r\psi$-related (i.e. $\mathcal{R}iem(P^r\psi) \circ A(\nabla_1) = A(\nabla_2) \circ P^r\psi$). The regularity means that A transforms smoothly parametrized families of classical linear connections into smoothly parametrized families of Riemannian structures.

For $r = 1$, P^1M is equivalent with the bundle LM of linear frames over M (we identify $j_0^1\varphi \in P^1M$ with $(T_0\varphi(\frac{\partial}{\partial x^i}|_0))_{i=1}^m \in LM$). In this case we have the following very important classical example of $\mathcal{M}f_m$-natural operator $Q \rightsquigarrow \mathcal{R}iem P^1$ presented in the proof of Theorem 1.5 in [1].

Example 2.1. Let ∇ be a classical linear connection on an m-manifold M and let $\omega = (\omega_k^j) : TP^1M \to gl(m)$ be its connection form. Let $\theta = (\theta^i) : TP^1M \to \mathbb{R}^m$ be the canonical form on P^1M. We put

$$g^\nabla(X^*, Y^*) = \sum_i \theta^i(X^*)\theta^i(Y^*) + \sum_{j,k} \omega_k^j(X^*)\omega_k^j(Y^*), \ X^*, Y^* \in T_uP^1M.$$

Then g^∇ is a Riemannian structure on P^1M, see the proof of Theorem 1.5 in [1]. Clearly, the correspondence $A^1 : Q \rightsquigarrow \mathcal{R}iem P^1$ given by $A^1(\nabla) = g^\nabla$ for all $\nabla \in \underline{Q}(M)$ is an $\mathcal{M}f_m$-natural operator.

Example 2.2. Let ∇ be a classical linear connection on M. Let (θ^i, ω_k^j) be the basis of 1-forms on P^1M, where $\theta = (\theta^i)$ and $\omega = (\omega_k^j)$ is as in Example 1 for ∇. Let $X^1(\nabla), ..., X^{L_1}(\nabla)$, where $L_1 = dim(P^1\mathbb{R}^m) = m + m^2$, be the dual (to (θ^i, ω_k^j)) basis of vector fields on P^1M. Then $\tilde{g}^\nabla = \sum_{s=1}^{L_1}(X^s(\nabla))^* \odot (X^s(\nabla))^*$ is a Riemannian structure on P^1M (the same as in Example 2.1).

From now on let $A_1, ..., A_{dim(g_m^r)}$ be the standard basis in $g_m^r = \mathcal{L}ie(G_m^r)$ (i.e. the basis $j_0^r(x^\alpha \frac{\partial}{\partial x^i}) \in (J_0^r T\mathbb{R}^m)_0 = g_m^r$ for $i = 1, ..., m, 1 \le |\alpha| \le r$).

Example 2.3. *Construction of an absolute parallelism on P^rM from a classical linear connection ∇ on M.* Let ∇ be a classical linear connection on an m-manifold M. Let $i = 1, ..., m$. We have a vector field $Y^i(\nabla)$ on P^rM defined as follows. Let $\sigma = j_0^r\varphi \in (P^rM)_x$, $x \in M$. Let $v^i = T\varphi(\frac{\partial}{\partial x^i}|_0) \in T_xM$. We extend v^i to the constant vector field $\tilde{v}^i$ on T_xM. Then on some neighborhood of x we have the vector field $V^i(\nabla) = (Exp_x^\nabla)_*\tilde{v}^i$, where $Exp_x^\nabla : T_xM \supset U_{0_x} \to \tilde{U}_x \subset M$ is the exponent of ∇. We define $Y^i(\nabla)_\sigma := \mathcal{P}^r(V^i(\nabla))_\sigma$, where $\mathcal{P}^rV$ is the flow lifting of a vector field V on M to P^rM (if $\{\varphi_t\}$ is the flow of V then $\{P^r\varphi_t\}$ is the flow of $\mathcal{P}^rV$). It is easy to see that $Y^i(\nabla)_\sigma$ projects onto v^i by the bundle projection $P^rM \to M$. So, it is a simple observation that $Y^i(\nabla), A_j^*$ for $i = 1, ..., m, j = 1, ..., dim(g_m^r)$ is an absolute parallelism on P^rM (canonically depending on ∇), where given $A \in \mathcal{L}ie(G_m^r)$ we denote the fundamental vector field on the principal bundle P^rM by A^*.

Example 2.4. Let ∇ be a classical linear connection on an m-manifold M. Let $(Y^i(\nabla), A_j^*)$ be the parallelism from Example 2.3. Let $\omega^s(\nabla)$ for $s = 1, ..., dim(P^r\mathbb{R}^m)$ be the dual basis of 1-form on P^rM. We put

$$\tilde{g}^{r,\nabla} := \sum_s \omega^s(\nabla) \odot \omega^s(\nabla).$$

Clearly, $\tilde{g}^{r,\nabla}$ is a Riemannian structure on P^rM. Clearly, the correspondence $A^{[r]} : Q \rightsquigarrow \mathcal{R}iem P^r$ given by $A^{[r]}(\nabla) = \tilde{g}^{r,\nabla}$ for all ∇ in question is an $\mathcal{M}f_m$-natural operator. One can observe easily that $A^1 = A^{[1]}$.

According to the global basis of vector fields $(Y^i(\nabla), A_j^*)$ on P^rM from Example 3, given $\nabla \in \underline{Q}(M)$ we have a canonical (in ∇) fibred diffeomorphism

(∗) $$I_\nabla : P^r M \times Met(\mathbb{R}^{L_r}) \to \mathcal{R}iem(P^r M)$$

covering $id_{P^r M}$ defined by the condition that the matrix of $I_\nabla(\sigma, G)$ in the basis $(Y^i(\nabla)(\sigma), A_j^*(\sigma))$ is the same as the one of G in the usual canonical basis of $\mathbb{R}^{L_r}$.

From now on we denote

$$Q^r = (\pi_1^r)^{-1}(l^o) \subset (P^r \mathbb{R}^m)_0,$$

where l^o is the usual canonical basis in $\mathbb{R}^m = T_0 \mathbb{R}^m$ and $\pi_1^r : P^r M \to P^1 M$ is the jet projection. Of course, Q^r is a submanifold in $(P^r \mathbb{R}^m)_0$.

For $s = 0, 1, ..., \infty$, let $Z^s = J_0^s(Q(\mathbb{R}^m))$ be the set of s-jets $j_0^s \nabla$ of all classical linear connections ∇ on $\mathbb{R}^m$ with

(∗∗) $$\sum_{j,k=1}^m \nabla^i_{jk}(x) x^j x^k = 0 \text{ for } i = 1, ..., m.$$

We inform that the condition (∗∗) means (it is equivalent to) that the usual coordinates $x^1, ..., x^m$ on $\mathbb{R}^m$ are ∇-normal with centre 0. If s is finite, then (from (∗∗)) Z^s is a finite dimensional manifold (diffeomorphic to a finite dimensional vector space). Z^∞ is a topological space with respect to the inverse limit topology given by the inverse system $... \to Z^{s+1} \to Z^s \to ... \to Z^0$ of jet projections.

Example 2.5. *General construction.* Let $\mu : Z^\infty \times Q^r \to Met(\mathbb{R}^{L_r})$, where $L_r = dim(P^r \mathbb{R}^m)$, be a map satisfying the following local finite determination property (a_r):

(a_r) For any $\rho \in Z^\infty$ and $\sigma \in Q^r$ we can find an open neighborhood $U \subset Z^\infty$ of ρ, an open neighborhood $V \subset Q^r$ of σ, a natural number s and a smooth map $f : \pi_s(U) \times V \to Met(\mathbb{R}^{L_r})$ such that $\mu = f \circ (\pi_s \times id_V)$ on $U \times V$, where $\pi_s : Z^\infty \to Z^s$ is the jet projection.

(A simple example of such μ is $\mu = f \circ (\pi_s \times id_{Q^r})$ for smooth $f : Z^s \times Q^r \to Met(\mathbb{R}^{L_r})$ for finite s.) Given a classical linear connection ∇ on an m-manifold M we define a Riemannian structure $A^{<\mu>}(\nabla)$ on $P^r M$ as follows. Let $\sigma \in (P^r M)_x$, $x \in M$. Choose a ∇-normal coordinate system ψ on M with center x such that $P^r \psi(\sigma) \in Q^r$. Of course, such ψ exists. Then $germ_x(\psi)$ is uniquely determined. We put

$$A^{<\mu>}(\nabla)_\sigma = \mathcal{R}iem(P^r(\psi^{-1}))(I_{\psi_* \nabla}(P^r \psi(\sigma), \mu(j_0^\infty(\psi_* \nabla), P^r \psi(\sigma)))).$$

Since $germ_x(\psi)$ is uniquely determined the definition $A^{<\mu>}(\nabla)_\sigma$ is correct. The family $A^{<\mu>} : Q \rightsquigarrow \mathcal{R}iem P^r$ is an $\mathcal{M}f_m$-natural operator.

3. The main result

Theorem 3.1. *Any $\mathcal{M}f_m$-natural operator $A : Q \rightsquigarrow \mathcal{R}iemP^r$ is $A^{<\mu>}$ for some uniquely determined (by A) function $\mu : Z^\infty \times Q^r \to Met(\mathbb{R}^{L_r})$ satisfying the property (a_r).*

Proof. Let $A : Q \rightsquigarrow \mathcal{R}iemP^r$ be an $\mathcal{M}f_m$-natural operator. We must define $\mu : Z^\infty \times Q^r \to Met(\mathbb{R}^{L_r})$ by

$$(\sigma, \mu(j_0^\infty \nabla, \sigma)) = I_\nabla^{-1}(A(\nabla)(\sigma)).$$

Then by the non-linear Peetre theorem [2], μ satisfies the property (a_r). Then by the definition of μ and $A^{<\mu>}$ we see that $A(\nabla)(\sigma) = A^{<\mu>}(\nabla)(\sigma)$ for any classical linear connection ∇ on $\mathbb{R}^m$ such that the identity map $id_{\mathbb{R}^m}$ is a ∇-normal coordinate system with center 0 and any $\sigma \in Q^r$. Then by the invariance of A and $A^{<\mu>}$ with respect to normal coordinates we deduce that $A = A^{<\mu>}$. □

Remark 3.1. Similar problems as the one of the present note have been studied in many papers, [3]–[7]. The method used in the present note is the same as in [5] and [6].

References

1. Kobayashi S., Nomizu K., *Foundations of Differential Geometry*, Moskow (Mir) 1981 (in Russian).
2. Kolář I., Michor P.W., Slovák J., *Natural operations in differential geometry*, Springer-Verlag, Berlin, 1993
3. Kowalski O., Sekizawa M., Differential geometry and its applications (Brno, 1986), 149–178, *Math. Appl. (East European Ser.)*, **27**, Dordrecht, 1987.
4. Kowalski O., Sekizawa M., *Arch. Math. Brno* **44** (2008), 139–147.
5. Kurek J., Mikulski W.M., *Lobachevskii J. Math.* **27** (2007), 41–46.
6. Mikulski W.M., *Demonstratio Math.* **40** (2007), 481–484.
7. Sekizawa M., *Monatsh. Math.* **105** (1988), 229–243

DISTRIBUTIONS ON THE COTANGENT BUNDLE FROM TORSION-FREE CONNECTIONS

J. Kurek

Institute of Mathematics, Maria Curie-Sklodowska University
pl. Marii Curie Skłodowskiej 1, 20–031 Lublin, Poland
E-mail: kurek@hektor.umcs.lublin.pl

W. M. Mikulski

Institute of Mathematics Jagiellonian University
Łojasiewicza 6, 30–348 Kraków, Poland
E-mail: Wlodzimierz.Mikulski@im.uj.edu.pl

We describe all $\mathcal{M}f_m$-natural operators $A : Q^o \rightsquigarrow Gr_qT^*$ lifting torsion free classical linear connections ∇ on m-manifolds M into q-dimensional distributions $A(\nabla) \subset TT^*M$ on the cotangent bundle T^*M.

Keywords: (Torsion free) classical linear connection, distribution, Grassmann bundle, natural operator

1. Introduction

Let m and q be arbitrary integers such that $m \geq 1$ and $0 \leq q \leq 2m$. We study the problem how a torsion-free classical linear connection ∇ on an m-dimensional manifold M induces canonically a smooth (C^∞) distribution $A(\nabla) \subset TT^*M$ on the cotangent bundle T^*M of M such that $dim(A(\nabla)_\omega) = q$ for any $\omega \in T^*M$. This problem is reflected in the concept of natural operators $A : Q^o \rightsquigarrow Gr_qT^*$ transforming torsion free classical linear connections ∇ on m-manifolds M (sections of the bundle $Q^oM \to M$ of torsion-free connections) into q-dimensional distributions $A(\nabla)$ on T^*M (sections of the Grassmann bundle $Gr_q(T^*M) \to T^*M$ of q-dimensional subspaces tangent to T^*M). The main result constitutes Theorem 2.1, where we describe all natural operators $A : Q^o \rightsquigarrow Gr_qT^*$ in question. The proof of Theorem 2.1 is based on a modification of the method from [2].

All manifolds and maps are assumed to be smooth (of class C^∞). Manifolds are assumed to be finite dimensional and without boundary.

2. The main result

Given an m-dimensional manifold M, let TM be the tangent bundle of M. Every embedding $f: M \to M_1$ of m-dimensional manifolds induces the corresponding tangent map $Tf: TM \to TM_1$ of f.

Given an m-dimensional manifold M, let $T^*M = (TM)^*$ be the cotangent bundle of M. Every embedding $f: M \to M_1$ of m-manifolds induces a vector bundle embedding $T^*f := (T(f^{-1}))^*: T^*M \to T^*M_1$ covering f, where Tf denotes the tangent map of f.

Given an n-dimensional manifold N, let Gr_pN be the Grassmann bundle of p-dimensional vector subspaces tangent to N. Every embedding $f: N \to N_1$ of n-manifolds induces a fibred map $Gr_pf: Gr_pN \to Gr_pN_1$ covering f given by $Gr_pf(V) = Tf(V)$ for any $V \in (Gr_pN)_x$ (i.e. for any p-dimensional subspace $V \subset T_xN$), $x \in N$.

A p-dimensional distribution on an n-dimensional manifold N is a smooth (C^∞) vector sub-bundle $D \subset TN$ of the tangent bundle of N such that $dim(D_x) = p$ for any point $x \in N$. Thus a p-dimensional distribution D is a smooth (C^∞) section of the Grassmann bundle Gr_pN of N.

Given an m-manifold M, let $QM := (id_{T^*M} \otimes \pi^1)^{-1}(id_{TM}) \subset T^*M \otimes J^1TM$ be the connection bundle of M, where $\pi^1 : J^1TM \to TM$ is the projection of the first jet prolongation $J^1TM = \{j_x^1 X | X \in \mathcal{X}(M), x \in M\}$ of the tangent bundle TM of M. Every embedding $f: M \to M_1$ induces (in obvious way) a fibred map $Qf: QM \to QM_1$ covering f.

There exists a sub-bundle Q^oM of QM such that the sections of Q^oM are the torsion free classical linear connections on M.[1,3] More precisely, if $\lambda^o : \mathbb{R}^m \to Q\mathbb{R}^m$ is the usual flat classical linear connection on $\mathbb{R}^m$ then $Q^oM = \{Qf(\lambda_0^o) \mid f : \mathbb{R}^m \to M$ is an embedding$\}$. Every embedding $f: M \to M_1$ of m-manifolds induces (by restriction of Qf) a fibred map $Q^of: Q^oM \to Q^oM_1$ covering f.

We denote the set of all classical linear connections on M by $\underline{Q}(M)$ and the set of all torsion free classical linear connections on M by $\underline{Q^o}(M)$.

Definition 2.1. ([1]) Let m and q be integers such that $m \geq 1$ and $0 \leq q \leq 2m$. An $\mathcal{M}f_m$-natural operator $A: Q^o \rightsquigarrow Gr_qT^*$ lifting torsion free classical linear connections ∇ on m-manifolds M into q-dimensional distributions $A(\nabla)$ on the cotangent bundle T^*M is a family of $\mathcal{M}f_m$-invariant regular operators (functions)

$$A: \underline{Q^o}(M) \to \underline{Gr_q}(T^*M)$$

from the set $\underline{Q}^o(M) \subset Q(M)$ of all torsion free classical linear connections on M (sections of $Q^oM \to M$) into the set $Gr_q(T^*M)$ of all q-dimensional distributions on T^*M (sections of $Gr_q(T^*M) \to T^*M$) for any $\mathcal{M}f_m$-object M, where $\mathcal{M}f_m$ is the category of m-dimensional manifolds and their embeddings. The invariance means that if $\nabla_1 \in \underline{Q}^o(M)$ and $\nabla_2 \in \underline{Q}^o(N)$ are f-related (i.e. $Q^of \circ \nabla_1 = \nabla_2 \circ f$) for some $\mathcal{M}f_m$-map $f : M \to N$ then $A(\nabla_1)$ and $A(\nabla_2)$ are T^*f-related (i.e. $Gr_q(T^*f) \circ A(\nabla_1) = A(\nabla_2) \circ T^*f$). The regularity means that A transforms smoothly parametrized families of torsion free classical linear connections into smoothly parametrized families of distributions.

We have the following $\mathcal{M}f_m$-natural operators $A : Q^o \rightsquigarrow Gr_qT^*$.

Example 2.1. $A^{[1]}(\nabla)_\omega = \{0\}$ for all $\omega \in (T^*M)_x$, $x \in M$, $\nabla \in \underline{Q}^o(M)$.

Example 2.2. $A^{[2]}(\nabla)_\omega = \ker(T_\omega \pi_M) = V_\omega T^*M$ for all $\omega \in (T^*M)_x$, $x \in M$, $\nabla \in \underline{Q}^o(M)$, where $\pi_M : T^*M \to M$ is the cotangent bundle projection.

Example 2.3. $A^{[3]}(D)_\omega = T_\omega T^*M$ for all $\omega \in (T^*M)_x$, $x \in M$, $\nabla \in \underline{Q}^o(M)$.

To present more complicated examples of $\mathcal{M}f_m$-natural operators $A : Q^o \rightsquigarrow Gr_qT^*$ we need a preparation.

Definition 2.2. An $\mathcal{M}f_m$-natural operator $B : Q^o \rightsquigarrow (T^*, T^* \otimes T^*)$ is an $\mathcal{M}f_m$-invariant family of regular operators $B : \underline{Q}^o(M) \rightsquigarrow C_M^\infty(T^*M, T^*M \otimes T^*M)$ for any m-manifold M, where $C_M^\infty(T^*M, T^*M \otimes T^*M)$ is the space of all (not necessarily vector) fibred maps $T^*M \to T^*M \otimes T^*M$ covering the identity map id_M.

Remark 2.1. We have the following $\mathcal{M}f_m$-natural operators $B^{[i]} : Q^o \rightsquigarrow (T^*, T^* \otimes T^*)$ for $i = 1, 2, 3$ given by $B^{[1]}(\nabla)(\omega) = \omega \otimes \omega$, $B^{[2]}(\nabla)(\omega) = sym((Ric_\nabla)_x)$ and $B^{[3]}(\nabla)(\omega) = alt((Ric_\nabla)_x)$ for $\nabla \in \underline{Q}^o(M)$, $\omega \in (T^*M)_x$, $x \in M$, where $sym((Ric_\nabla)_x)$ is the symmetrization of the Ricci tensor field Ric_∇ of ∇ at x and $alt((Ric_\nabla)_x)$ is the alternation of $(Ric_\nabla)_x$. (We remark that $R_\nabla(X,Y,Z) = R_\nabla(X,Y)Z = \nabla_X(\nabla_Y Z) - \nabla_Y(\nabla_X Z) - \nabla_{[X,Y]}Z$ and $Ric_\nabla(X,Y) = tr\{Z \to R_\nabla(Z,X)Y\}$.) Using methods from[1] (in particular Example 28.7) one can standardly show that any $\mathcal{M}f_m$-natural operator $B : Q^o \rightsquigarrow (T^*, T^* \otimes T^*)$ is a linear combination of the natural operators $B^{[i]}$ for $i = 1, 2, 3$ with real coefficients.

Let $B : Q^o \rightsquigarrow (T^*, T^* \otimes T^*)$ be an $\mathcal{M}f_m$-natural operator. Then we have the corresponding $\mathcal{M}f_m$-natural operator $A^{[B]} : Q^o \rightsquigarrow Gr_m T^*$ defined as follows.

Example 2.4. $A^{[B]}(\nabla)_\omega = \{v_\omega^{\nabla^*} + \frac{d}{dt}|_{t=0}(\omega + t < B(\nabla)(\omega), v >) \mid v \in T_x M\}$, $\nabla \in \underline{Q}^o(M)$, $\omega \in (T^*M)_x$, $x \in M$, where $v_\omega^{\nabla^*} \in H_\omega^{\nabla^*} \subset T_\omega T^* M$ is the unique ∇^*-horizontal vector over v with respect to the cotangent bundle projection (i.e. $v_\omega^{\nabla^*}$ is the ∇^*-horizontal lift of v at ω) and where $T_\omega T^* M = V_\omega T^* M \oplus H_\omega^{\nabla^*}$ is the decomposition corresponding to linear connection ∇^* on $T^*M \to M$ dual to ∇ (i.e. H^{∇^*} is the ∇^*-horizontal distribution on T^*M). (The pairing $< \cdot, \cdot >$ is the contraction C_1^1 as in Section 1.)

The main result in this paper is the following theorem.

Theorem 2.1. *All $\mathcal{M}f_m$-natural operators $A : Q^o \rightsquigarrow Gr_q T^*$ are described in Examples 2.1–2.4.*

3. Proof of the main result

Lemma 3.1. *Let $A : Q^o \rightsquigarrow Gr_q T^*$ be an $\mathcal{M}f_m$-natural operator such that $A(\nabla) \subset VT^*M = ker(T\pi_M)$ for any $\nabla \in \underline{Q}^o(M)$. Then either $A = A^{[1]}$ or $A = A^{[2]}$.*

Proof. From now on let $\nabla^o \in \underline{Q}^o(\mathbb{R}^m)$ be the usual flat classical linear connection on $\mathbb{R}^m$. We may assume that $A(\nabla^o)_\theta \neq \{0\}$ (as if $A(\nabla^o)_\theta = \{0\}$ then $q = 0$ and then $A = A^{[1]}$). Consider a non-zero vector $v \in A(\nabla^o)_\theta$. Then using the invariance of A with respect to linear isomorphisms (preserving ∇^o) we see that $V_\theta T^* \mathbb{R}^m \subset A(\nabla^o)_\theta$ (if $w \in V_\theta T^* \mathbb{R}^m$, $w \neq 0$, then there exists a linear isomorphism ψ (preserving ∇^o) such that $w = TT^*\psi(v)$, and then $w \in A(\nabla^o)_\theta$ because of the invariance of A with respect to ψ). Then $A(\nabla^o)_\theta = V_\theta T^* \mathbb{R}^m$ (as by the assumption $A(\nabla^o)_\theta \subset V_\theta T^* \mathbb{R}^m$). Then $q = m$. Consequently $A(\nabla)_\omega = V_\omega T^* M$ as $A(\nabla)_\omega \subset V_\omega T^* M$ (because of the assumption of the lemma) and $dim(A(\nabla)_\omega) = q = m = dim(V_\omega T^* M)$. Hence $A = A^{[2]}$. □

Proof of Theorem 2.1. Let $A : Q^o \rightsquigarrow Gr_q T^*$ be an $\mathcal{M}f_m$-natural operator. By the $\mathcal{M}f_m$-invariance of A and by Lemma 3.1 we may assume that $A(\nabla)_\omega \setminus V_\omega T^* \mathbb{R}^m \neq \emptyset$ for some $\nabla \in \underline{Q}^o(\mathbb{R}^m)$ and some $\omega \in (T^*\mathbb{R}^m)_0$, $\omega \neq 0$. Then using the invariance of A we may additionally assume that $\omega = d_0 x^1$

and $\nabla_0 = \nabla_0^o$, where $x^1, ..., x^m$ are the usual coordinates on $\mathbb{R}^m$. Let $v \in A(\nabla)_{d_0 x^1} \setminus V_{d_0 x^1} T^* \mathbb{R}^m$. Then standardly one can show that there exist a constant vector field $Y^o \neq 0$ and a vector field Y^1 with $Y^1(0) = 0$ such that $T^*(Y)_{d_0 x^1} = v$, where $Y = Y^o + Y^1$. (Here $T^*(X)$ means the flow lifting of a vector field $X \in \mathcal{X}(M)$ to T^*M.) Then there exist a local diffeomorphism ψ such that $j_0^1 \psi = id$ and $\psi_* Y = Y^o$ near 0. Then using the invariance of A with respect to ψ we may additionally assume $v = T^*(Y^o)_{d_0 x^1}$. Then using the invariance of A with respect to the homotheties $a_t = \frac{1}{t} id_{\mathbb{R}^m}$ for $t > 0$ and putting $t \to 0$ we have $T^*(Y^o)_\theta \in A(\nabla^o)_\theta \setminus V_\theta T^* \mathbb{R}^m$. Then using the invariance of A with respect to linear isomorphisms we deduce that $T^*(Z)_\theta \in A(\nabla^o)_\theta$ for any constant vector field $Z \in \mathcal{X}(\mathbb{R}^m)$. Consequently $H_\theta^{(\nabla^o)^*} \subset A(\nabla^o)_\theta$.

If $q > m$, then (from $H_\theta^{(\nabla^o)^*} \subset A(\nabla^o)_\theta$) it follows that there exists $u \in V_\theta T^* \mathbb{R}^m \bigcap A(\nabla^o)_\theta$, $u \neq 0$. Then $V_\theta T^* \mathbb{R}^m \subset A(\nabla^o)_\theta$ (as if $v \in V_\theta T^* \mathbb{R}^m$, $v \neq 0$, then there is a linear isomorphism ψ such that $v = TT^* \psi(u)$, and then $v \in A(\nabla^o)_\theta$ because of the invariance of A with respect to ψ). Then $A(\nabla^o)_\theta = T_\theta T^* \mathbb{R}^m$, and then $q = 2m$. Consequently $A = A^{[3]}$.

If $q = m$ and $V_\omega T^* \mathbb{R}^m \bigcap A(\nabla)_\omega \neq \{0\}$ for some $\nabla \in \underline{Q}^o(\mathbb{R}^m)$ with $\nabla_0 = \nabla_0^o$ and some $\omega \in (T^* \mathbb{R}^m)_0$, then using the invariance of A with respect to the homotheties $\frac{1}{t} id_{\mathbb{R}^m}$ for $t > 0$ and then putting $t \to 0$ we deduce (because of the regularity of A) that $V_\theta T^* \mathbb{R}^m \bigcap A(\nabla^o)_\theta \neq \{0\}$. Then (since $H_\theta^{(\nabla^o)^*} \subset A(\nabla^o)_\theta$) we have $q > m$. Contradiction.

So, we may the assumption $q = m$ and $V_\omega T^*M \bigcap A(\nabla)_\omega = \{0\}$ for any $\nabla \in \underline{Q}^o(M)$ and any $\omega \in (T^*M)_x$, $x \in M$. Then we have the decomposition $T_\omega T^*M = V_\omega T^*M \oplus A(\nabla)_\omega$ for any ∇ and ω as above. Then we can define an $\mathcal{M}f_m$-natural operator $B = B^{(A)} : Q^o \rightsquigarrow (T^*, T^* \otimes T^*)$ postulating that $< -B(\nabla)(\omega), v > \in (T^*M)_x = V_\omega T^*M$ is the $V_\omega T^*M$-component of the ∇^*-horizontal lift $v_\omega^{\nabla^*}$ of $v \in T_x M$ at $\omega \in (T^*M)_x$ with respect to the decomposition $T_\omega T^*M = V_\omega T^*M \oplus A(\nabla)_\omega$, where $\nabla \in \underline{Q}^o(M)$, $v \in T_x M$, $\omega \in (T^*M)_x$, $x \in M$. Clearly, $A = A^{[B]}$ (where $A^{[B]}$ is described in Example 2.4). □

References

1. I. Kolář, P. W. Michor and J. Slovák, *Natural Operations in Differential Geometry*, Springer-Verlag Berlin 1993.
2. W. M. Mikulski, *Ann. Polon Math.* **93** (3)(2008), 211-215.
3. M. Paluszny and A. Zajtz, Foundations of the geometry of natural bundles, Lect. Notes Univ. Caracas, 1984.

ON THE GEODESICS OF THE ROTATIONAL SURFACES IN THE BIANCHI-CARTAN-VRANCEANU SPACES

P. Piu* and M. M. Profir**

Dipartimento di Matematica e Informatica, Università degli Studi di Cagliari
Via Ospedale 72,09124 Cagliari, Italy
* E-mail: piu@unica.it
** E-mail: profirmanuela@yahoo.com

This paper is devoted to the problem of finding the geodesics of the rotational surfaces in the Bianchi-Cartan-Vranceanu spaces. We find a Clairaut-like relation and calculated the geodesics of the "cylinder".

Keywords: Geodesics, rotational surfaces, Bianchi-Cartan-Vranceanu spaces

Mathematics Subject Classification 2000: 58E20, 53A05

1. Introduction

We consider a two-parameter family of three-dimensional Riemannian manifolds $(M, ds^2_{\ell,m})$, where the metrics have the expression

$$ds^2_{\ell,m} = \frac{dx^2 + dy^2}{[1 + m(x^2 + y^2)]^2} + \left(dz + \frac{\ell}{2}\frac{ydx - xdy}{[1 + m(x^2 + y^2)]}\right)^2, \quad (1)$$

with $\ell, m \in \mathbb{R}$. The underlying differentiable manifolds M are $\mathbb{R}^3$ if $m \geq 0$, and $M = \{(x, y, z) \in \mathbb{R}^3 \ : \ x^2 + y^2 < -\frac{1}{m}\}$ otherwise. These metrics (B-C-V metrics) have been known for a long time. They can be found in the classification of three-dimensional homogeneous metrics given by L. Bianchi in 1897 (see[1]); later, they appeared, in form (1), in É. Cartan (see[2]) and in G. Vranceanu (see[8]). Their geometric interest lies in the following fact: *the family of metrics (1) includes all three-dimensional homogeneous metrics whose group of isometries has dimension 4 or 6, except for those of constant negative sectional curvature.* The group of isometries of these spaces contains a subgroup isomorphic to the group $SO(2)$, and the surfaces invariant by the action of $SO(2)$ are called *rotational surfaces*. In,[3,4] and,[7] R Caddeo, P. Piu, A. Ratto, and P. Tomter have studied rotational surfaces of the Heisenberg group $\mathbb{H}_3$ with constant (mean or Gauss) curvature. The geodesics of the B-C-V spaces were studied by A. Serra and P. Sitzia in their degree thesis (see,[56]). In this note we write down the differential equations of the geodesics and we find the conditions that meridians and parallels must satisfy in order to be geodesics. Then we determine the geodesics of the rotational cylinders and find a Clairaut-like relation. It is maybe usufull to recall the spaces that correspond to the different values of ℓ and m.

- If $\ell = 0$, then M is the product of a surface S with constant Gaussian curvature $4m$ and the real line $\mathbb{R}$.
- If $4m - \ell^2 = 0$, then M has non negative constant sectional curvature.
- If $\ell \neq 0$ and $m > 0$, then M is locally $SU(2)$.
- If $\ell \neq 0$ and $m < 0$, then M is locally $\widetilde{SL}(2,\mathbb{R})$,
- If $\ell \neq 0$ and $m = 0$ we get a left invariant metric on the Heisenberg Lie group $\mathbb{H}_3$.

2. Geodesics on surfaces of revolution

First of all, we want to obtain the differential equations of the geodesics for the $SO(2)$-invariant surfaces of the B-C-V spaces and point out the analogies with the Euclidean case. The metrics (1) are invariant with respect to the rotations around the z-axis and this fact leads to the study of the rotational surfaces, locally parametrized by

$$X(u,v) = (f(u)\cos v, f(u)\sin v, g(u))$$

where $0 \leq v < 2\pi$ and f, g are real functions with $f > 0$. The metric induced by (1) on the surface X is

$$ds^2 = \frac{f'(u)^2 du^2 + f(u)^2 dv^2}{[1+mf(u)^2]^2} + \left(g'(u)du - \frac{\ell}{2}\frac{f(u)^2 dv}{1+mf(u)^2}\right)^2. \quad (2)$$

Therefore the coefficients of the first fundamental form of the surface are

$$E = \frac{f'(u)^2}{[1+mf(u)^2]^2} + g'(u)^2, \quad F = -\frac{\ell f(u)^2 g'(u)}{2(1+mf(u)^2)}, \quad G = \frac{4f(u)^2 + \ell^2 f(u)^4}{4[1+mf(u)^2]^2}.$$

In order to obtain the geodesics of the rotational surfaces we use the Euler-Lagrange equations. The Lagrangian associated to the metric (2) is

$$L = \frac{1}{2}(E\dot{u}^2 + 2F\dot{u}\dot{v} + G\dot{v}^2)$$

and the corresponding Euler-Lagrange equations are

$$\begin{cases} \frac{d}{dt}\left[2\left(\frac{f'(u)^2}{[1+mf(u)^2]^2} + g'(u)^2\right)\dot{u} - \frac{\ell f(u)^2 g'(u)}{1+mf(u)^2}\dot{v}\right] = E'(u)\dot{u}^2 + 2F'(u)\dot{u}\dot{v} + G'(u)\dot{v}^2 \\ \frac{d}{dt}\left[2\frac{4f(u)^2 + \ell^2 f(u)^4}{4[1+mf(u)^2]^2}\dot{v} - \frac{\ell f(u)^2 g'(u)}{1+mf(u)^2}\dot{u}\right] = 0. \end{cases}$$

Suppose that the generating curve $\alpha(u) = (f(u), 0, g(u))$ of X is parametrized by arc length, in which case $E = 1$. Then the equations of the geodesics become

$$\begin{cases} \frac{d}{dt}\left[2\dot{u} - \frac{\ell f(u)^2 g'(u)}{1+mf(u)^2}\dot{v}\right] = 2F'(u)\dot{u}\dot{v} + G'(u)\dot{v}^2 \\ \frac{d}{dt}\left[\frac{4f(u)^2 + \ell^2 f(u)^4}{2[1+mf(u)^2]^2}\dot{v} - \frac{\ell f(u)^2 g'(u)}{1+mf(u)^2}\dot{u}\right] = 0. \end{cases} \quad (3)$$

A parallel $u = u_0$, will be a geodesic if u_0 is a solution of

$$\begin{cases} \frac{d}{dt}\left[-\frac{\ell f(u)^2 g'(u)}{1+mf(u)^2}\dot{v}\right] = G'(u)\dot{v}^2 \\ \frac{d}{dt}\left[\frac{4f(u)^2+\ell^2 f(u)^4}{2[1+mf(u)^2]^2}\dot{v}\right] = 0 \end{cases}$$

that is

$$\begin{cases} \frac{\ell f(u_0)^2 g'(u_0)}{1+mf(u_0)^2}\ddot{v} = -G'(u_0)\dot{v_0}^2 \\ \frac{4f(u_0)^2+\ell^2 f(u_0)^4}{2[1+mf(u_0)^2]^2}\dot{v} = G(u_0)\dot{v} = const. \end{cases}$$

and then $\dot{v} = const$. If we substitute $\dot{u} = \ddot{v} = 0$ in the first equation we have

$$G'(u_0)\dot{v}^2 = 0$$

and therefore ($\dot{v} \neq 0$)

$$G'(u_0) = \frac{f(u_0)f'(u_0)[2+\ell^2 f(u_0)^2 - 2mf(u_0)^2]}{[1+mf(u_0)^2]^3} = 0.$$

It follows that the parallels are geodesics if

$$\frac{f'(u_0)[2+\ell^2 f(u_0)^2 - 2mf(u_0)^2]}{[1+mf(u_0)^2]^3} = 0. \quad (4)$$

Thus we have:

- the only parallels which are geodesics of the rotational surfaces in $\mathbb{H}_3$, $\widetilde{SL}(2,\mathbb{R})$, $\mathbb{H}^2 \times \mathbb{R}$ are, just like in the Euclidean case, those generated by the rotation of a point of the generating curve where the tangent is parallel to the axis of revolution ($f' = 0$). [For these spaces we have $\ell^2 \geq 2m$] ;
- for the rotational surfaces of the product manifold $\mathbb{S}^2 \times \mathbb{R}$ the parallels which are geodesics have $f' = 0$ or $f(u_0) = \sqrt{\frac{2}{2m-\ell^2}}$. In this case we have $\ell^2 < 2m$,
- for the rotational surfaces of $SU(2)$, besides the parallels with $f' = 0$, there are the parallels for which $f(u_0) = \sqrt{\frac{2}{2m-\ell^2}}$.

A meridian $v = v_0$ is a geodesic if it satisfies the equations

$$\begin{cases} \dot{u} = a \\ \frac{\ell f(u)^2 g'(u)}{1+mf(u)^2} \dot{u} = b, \end{cases} \quad a, b \in \mathbb{R}.$$

As $E = 1$, we obtain for $g'(u)$ the expression

$$g'(u) = \sqrt{1 - \frac{f'(u)^2}{[1+mf(u)^2]^2}}$$

and we conclude that a meridian is a geodesic if

$$\frac{\ell f(u)^2 \sqrt{[1+mf(u)^2]^2 - f'(u)^2}}{[1+mf(u)^2]^2} = const.$$

It follows that

- for the rotational surfaces of the product manifold $\mathbb{S}^2 \times \mathbb{R}$ and $\mathbb{H}^2 \times \mathbb{R}$, ($\ell = 0$), like in the Euclidean case, all the meridians are geodesics;
- all the meridians of the cylinders $f(u) = const.$ are geodesics;
- if $\ell \neq 0$ the meridians are geodesics for $m \gtreqless 0$ if the function f is

$$f(u) = \frac{\tan(\sqrt{m}u + c)}{\sqrt{m}}, \quad f(u) = u, \quad f(u) = \frac{\tanh(\sqrt{-m}u + c)}{\sqrt{-m}}$$

respectively, or if f is a solution of the equation

$$(2f'(u) + 4mf(u)^2 f'(u) + 2m^2 f(u)^4 f'(u) - 2f'(u)^3 \\ + 2mf(u)^2 f'(u)^3 - f(u)f'(u)f''(u) - mf(u)^3 f'(u)f''(u) = 0.$$

3. The Clairaut's relation

If we consider a geodesic γ parametrized by arc length and ω is the angle between γ and a parallel, by using $\dot{\gamma} = \dot{u}X_u + \dot{v}X_v$, we have:

$$\cos \omega = \frac{\langle \dot{\gamma}, X_v \rangle}{\|X_v\|} = \frac{1}{\sqrt{G}} \left[-\frac{\ell f(u)^2 g'(u)}{2(1+mf(u)^2)} \dot{u} + \frac{4f(u)^2 + \ell^2 f(u)^4}{4[1+mf(u)^2]^2} \dot{v} \right]$$

The Clairaut's relation will be:

$$\sqrt{\frac{4f(u)^2 + \ell^2 f(u)^4}{4[1+mf(u)^2]^2}} \cos \omega = const.$$

As in the case of the surfaces of the Euclidean space, we can consider $f(u)$ as the distance from the z-axis, $f(u) = \sqrt{x^2 + y^2} = r$.

4. Geodesics on the cylinder

If we consider the "cylinder" parametrized by

$$S(u,v) = (a\cos v, a\sin v, u), \quad a \in \mathbb{R}, \tag{5}$$

the Euler-Lagrange equations of the geodesics become

$$\begin{cases} \dot{u} - \frac{\ell a^2}{2(1+ma^2)}\dot{v} = const. \\ -\frac{\ell a^2}{2(1+ma^2)}\dot{u} + \frac{4a^2+\ell^2 a^4}{4(1+ma^2)^2}\dot{v} = const. \end{cases} \tag{6}$$

and thus we have

$$\frac{4a^2}{4(1+ma^2)^2}\dot{v} = const.$$

This implies that $\dot{u} = const.$, $\dot{v} = const.$ and we have

Proposition 4.1. *The geodesics of the cylinder are the curves of equation*

$$\gamma(s) = (a\cos(As+B), a\sin(As+B), Cs+D),$$

that includes:

- *the meridians,*
- *the parallels,*
- *the helices, that is the curves with constant geodesic curvature and geodesic torsion, analogous of the helices of $\mathbb{R}^3$.*

References

1. L. Bianchi, *Mem. Soc. It. delle Scienze (dei XL) (b)* **11** (1897), 267–352.
2. É. Cartan, *Leçons sur la géométrie des espaces de Riemann,* Gauthier-Villars Paris, II Ed. (1946).
3. R. Caddeo, P. Piu, A. Ratto, *Boll. Un. Mat. Ital. B (7)* **10** (1996), 341–357.
4. R. Caddeo, P. Piu, A. Ratto, *Manuscripta Math.* **87** (1995), 1–12.
5. A. Serra, *Sulla funzione densità di volume di una varietà riemanniana,* Tesi di Laurea, Università di Cagliari, Luglio 1988.
6. P. Sitzia, *Sulle geodetiche degli spazi biarmonici tridimensionali,* Tesi di Laurea, Università di Cagliari, Novembre 1988.
7. P. Tomter, Differential geometry: partial differential equations on manifolds (Los Angeles, CA, 1990), 485–495, *Proc. Symp. Pure Math.* **54** Part I, Amer. Math. Soc., Providence, RI, 1993.
8. G. Vranceanu, *Leçons de Géométrie Différentielle,* Ed. Ac. Rep. Pop. Roum., Vol. I, Bucarest (1957).

COTANGENT BUNDLES WITH GENERAL NATURAL KÄHLER STRUCTURES OF QUASI-CONSTANT HOLOMORPHIC SECTIONAL CURVATURES

S. L. Druţă

Faculty of Mathematics, University "Al. I. Cuza"
Iaşi, 700506, Romania
E-mail: simonadruta@yahoo.com

We prove that the only general natural Kählerian cotangent bundles of quasi-constant holomorphic sectional curvatures, are those of constant holomorphic sectional curvature.

Keywords: cotangent bundle; Riemannian metric; general natural lift; quasi-constant holomorphic sectional curvature.

Mathematics Subject Classification 2000: Primary 53C55, 53C15, 53C05

1. Introduction

The similitude between the geometry of the tangent and cotangent bundles of a Riemannian manifold (M, g), may be explained by their duality. Anyway, there are some fundamental differences between them, due to the different construction of lifts to T^*M, particularly of the natural ones (see Refs. 5, 7, 8), which cannot be defined just like in the case of TM (see Ref. 15). The possibility to consider vertical, complete and horizontal lifts on T^*M leads to interesting geometric structures, studied in some recent papers like Refs. 10, 12, 13, 14.

In the present paper we study the conditions under which the general natural Kählerian structures on T^*M, determined in Ref. 3, have quasi-constant holomorphic sectional curvatures. We prove that the general natural Kählerian cotangent bundles of quasi-constant holomorphic sectional curvatures are only those of constant holomorphic sectional curvature. In Ref. 2 it was proved a similar result for the natural Kählerian structures of diagonal type on TM.

Let (M, g) be a smooth Riemannian manifold of the dimension n. We denote its cotangent bundle by $\pi : T^*M \to M$, and recall the splitting of the tangent bundle to T^*M into the vertical distribution $VT^*M = \text{Ker } \pi_*$ and the horizontal one, determined by the Levi Civita connection $\dot{\nabla}$ of g:

$$TT^*M = VT^*M \oplus HT^*M.$$

If $(\pi^{-1}(U), q^1, \ldots, q^n, p_1, \ldots, p_n)$ is a local chart on T^*M, induced from the local chart $(U, x^1, \ldots, x^n)$ on M, the local vector fields $\{\frac{\partial}{\partial p_i}\}_{i=1\ldots n}$ on $\pi^{-1}(U)$ define a local frame for VT^*M over $\pi^{-1}(U)$ and the local vector fields $\{\frac{\delta}{\delta q^j}\}_{j=1\ldots n}$ define a local frame for HT^*M over $\pi^{-1}(U)$, where

$$\frac{\delta}{\delta q^i} = \frac{\partial}{\partial q^i} + \Gamma^0_{ih}\frac{\partial}{\partial p_h}, \quad \Gamma^0_{ih} = p_k \Gamma^k_{ih},$$

and $\Gamma^k_{ih}(\pi(p))$ are the Christoffel symbols of g.

The set of vector fields $\{\frac{\delta}{\delta q^i}, \frac{\partial}{\partial p_j}\}_{i,j=1,\ldots,n}$, denoted by $\{\delta_i, \partial^j\}_{i,j=1,\ldots,n}$, defines a local frame on T^*M, adapted to the direct sum decomposition.

The M-tensor fields on the cotangent bundle may be introduced in the same manner as the M-tensor fields were introduced Ref. 9 on the tangent bundle of a Riemannian manifold.

Roughly speaking, the natural lifts have coefficients as functions of the density energy (see Ref. 5 by Janyška, and Ref. 11 by Oproiu), given as

$$t = \frac{1}{2}\|p\|^2 = \frac{1}{2}g^{-1}_{\pi(p)}(p,p) = \frac{1}{2}g^{ik}(x)p_i p_k, \quad p \in \pi^{-1}(U).$$

We have $t \in [0, \infty)$ for all $p \in T^*M$.

In Ref. 3, the present author considered the real valued smooth functions $a_1, a_2, a_3, a_4, b_1, b_2, b_3, b_4$ on $[0, \infty) \subset \mathbf{R}$ and studied a general natural tensor of type $(1, 1)$ on T^*M, defined by the relations

$$\begin{cases} JX^H_p = a_1(t)(g_X)^V_p + b_1(t)p(X)p^V_p + a_4(t)X^H_p + b_4(t)p(X)(p^\sharp)^H_p, \\ J\theta^V_p = a_3(t)\theta^V_p + b_3(t)g^{-1}_{\pi(p)}(p, \theta)p^V_p - a_2(t)(\theta^\sharp)^H_p - b_2(t)g^{-1}_{\pi(p)}(p, \theta)(p^\sharp)^H_p, \end{cases}$$

in every point p of the induced local card $(\pi^{-1}(U), \Phi)$ on T^*M, $\forall X \in \mathcal{X}(M), \forall \theta \in \Lambda^1(M)$, where g_X is the 1-form on M defined by $g_X(Y) = g(X, Y)$, $\forall Y \in \mathcal{X}(M)$, $\theta^\sharp = g_\theta^{-1}$ is a vector field on M defined by $g(\theta^\sharp, Y) = \theta(Y)$, $\forall Y \in \mathcal{X}(M)$, the vector $p^\sharp$ is tangent to M in $\pi(p)$, p^V is the Liouville vector field on T^*M, and $(p^\sharp)^H$ is the geodesic spray on T^*M.

In the same paper, Ref. 3, the author defined a Riemannian metric G of general natural lift type, given by the relations

$$\begin{cases} G_p(X^H, Y^H) = c_1(t)g_{\pi(p)}(X, Y) + d_1(t)p(X)p(Y), \\ G_p(\theta^V, \omega^V) = c_2(t)g^{-1}_{\pi(p)}(\theta, \omega) + d_2(t)g^{-1}_{\pi(p)}(p, \theta)g^{-1}_{\pi(p)}(p, \omega), \\ G_p(X^H, \theta^V) = G_p(\theta^V, X^H) = c_3(t)\theta(X) + d_3(t)p(X)g^{-1}_{\pi(p)}(p, \theta), \end{cases} \quad (1)$$

$\forall\ X, Y \in \mathcal{X}(M),\ \forall\ \theta, \omega \in \Lambda^1(M), \forall\ p \in T^*M$, and she proved that (T^*M, G, J) is an almost Hermitian manifold, if and only if the coefficients c_1, c_2, c_3, are proportional to a_1, a_2, a_3, with the proportionality factor $\lambda > 0$, and $c_1 + 2td_1$, $c_2 + 2td_2$, $c_3 + 2td_3$ are related to the coefficients $a_1 + 2tb_1$, $a_2 + 2tb_2$, $a_3 + 2tb_3$, by another proportionality relations, the new proportionality factor being $\lambda + 2t\mu > 0$. Then, the almost Hermitian structure (T^*M, G, J) is Kählerian if and only if $\mu = \lambda'$, and the almost complex structure J is integrable.

2. The quasi-constant holomorphic sectional curvatures of the cotangent bundles with general natural lifted metrics

A Kählerian manifold (M, g, J), endowed with a unit vector field ξ, is said to be of quasi-constant holomorphic sectional curvatures (see Refs. 1 and 6), if for any holomorphic plane $span\{X, JX\}$, generated by the unit vector $X \in T_pM$, $p \in M$, the sectional curvature $R(X, JX, JX, X)$ depends only on the point p and on the angle between the holomorphic plane and the unit vector field ξ. In Refs. 1 and 6 it was shown that a Kählerian manifold (M, g, J, ξ) is of quasi-constant holomorphic sectional curvature if and only if the curvature tensor field R of the Levi-Civita connection ∇, satisfy

$$R - k_0 R_0 - k_1 R_1 - k_2 R_2 = 0, \tag{2}$$

where k_0, k_1, k_2 are smooth functions on M and R_0, R_1, R_2 are certain tensor fields of curvature type on M, given by

$$\begin{aligned}R_0(X,Y)Z = \tfrac{1}{4}\{&g(Y,Z)X - g(X,Z)Y \\ &+ g(JY,Z)JX - g(JX,Z)JY + 2g(X,JY)JZ\},\end{aligned} \tag{3}$$

$$\begin{aligned}R_1(X,Y)Z = &U(X,Y,Z) - U(Y,X,Z) \\ &+ U(JX, JY, Z) - U(JY, JX, Z),\end{aligned} \tag{4}$$

$$R_2(X,Y)Z = \{\eta(X)\eta(JY) - \eta(JX)\eta(Y)\}\{\eta(JZ)\xi + \eta(Z)J\xi\}, \tag{5}$$

where η is a 1-form defined by $\eta(X) = g(X, \xi)$, and U is an auxiliar (1,3)-tensor field, defined by

$$\begin{aligned}U(X,Y,Z) = \tfrac{1}{8}\{&\eta(Y)\eta(Z)X + \eta(X)\eta(JZ)JY + \eta(X)\eta(JY)JZ+ \\ &+ g(Y,Z)\eta(X)\xi + g(X,JZ)\eta(Y)J\xi+ \\ &+ \tfrac{1}{2}g(X,JY)\eta(JZ)\xi + \tfrac{1}{2}g(X,JY)\eta(Z)J\xi\}.\end{aligned}$$

Let us consider on the cotangent bundle T^*M, the general natural Kählerian structure (G, J), and the Liouville vector field $C = p_i \partial^i$. We mention that on the bundle T^*M_0, of non-zero vector fields, we may work with C instead of the unitary vector field ξ, since C is non-null on T^*M_0, $\frac{1}{\|C\|}C$ is unitary, and the scalar factors may be incorporated into k_1, k_2. Hence we may study the property of (T^*M_0, G, J, C) to have quasi-constant holomorphic sectional curvatures (i.e we find the condition under which the curvature tensor field of (T^*M_0, G, J, C) may be written in the form (2)).

By using the local adapted frame $\{\delta_i, \partial^j\}_{i,j=1,...,n}$ on T^*M, we obtain, after a standard straightforward computation, the horizontal and vertical components of the curvature tensor field R of (T^*M, G). For example:

$$R(\partial^i, \delta_j)\delta_k = PQQQ^i{}_{jk}{}^h \delta_h + PQQP^i{}_{jkh} \partial^h,$$

where the M-tensor fields which appear as coefficients, have quite long expressions, presented in Ref. 4. We mention that we use the character Q on a certain position to indicate that the argument on that position was a horizontal vector field and, similarly, we used the character P for vertical vector fields.

Analogously, from (3), (4), and (5) we obtain after standard straightforward computations, the components of R_0, R_1, R_2.

The next lemma will be useful in the study of the conditions under which (T^*M_0, G, J, C) has quasi-constant holomorphic sectional curvatures.

Lemma 2.1. *If $\alpha_1, \ldots, \alpha_{10}$ are smooth functions on T^*M such that*

$$\alpha_1 \delta_i^h g_{jk} + \alpha_2 \delta_j^h g_{ik} + \alpha_3 \delta_k^h g_{ij} + \alpha_4 \delta_k^h p_i p_j + \alpha_5 \delta_j^h p_i p_k + \alpha_6 \delta_i^h p_j p_k + \\ \alpha_7 g_{jk} p_i g^{0h} + \alpha_8 g_{ik} p_j g^{0h} + \alpha_9 g_{ij} p_k g^{0h} + \alpha_{10} p_i p_j p_k g^{0h} = 0, \quad (6)$$

then $\alpha_1 = \cdots = \alpha_{10} = 0$.

We may prove that all the components in (2) are of the form (6). Using lema 2.1, we get for the proportionality factor λ the expression obtained in Ref. 4, and then we may prove that k_1 and k_2 vanish. Thus we may state:

Theorem 2.1. *The bundle T^*M_0 of non-zero cotangent vectors to M, endowed with a general natural Kähler structure (G, J), and with the Liouville vector field C, has quasi-constant holomorphic sectional curvatures, if and only if (T^*M, G, J) is of constant holomorphic sectional curvature.*

Remark. The theorem 2.1 may be reformulated as a generalisation of Shur's theorem: if the sectional curvature of any holomorphic plane generated by the unit nonzero cotangent vector $P \in T_p^*M$ depends only on the angle with the Liouville vector field and on the point $p \in M$, then T^*M has constant holomorphic sectional curvature.

Acknowledgements

The author wants to acknowledge the support of professor V. Oproiu throughout this work and throughout the PhD period. The research was partially supported by the Grant TD 158/2007, CNCSIS, Ministerul Educaţiei şi Cercetării, România

References

1. C.L. Bejan, and M. Benyounes, *J. Geom.* **88** (2008), 1-14.
2. C.L. Bejan, and V. Oproiu, *Balkan J. Geom. Applic.* **11** (2006), 11-22.
3. S.L. Druţă, Cotangent Bundles with General Natural Kähler Structures, to appear in *Rév. Roum. Math. Pur. Appl.*
4. S.L. Druţă, Quasi-constant holomorphic sectional curvatures of tangent bundles with general natural Kähler structures, submited.
5. J. Janyška, *Arch. Math. (Brno)* **37** (2001), 143-160.
6. G. Ganchev, and V. Mihova, *Central European J. Math.* **6** (2008) 43-75.
7. I. Kolář, P. Michor, and J. Slovak, *Natural Operations in Differential Geometry*, Springer Verlag, Berlin, 1993.
8. O. Kowalski, and M. Sekizawa, *Bull. Tokyo Gakugei Univ. (4)* **40** (1988), 1-29.
9. K.P. Mok, E.M. Patterson, and Y.C. Wong, *Trans. Amer. Math. Soc.* **234** (1977), 253-278.
10. M. Munteanu, *An. Şt. Univ. "Al. I. Cuza" Iaşi, Math.* **44** (1998), 125-136.
11. V. Oproiu, *Math. J. Toyama Univ.* **22** (1999), 1-14.
12. V. Oproiu, and D.D. Poroşniuc, *An. Ştiinţ. Univ. Al. I. Cuza, Iaşi , s.I, Mathematics* **49** (2003), 399-414.
13. V. Oproiu, and D.D. Poroşniuc, *Publ. Math. Debrecen* **66** (2005), 457-478.
14. D.D. Poroşniuc, *Balkan J. Geom. Appl.* **9** (2004), 94-103.
15. K. Yano, and S. Ishihara, *Tangent and Cotangent Bundles*, M. Dekker Inc., New York, 1973.

POLYNOMIAL TRANSLATION WEINGARTEN SURFACES IN 3-DIMENSIONAL EUCLIDEAN SPACE

M. I. Munteanu* and A. I. Nistor**

Department of Mathematics, Al.I. Cuza University
Bd. Carol I, no. 11, 700506 – Iasi, Romania
* E-mail: marian.ioan.munteanu@gmail.com
www.math.uaic.ro\ ~munteanu
** E-mail: ana.irina.nistor@gmail.com

In this paper we will classify those translation surfaces in $\mathbb{E}^3$ involving polynomials which are Weingarten surfaces.

Keywords: W-surfaces, translation surfaces, minimal surfaces, constant mean curvature (CMC), constant Gaussian curvature

Mathematics Subject Classification 2000: 53A05, 53A10

1. Preliminaries

The surfaces of constant mean curvature, H-surfaces and those of constant Gaussian curvature, K-surfaces in the Euclidean 3-dimensional space, $\mathbb{E}^3$, have been studied extensively. One interesting class of surfaces in $\mathbb{E}^3$ is that of *translation surfaces*, which can be parametrized, locally, as $r(u,v) = (u, v, f(u)+g(v))$, where f and g are smooth functions. This type of surfaces are important either because they are interesting themselves or because they furnish counterexamples for some problems (e.g. it is a known fact that a minimal surface has vanishing second Gaussian curvature but not conversely — see for details Ref. 1). We call *polynomial translation surfaces* (in short, *PT surfaces*) those translation surfaces for which f and g are polynomials. For technological applications in which different surfaces are needed (such as Computed Aided Manufacturing) polynomial forms are preferred since they may be incorporated into the CAD software in order to be easily processed by numerical computations.

Scherk's surface, obtained in 1834 by H. Scherk[2], is the only non flat minimal surface, that can be represented as a translation surface. More precisely, we have

Theorem A. *Let S be a translation minimal surface in 3-dimensional Euclidean space. Then S is an open part of $\mathbb{E}^3$ or it is congruent to the following surface*

$$z = \frac{1}{a} \log \left| \frac{\cos(ax)}{\cos(ay)} \right|, \quad a \neq 0.$$

Other interesting results concerning translation surfaces having either constant mean curvature or constant Gaussian curvature are the following:

Theorem B. *Let S be a translation surface with constant Gauss curvature K in 3-dimensional Euclidean space. Then S is congruent to a cylinder, so it is flat ($K = 0$)* (Theorem 1 in Ref. 3).

Theorem C. *Let S be a translation surface with constant mean curvature $H \neq 0$ in 3-dimensional Euclidean space $\mathbb{E}^3$. Then S is congruent to the following surface*

$$z = \frac{\sqrt{1+a^2}}{2H} \sqrt{1 - 4H^2 x^2} + ay, \quad a \in \mathbb{R}.$$

(Theorem 2, statement (1) in Ref. 3). Cf. also References in op. cit.

A surface S is called a *Weingarten surface* (cf. Ref. 4) if there is some (smooth) relation $W(\kappa_1, \kappa_2) = 0$ between its two principal curvatures κ_1 and κ_2, or equivalently, if there is a (smooth) relation $U(K, H) = 0$ between its mean curvature H and its Gaussian curvature K.

In this paper we study those PT surfaces that are Weingarten surfaces. We abbreviate by a *WPT surfaces*. We give the following classification theorem:

Theorem 1.1. *Let S be a WPT surface in 3-dimensional Euclidean space. Then*

(i) S is a cylinder, case in which $K = 0$;

(ii) S is a paraboloid of revolution, case in which the mean curvature H and the Gaussian curvature K are positive everywhere and related by the formula

$$8aH^2 = \sqrt{K}(2a + \sqrt{K})^2 \tag{1}$$

where a is a positive constant.

2. Weingarten translation surfaces

Let $r: S \to \mathbb{R}^3$ be an isometrical immersion of a translation surface of type

$$r(u, v) = (u, v, f(u) + g(v)) \tag{2}$$

where f and g are smooth functions. The first fundamental form **I** and the second fundamental form **II** have particular expressions, namely

$$\mathbf{I} = \left(1 + f'(u)^2\right) du^2 + 2f'(u)g'(v) du\, dv + \left(1 + g'(v)^2\right) dv^2$$

$$\mathbf{II} = \frac{1}{\sqrt{\Delta}} \left(f''(u)\, du^2 + g''(v)\, dv^2\right)$$

where $\Delta = 1 + f'(u)^2 + g'(v)^2$. Let us denote f' by α and g' by β. Hence, the mean curvature H and the Gaussian curvature K can be written as

$$H = \frac{\left(1 + \beta(v)^2\right)\alpha'(u) + \left(1 + \alpha(u)^2\right)\beta'(v)}{2\left[1 + \alpha(u)^2 + \beta(v)^2\right]^{\frac{3}{2}}} \qquad (3)$$

$$K = \frac{\alpha'(u)\beta'(v)}{\left[1 + \alpha(u)^2 + \beta(v)^2\right]^2}. \qquad (4)$$

The existence of a Weingarten relation $U(H,K) = 0$ means that curvatures H and K are functionally related, and since H and K are differentiable functions depending on u and v, this implies the *Jacobian condition* $\partial(H,K)/\partial(u,v) = 0$. (See e.g. Ref. 5.) More precisely the following condition

$$\frac{\partial H}{\partial u}\frac{\partial K}{\partial v} - \frac{\partial H}{\partial v}\frac{\partial K}{\partial u} = 0 \qquad (5)$$

needs to be satisfied. Conversely, if the above condition is satisfied, then the curvatures must be functionally related and thus, by definition, the surface is Weingarten. The Jacobian condition characterizes W-surfaces and it is used to identify them when an explicit Weingarten relation cannot be found.

In our case, the Jacobian condition yields the following relation

$$\begin{aligned}
0 = {} & 8\alpha(u)\beta(v)\alpha'(u)^3\beta'(v)^2 - 8\alpha(u)\beta(v)\alpha'(u)^2\beta'(v)^3 \\
& -3\beta(v)\alpha'(u)\beta'(v)^2\alpha''(u) + 3\alpha(u)\alpha'(u)^2\beta'(v)\beta''(v) \\
& -2\alpha(u)^2\beta(v)\alpha'(u)\beta'(v)^2\alpha''(u) + 2\alpha(u)\beta(v)^2\alpha'(u)^2\beta'(v)\beta''(v) \\
& -3\beta(v)^3\alpha'(u)\beta'(v)^2\alpha''(u) + 3\alpha(u)^3\alpha'(u)^2\beta'(v)\beta''(v) \\
& -3\alpha(u)\alpha'(u)^3\beta''(v) + 3\beta(v)\beta'(v)^3\alpha''(u) \\
& +3\alpha(u)^2\beta(v)\beta'(v)^3\alpha''(u) - 3\alpha(u)\beta(v)^2\alpha'(u)^3\beta''(v) \\
& +\alpha'(u)\alpha''(u)\beta'(v) - \beta'(v)\alpha''(u)\beta''(v) \\
& +\alpha(u)^2\alpha'(u)\alpha''(u)\beta''(v) - \beta(v)^2\beta'(v)\alpha''(u)\beta''(v) \\
& +2\beta(v)^2\alpha'(u)\alpha''(u)\beta''(v) - 2\alpha(u)^2\beta'(v)\alpha''(u)\beta''(v) \\
& +\alpha(u)^2\beta(v)^2\alpha'(u)\alpha''(u)\beta''(v) - \alpha(u)^2\beta(v)^2\beta'(v)\alpha''(u)\beta''(v) \\
& -\alpha(u)^4\beta'(v)\alpha''(u)\beta''(v) + \beta(v)^4\alpha'(u)\alpha''(u)\beta''(v).
\end{aligned} \qquad (6)$$

At this point we will consider α and β to be polynomials of degree m and n respectively. More precisely we shall consider

$$\alpha = a_m u^m + a_{m-1} u^{m-1} + \ldots \quad \text{and} \quad \beta = b_n v^n + b_{n-1} v^{n-1} + \ldots$$

where a_m and b_n are different from 0. Replacing α and β in Eq. (6) we obtain a polynomial expression in u and v vanishing identically. This means that all the coefficients are 0.

Let us distinguish several cases:

Case 1: $m, n \geq 2$, i.e. $\alpha'' \neq 0$ and $\beta'' \neq 0$.

a. Suppose $m > n (\geq 2)$.

The dominant term corresponds to $u^{5m-2} v^{2n-3}$ and it comes from

$$3\alpha(u)^3 \alpha'(u)^2 \beta'(v) \beta''(v) - \alpha(u)^4 \beta'(v) \alpha''(u) \beta''(v)$$

having the coefficient

$$a_m^5 b_n^2 \left(3m^2 n^2 (n-1) - mn^2 (m-1)(n-1) \right).$$

This cannot vanish since $a_m, b_n \neq 0$ and $m > n \geq 2$.

The subcase $n > m \geq 2$ can be treated in similar way.

b. Suppose $m = n \geq 2$.

In the same manner, this case cannot occur.

Case 2: $m > n = 1$.

In this case β can be expressed as $\beta(v) = av + b$, with a and b real constants, $a \neq 0$. We rewrite the Jacobian condition in the following way

$$\begin{aligned}0 = {} & 8a^2 \alpha(u) \beta(v) \alpha'(u)^3 - 8a^3 \alpha(u) \beta(v) \alpha'(u)^2 - 3a^2 \beta(v) \alpha'(u) \alpha''(u) \\ & - 2a\alpha(u)^2 \beta(v) \alpha'(u) \alpha''(u) - 3a^2 \beta(v)^3 \alpha'(u) \alpha''(u) \\ & + 3a^3 \beta(v) \alpha''(u) + 3a^3 \alpha(u)^2 \beta(v) \alpha''(u). \end{aligned} \quad (7)$$

Let us analyze the terms in u having maximum degree, namely $u^{4m-3} v$. This comes from $-2a\alpha(u)^2 \beta(v) \alpha'(u) \alpha''(u)$ and yields the relation $2a^2 a_m^4 m^2 (m-1) = 0$ which cannot hold in this case.

The case $n > m = 1$ can be treated in similar way.

Case 3: $m = n = 1$

In this case α and β can be expressed as $\alpha(u) = Au + B$ and $\beta(v) = av + b$, with A, B, a and b real constants, $A, a \neq 0$. The Jacobian condition becomes

$$8\alpha(u)\beta(v)\alpha'(u)^3 \beta'(v)^2 - 8\alpha(u)\beta(v)\alpha'(u)^2 \beta'(v)^3 = 0. \quad (8)$$

Using the same technique as above one gets $A = a$. So, the parametrization of the surface can be written (after a possible translation in $\mathbb{E}^3$) in the form

$$r(u, v) = \left(u, v, a(u - u_0)^2 + a(v - v_0)^2 \right) \quad (9)$$

where $u_0, v_0 \in \mathbb{R}$. This is a paraboloid of revolution, and its curvatures H and K are both everywhere positive and they are related by the relation Eq. (1).

Case 4: $m = 0$ (or $n = 0$).

In this case α (or β) is constant and the Jacobian condition is automatically satisfied. So, the parametrization of the surface can be written in the form

$$r(u,v) = (u, v, au + g(v)) \qquad (10)$$
$$r(u,v) = (u, v, f(u) + av) \qquad (11)$$

where f and g are arbitrary polynomials, $a \in \mathbb{R}$ (it can also vanishes). These two surfaces are cylinders and they are obviously flat.

It is interesting to remark that in the case when $f(u) = au + b$ (or $g(v) = av + b$) (with a and b real constants) and the other function is not polynomial we still obtain a cylinder, hence a flat surface.

Acknowledgements

The first author was supported by grant ID_ 398/2007–2010 , ANCS, Romania. The second author was partially supported by grant CEEX – ET n. 5883/2006–2008, ANCS, Romania.

References

1. D. E. Blair and Th. Koufogiorgos, *Monat. Math.* **113** (4) (1992) 177–181.
2. H. F. Scherk, *J. Reine Angew. Math.* **13** (1834) 185–208.
3. H. Liu, *J. Geom.* **64** (1999) 141–149.
4. J. Weingarten, *J. Reine Angew. Math.* **59** (1861) 382–393.
5. B. van Brunt and K. Grant, *Comput. Aided Geom. Design* **13** (1996), 569–582.

G-STRUCTURES DEFINED ON PSEUDO-RIEMANNIAN MANIFOLDS

IGNACIO SÁNCHEZ-RODRÍGUEZ

Department of Geometry and Topology, University of Granada
Granada, E-18071, Spain
E-mail: ignacios@ugr.es
www.ugr.es/local/ignacios

Concepts and techniques from the theory of G-structures of higher order are applied to the study of certain structures (volume forms, conformal structures, linear connections and projective structures) defined on a pseudo-Riemannian manifold. Several relationships between the structures involved have been investigated. The operations allowed on G-structures, such as intersection, inclusion, reduction, extension and prolongation, were used for it.

Keywords: Higher order G-structures, pseudo-Riemannian conformal structure, projective differential geometry, general relativity

1. Introduction and motivation

A *differential structure* of a manifold M is a C^∞ maximal atlas and, indeed, the charts of the atlas make up the *primordial* structure. The idea of a *geometrical structure* can be realized by the concept of G-structure when choosing the allowable meaningful classes of charts.

General relativity is a physical theory, which is heavily based on differential geometry. The fundamental mathematical tools used by this theory to explain and to handle gravity are the geometrical structures. The space-time is described by a 4-dimensional manifold with a *Lorentzian metric field*, and the theory put the matter on space-time, being mainly represented by curves in the manifold or by the overall stress-energy tensor.

The law of inertia in the space-time is translated into a *projective structure* on the manifold, which is provided by the geodesics of the metric in keeping with the equivalence principle. Furthermore, the space-time in general relativity is a dynamical entity because the metric field is subject to the Einstein field equations, which almost equate Ricci curvature with stress-energy of matter.

Other main structures are the *volume form*, that is used to get action functionals by integration over the manifold, and the *Lorentzian conformal structure*, that gives an account of light speed invariance. Different approaches to gravity try to *separate* the geometry into independent compounds to promote the understanding about physical interpretation of geometric variables.

The theory of G-structures of higher order is possibly the more natural framework for studying the interrelations involved among the relevant structures. In a pseudo-Riemannian manifold there are defined unambiguously the following structures: *volume form, conformal structure, pseudo-Riemannian metric, symmetric linear connection and projective structure*. Volume, conformal and metric structures are G-structures of first order, but each of them lead to a *prolonged* second order structure. Symmetric linear connection and projective structure are *inherently* G-structures of second order. We will try to clarify this unified description.

2. The bundle of r-frames

A differentiable manifold M is a set of points with the property that we can cover it with the charts of a C^∞ n-dimensional maximal atlas $\mathcal{A}$. The *bundle of r-frames* $\mathcal{F}^r M$ is a quotient set over $\mathcal{A}$. Every class-point, an r-frame, collect the charts with equal origin of coordinates which produce identical r-th order Taylor series expansion of functions (see Refs. 1,2).

An r-frame is an r-jet at 0 of inverses of charts of M; two charts are in the same r-jet if they have the same partial derivatives up to r-th order at the same origin of coordinates. Every $\mathcal{F}^r M$ is naturally equipped with a *principal bundle* structure with respect to the group G_n^r of r-jets at 0 of diffeomorphisms of $\mathbb{R}^n$, $j_0^r \phi$, with $\phi(0) = 0$.

The group of the *bundle of 1-frames* is $\mathrm{GL}(n, \mathbb{R}) \cong G_n^1$. Its natural representation on $\mathbb{R}^n$ gives an *associated bundle* coinciding with the tangent bundle TM. In the end, we identify $\mathcal{F}^1 M$ with the *linear frame bundle LM*. Other representations of G_n^1 on subspaces of the tensorial algebra over $\mathbb{R}^n$ give associated bundles whose sections are the well-known tensor fields.

The *bundle of 2-frames* $\mathcal{F}^2 M$ is somehow more complicated. Every 2-frame is characterized by a *torsion-free transversal n-subspace* $H_l \subset T_l LM$. It happens that the chart's first partial derivatives fix $l \in LM$ and the second partial derivatives give the 'inclination' of that n-subspace. The group G_n^2 is isomorphic to $G_n^1 \rtimes S_n^2$, a semidirect product, with S_n^2 the additive group of symmetric bilinear maps of $\mathbb{R}^n \times \mathbb{R}^n$ into $\mathbb{R}^n$; the multiplication rule is $(a, s)(b, t) := (ab, b^{-1} s(b, b) + t)$, for $a, b \in G_n^1$, $s, t \in S_n^2$.

Let $\mathfrak{g}$ denote the Lie algebra of $G \subset G_n^1$, then the named *first prolongation of* $\mathfrak{g}$ is defined by $\mathfrak{g}_1 := S_n^2 \cap L(\mathbb{R}^n, \mathfrak{g})$. We obtain that $G \rtimes \mathfrak{g}_1$ is a subgroup of $G_n^1 \rtimes S_n^2 \cong G_n^2$. This will be used in Sec. 4.

3. First order G-structures

An *r-th order G-structure* of M is a *reduction* (see [3, p. 53]) of $\mathcal{F}^r M$ to a subgroup $G \subset G_n^r$. First order G-structures are just called G-structures. Let us see some of them.

Let's define a *volume on* M as a first order G-structure, with $G = \mathrm{SL}_n^\pm := \{a \in G_n^1 : |\det(a)| = 1\}$. If M is orientable, a volume on M has two components: two $\mathrm{SL}(n,\mathbb{R})$-structures for two equal, except sign, *volume n-forms*. For a general M, a volume corresponds to an *odd type n-form*.

From *bundle theory*,[3] $\mathrm{SL}_n^\pm$-structures are sections of $\mathcal{V}M$, the *associated bundle* to LM and the action of G_n^1 on $G_n^1/\mathrm{SL}_n^\pm$, and they correspond to G_n^1-*equivariant functions* f of LM to $G_n^1/\mathrm{SL}_n^\pm$. The isomorphisms $G_n^1/\mathrm{SL}_n^\pm \simeq H_n := \{k\,I_n : k > 0\} \simeq \mathbb{R}^+$ allow to write $f \colon LM \to \mathbb{R}^+$; the equivariance condition is $f(la) = |\det a|^{\frac{-1}{n}} f(l)$, for $l \in LM$, $a \in G_n^1$.

Theorem 3.1. *We have the bijections:*

$$\text{Volumes on } M \longleftrightarrow \operatorname{Sec} \mathcal{V}M \longleftrightarrow C^\infty_{equi}(LM, \mathbb{R}^+)$$

Analogous bijective diagram can be obtained for every reduction of a principal bundle. The Lie algebra of $\mathrm{SL}_n^\pm$ is $\mathfrak{sl}(n,\mathbb{R})$ and its first prolongation is $\mathfrak{sl}(n,\mathbb{R})_1 = \{s \in S_n^2 : \sum_k s_{ik}^k = 0\}$; it's a Lie algebra of infinite type.

We define a *pseudo-Riemannian metric* as an $O_{q,n-q}$-structure, with $O_{q,n-q} := \left\{a \in G_n^1 : a^t \eta a = \eta := \begin{pmatrix} -I_q & 0 \\ 0 & I_{n-q} \end{pmatrix} \right\}$. As in Th.3.1, we obtain bijections between the metrics and the sections of the associated bundle with typical fiber $G_n^1/O_{q,n-q}$, and also with the equivariant functions of LM in $G_n^1/O_{q,n-q}$. The first prolongation of $\mathfrak{o}_{q,n-q}$ is $\mathfrak{o}_{q,n-q\,1} = 0$; a consequence of this fact is the uniqueness of the *Levi-Civita connection*.

A (pseudo-Riemannian) *conformal structure* is a $CO_{q,n-q}$-structure, with $CO_{q,n-q} := O_{q,n-q} \cdot H_n$ (direct product); this definition is equivalent to consider a class of metrics related by a positive factor, and in the Lorentzian case, $q = 1$, a conformal structure is characterized by the *field of null cones*. The first prolongation of $\mathfrak{co}_{q,n-q}$ is $\mathfrak{co}_{q,n-q\,1} = \{s \in S_n^2 : s_{jk}^i = \delta_j^i \mu_k + \delta_k^i \mu_j - \sum_s \eta^{is} \eta_{jk} \mu_s,\ \mu = (\mu_i) \in \mathbb{R}^{n*}\} \simeq \mathbb{R}^{n*}$. The named second prolongation $\mathfrak{co}_{q,n-q\,2}$ is equal to 0 (i. e., $\mathfrak{co}_{q,n-q}$ is of finite type 2); this deals with the existence and uniqueness of the normal Cartan connection but we do not deal with this here (see [2, §§VI.4.2, VII.3]).

Volumes on M and conformal structures are *extensions* (see [4, p. 202]) of pseudo-Riemannian metrics because of the inclusion of $O_{q,n-q}$ in $SL_n^{\pm}$ and $CO_{q,n-q}$. Reciprocally:

Theorem 3.2. *A pseudo-Riemannian metric field on M is given by a pseudo-Riemannian conformal structure and a volume on M.*

This statement is proved in Ref. 5 by the fact that $G_n^1 = SL_n^{\pm} \cdot CO_{q,n-q}$ and $O_{q,n-q} = SL_n^{\pm} \cap CO_{q,n-q}$ imply that volume and conformal G-structures intersect in $O_{q,n-q}$-structures.

4. Second order G-structures

A *symmetric linear connection* (SLC) on M is a distribution on LM of torsion-free transversal n-subspaces, which is invariant by the action of G_n^1. Identifying $G_n^1 \simeq G_n^1 \rtimes 0 \subset G_n^2$, we can define an SLC on M as a second order G_n^1-structure. From bundle theory, as in Th.3.1, every SLC ∇ is a section of the associated bundle to $\mathcal{F}^2 M$ and the action of G_n^2 on $G_n^2/G_n^1 \simeq S_n^2$, and corresponds to an equivariant function $f^\nabla: \mathcal{F}^2 M \to S_n^2$, verifying $f^\nabla(z(a,s)) = a^{-1} f^\nabla(z)(a,a) + s$, for $z \in \mathcal{F}^2 M$, $a \in G_n^1$, $s \in S_n^2$.

Let P be a first order G-structure on M; a *symmetric connection on P* is a distribution on P of torsion-free transversal n-subspaces, which is invariant by the action of G, thereby producing a second order G-structure, whose G_n^1-extension is an SLC on M. Reciprocally, a second order G-structure determines a first order G-structure and a symmetric connection on it.

Noteworthy examples of this are: **i)** A pseudo-Riemannian metric and its *Levi-Civita connection* are given by a second order $O_{q,n-q}$-structure. **ii)** An *equiaffine structure* on M is a SLC with a parallel volume; it is given by a second order $SL_n^{\pm}$-structure. **iii)** A *Weyl structure* is a conformal structure with a SLC compatible; it is given by a second order $CO_{q,n-q}$-structure.

The following result is an important theorem, arisen from the Weyl's 'Raumproblem', studied by Cartan and others. The theorem is proved in Ref. 6, with a correction revealed in Ref. 7.

Theorem 4.1. *Let G be a subgroup of G_n^1, with $n \geq 3$. Any first order G-structure admits a symmetric connection if and only if $\mathfrak{g}$ is one of these: $\mathfrak{sl}(n,\mathbb{R})$, $\mathfrak{o}_{q,n-q}$, $\mathfrak{co}_{q,n-q}$, $\mathfrak{gl}_{n,W}$ (algebra of endomorphisms with an invariant 1-dimensional subspace W), $\mathfrak{gl}_{n,W,c}$ (certain subalgebra of the last one, for each $c \in \mathbb{R}$) or, for $n=4$, $\mathfrak{csp}(2,\mathbb{R})$.*

A G-structure P can admit many torsion-free transversal n-subspaces in $T_l P$, for every $l \in P$. We have the following result (see [2, p.150-155]):

Theorem 4.2. *If a first order G-structure P admits a symmetric connection, the set P^2 of 2-frames corresponding with torsion-free transversal n-subspaces included in TP is a second order $G \rtimes \mathfrak{g}_1$-structure.*

We name P^2 the *(first holonomic) prolongation of P*. Necessarily, a second order $G \rtimes \mathfrak{g}_1$-structure is the prolongation of a G-structure.

A *(differential) projective structure* is a set of SLCs which have the same geodesics up to reparametrizations; we can define it as a second order $G_n^1 \rtimes \mathfrak{p}$-*structure* with $\mathfrak{p} := \{s \in S_n^2 : s_{jk}^i = \delta_j^i \mu_k + \mu_j \delta_k^i, \ \mu = (\mu_i) \in \mathbb{R}^{n*}\} \simeq \mathbb{R}^{n*}$. Considering geometrical structures on second order, with the same techniques than in the Th.3.2 for the first order (see Ref. 5), we obtain:

Theorem 4.3. *A projective structure and a volume on M give an SLC belonging to the former and making the volume parallel.*

Hence, a volume select a class of affine parametrizations for the paths of a projective structure. Contrarily, a projective structure and the prolongation of a conformal structure not always intersect; if they intersect we get a Weyl structure.

5. Concluding remarks

The study of integrability conditions of higher order, like curvatures, with respect to the interrelations of these G-structures is probably the natural next step following this work.

The geometrical structures described herein can be considered components of the space-time geometry.[8] In this context, the named *causal set theory* make a conceptual separation between volume and conformal structures, and Stachel[9] proposes an approach, similar to the metric-affine variational principle, using conformal and projective structures as independent variables. In this line of thought, and from the above results, I suggest considering the volume on space-time as a set of independent dynamical variables to make a variational analysis.

Acknowledgments

This work has been partially supported by the *Andalusian Government* P.A.I.: FQM-324.

References

1. S. Kobayashi, *Transformation Groups in Differential Geometry* (Springer, Heidelberg, 1972).

2. I. Sánchez Rodríguez, Conexiones en el fibrado de referencias de segundo orden. Conexiones conformes, PhD thesis, Universidad Complutense de Madrid, (Madrid, Spain, 1994), pp. vii + 210, http://www.ugr.es/~ignacios/.
3. S. Kobayashi and K. Nomizu, *Foundations of Differential Geometry, Vol. I* (John Wiley - Interscience, New York, 1963).
4. W. Greub, S. Halperin and R. Vanstone, *Connections, Curvature, and Cohomology, Vol. II* (Academic Press, New York, 1973).
5. I. Sánchez Rodríguez, On the intersection of geometrical structures, in *Lorentzian Geometry - Benalmádena 2001*, eds. M. A. Cañadas Pinedo, M. Gutiérrez and A. Romero (R.S.M.E., Spain, 2003) pp. 239–246.
6. S. Kobayashi and T. Nagano, *J. Math. Soc. Japan* **17**, 84 (1965).
7. H. Urbantke, *Internat. J. Theoret. Phys.* **28**, 1233 (1989).
8. I. Sánchez-Rodríguez, in *AIP Conf. Proc.* **1023**, 202 (2008).
9. J. Stachel, Prolegomena to any Future Quantum Gravity, in *Approaches to Quantum Gravity*, ed. D. Oriti (C.U.P., Cambridge, 2009), http://arxiv.org/pdf/gr-qc/0609108.

LIST OF PARTICIPANTS

VIII International Colloquium on Differential Geometry

Judit Abardia Bochaca (Barcelona, Spain)
Rui Albuquerque (Évora, Portugal)
Fernando Alcalde Cuesta (Santiago, Spain)
Ahmet Altundag (Istanbul, Turkey)
Jesús Antonio Álvarez López (Santiago, Spain)
Vladica Andrejic (Belgrade, Serbia)
Vestislav Apostolov (Montréal, Canada)
Teresa Arias-Marco (Extremadura, Spain)
Taro Asuke (Tokyo, Japan)
Adina Balmus (Iasi, Romania)
Gianluca Bande (Cagliari, Italy)
Miguel Bermúdez (Jussieu, France)
Jürgen Berndt (Cork, Ireland)
Ajit Bhand (Oklahoma, USA)
Andrzej Bis (Lódz, Poland)
Dorota Blachowska (Lódz, Poland)
Agustín Bonome Dopico (Santiago, Spain)
Charles Boubel (Strasbourg, France)
Miguel Brozos Vázquez (Santiago, Spain)
Esteban Calviño Louzao (Santiago, Spain)
Xosé Manuel Carballés Vázquez (Santiago, Spain)
Regina Castro Bolaño (Santiago, Spain)
Wen-Haw Chen (Tunghai, Taiwan)
Luis A. Cordero (Santiago, Spain)
Maciej Czarnecki (Lodz, Poland)
Marcos Dajczer (IMPA, Brazil)
Francoise Dal'Bo (Rennes, France)
Hamidou Dathe (Dakar, Senegal)
Bertrand Deroin (Paris, France)

José Carlos Díaz-Ramos (Cork, Ireland)
Mirjana Djoric (Belgrade, Serbia)
Christopher T.J. Dodson (Manchester, UK)
Demetrio Domínguez (Cantabria, Spain)
Miguel Domínguez Vázquez (Santiago, Spain)
María Josefina Druetta (Córdoba, Argentina)
Simona-Luiza Druta (Iasi, Romania)
Zdenek Dusek (Olomouc, Czech Republic)
Manuel Fernández López (Santiago, Spain)
Eduardo García Río (Santiago, Spain)
Nicholas Gardner (Chicago, USA)
Sandra Gavino Fernández (Santiago, Spain)
Catalin Gherghe (Bucharest, Romania)
Peter B. Gilkey (Oregon, USA)
Antonio Mariano Gómez Tato (Santiago, Spain)
María del Rosario González Dorrego (Madrid, Spain)
Pablo González Sequeiros (Santiago, Spain)
Michel Goze (Mulhouse, France)
Georges Habib (Leipzig, Germany)
Gilbert Hector (Lyon, France)
Luis Hernández-Lamoneda (Guanajuato, Mexico)
Luís María Hervella Torrón (Santiago, Spain)
Steven Hurder (Chicago, USA)
Stere Ianus (Bucharest, Romania)
Adrian Mihai Ionescu (Bucharest, Romania)
Radu Iordanescu (Bucharest, Romania)
Stefan Ivanov (Sofia, Bulgary)
Wlodzimierz Jelonek (Kracow, Poland)
Vadim Kaimanovich (Bremen, Germany)
Hyunsuk Kang (Birmingham, UK)
Sebastian Klein (Cork, Ireland)
Victor Kleptsyn (Rennes, France)
Fabian Kopei (Münster, Germany)
Yuri Kordyukov (Moscow, Russia)
Jan Kurek (Lublin, Poland)
Rémi Langevin (Dijon, France)
Nina Lebedeva (Münster, Germany)
Manuel de León (ICMAT-CSIC, Spain)
Santiago López de Medrano (UNAM, Mexico)

Álvaro Lozano Rojo (País Vasco, Spain)
Alexander Lytchak (Bonn, Germany)
Marta Macho Stadler (País Vasco, Spain)
Enrique Macías-Virgós (Santiago, Spain)
David Marín (Barcelona, Spain)
Michael Markellos (Patras, Greece)
Xosé María Masa Vázquez (Santiago, Spain)
Shigenori Matsumoto (Tokyo, Japan)
Carlos Meniño (Santiago, Spain)
Eugenio Merino Gayoso (A Coruña, Spain)
Dimitrios Michalis (Patras, Greece)
Wlodzimierz Mikulski (Krakow, Poland)
Yoshihiko Mitsumatsu (Chuo, Japan)
Raluca Mocanu (Bucharest, Romania)
Stefano Montaldo (Cagliari, Italy)
Witold Mozgawa (Lublin, Poland)
Olaf Müller (México, Mexico)
Marian Ioan Munteanu (Iasi, Romania)
Antonio M. Naveira (Valencia, Spain)
Stana Nikcevic Simic (Belgrade, Serbia)
Ana Irina Nistor (Iasi, Romania)
Tadeo Noda (Chiba, Japan)
Hiraku Nozawa (Tokyo, Japan)
José Antonio Oubiña Galiñanes (Santiago, Spain)
Rui Pacheco (Covilha, Portugal)
Daniel Peralta-Salas (Madrid, Spain)
Ana Pereira do Vale (Braga, Portugal)
María José Pereira Sáez (Santiago, Spain)
María Pérez Fernández de Córdoba (Santiago, Spain)
Paola Piu (Cagliari, Italy)
Ana María Porto F. Silva (Madrid, Spain)
M. Ángeles de Prada (País Vasco, Spain)
Zoran Rakic (Belgrade, Serbia)
Elisabeth Remm (Mulhouse, France)
Jean Renault (Orléans, France)
Agustí Reventós Tarrida (Barcelona, Spain)
Ángel Manuel Rey Roca (Santiago, Spain)
Manuela Beatriz Rodríguez Moreiras (Santiago, Spain)
Julien Roth (Nice-Sophia Antipolis, France)

Vladimir Rovenski (Haifa, Israel)
José Ignacio Royo Prieto (País Vasco, Spain)
Tomasz Rybicki (Krakow, Poland)
Isabel Salavessa (Lisboa, Portugal)
Eliane Salem (Paris, France)
Marcos Salvai (Córdoba, Argentina)
Ignacio Sánchez-Rodríguez (Granada, Spain)
María Sancosmed Alvarez (Barcelona, Spain)
Luis Sanguiao Sande (Santiago, Spain)
Esperanza Sanmartín Carbón (Vigo, Spain)
Paul A. Schweitzer (Rio de Janeiro, Brazil)
Javier Seoane Bascoy (Santiago, Spain)
Fatma Muazzez Simsir (Amkara, Turkey)
Vladimir-Claudiu Slesar (Craiova, Romania)
Pedro Suárez (Barry, USA)
Ali Suri (Isfahan, Iran)
Ana Dorotea Tarrío Tobar (A Coruña, Spain)
Dirk Toeben (Cologne, Germany)
Ruy Tojeiro (São Carlos, Brazil)
Takashi Tsuboi (Tokyo, Japan)
Wilderich Tuschmann (Kiel, Germany)
María Elena Vázquez Abal (Santiago, Spain)
Ramón Vázquez Lorenzo (Santiago, Spain)
Alberto Verjovsky (UNAM, Mexico)
Silvia Vilariño Fernández (Santiago, Spain)
Rodica Cristina Voicu (Bucharest, Romania)
Pawel G. Walczak (Lódz, Poland)
Szymon Walczak (Lódz, Poland)